Rock Dynamics

ISRM Book Series
Series editor: Xia-Ting Feng
Institute of Rock and Soil Mechanics, Chinese Academy of Sciences, Wuhan, China

ISSN : 2326-6872
eISSN: 2326-778X

Volume 3

International Society for Rock Mechanics

ISRM

Rock Dynamics

Ömer Aydan

Department of Civil Engineering, University of the Ryukyus, Nishihara, Okinawa, Japan

CRC Press
Taylor & Francis Group
Boca Raton London New York

CRC Press is an imprint of the
Taylor & Francis Group, an **informa** business

A BALKEMA BOOK

Published by:
CRC Press/Balkema

P.O. Box 447, 2300 AK Leiden, The Netherlands
e-mail: Pub.NL@taylorandfrancis.com
www.crcpress.com – www.taylorandfrancis.com

First issued in paperback 2021

ISBN-13: 978-1-03-209689-6 (pbk)
ISBN-13: 978-1-138-03228-6 (hbk)

Publisher's Note
The publisher has gone to great lengths to ensure the quality of this reprint but points out that some imperfections in the original copies may be apparent.

Visit the Taylor & Francis Web site at
http://www.taylorandfrancis.com

and the CRC Press Web site at
http://www.crcpress.com

Library of Congress Cataloging-in-Publication Data

Names: Aydan, Ömer, author.
Title: Rock dynamics / Ömer Aydan, Department of Civil Engineering,
    University of the Ryukyus, Nishihara, Okinawa, Japan.
Description: Leiden, The Netherlands : CRC Press/Balkema, [2017] | Series:
    ISRM book series, ISSN 2326-6872 ; volume 3 | Includes bibliographical
    references and index.
Identifiers: LCCN 2016058563 (print) | LCCN 2016059918 (ebook) | ISBN
    9781138032286 (hardcover : alk. paper) | ISBN 9781315391304 (eBook) | ISBN
    9781315391304 (ebook)
Subjects: LCSH: Rock mechanics.
Classification: LCC TA706 .A929 2017 (print) | LCC TA706 (ebook) | DDC
    624.1/513—dc23
LC record available at https://lccn.loc.gov/2016058563

Typeset by MPS Limited, Chennai, India

Table of contents

About the author

Born in 1955, Professor Aydan studied Mining Engineering at the Technical University of Istanbul, Turkey (B.Sc., 1979), Rock Mechanics and Excavation Engineering at the University of Newcastle upon Tyne, UK (M.Sc., 1982), and finally received his Ph.D. in Geotechnical Engineering from Nagoya University, Japan in 1989. Prof. Aydan worked at Nagoya University as a research associate (1987–1991), and then at the Department of Marine Civil Engineering at Tokai University, first as Assistant Professor (1991–1993), then as Associate Professor (1993–2001), and finally as Professor (2001–2010). He then became Professor of the Institute of Oceanic Research and Development at Tokai University, and is currently Professor at the University of Ryukyus, Department of Civil Engineering & Architecture, Nishihara, Okinawa, Japan. He has furthermore played an active role on numerous ISRM, JSCE, JGS, SRI and Rock Mech. National Group of Japan committees, and has organized several national and international symposia and conferences. Professor Aydan has received the 1998 Matsumae Scientific Contribution Award, the 2007 Erguvanlı Engineering Geology Best Paper Award, the 2011 Excellent Contributions Award from the International Association for Computer Methods in Geomechanics and Advances, the 2011 Best Paper Award from the Indian Society for Rock Mechanics and Tunnelling Technology and was awarded the 2013 Best Paper Award at the 13th Japan Symposium on Rock Mechanics and 6th Japan-Korea Joint Symposium on Rock Engineering. He was also made Honorary Professor in Earth Science by Pamukkale University in 2008 and received the 2005 Technology Award, the 2012 Frontier Award and the 2015 Best Paper Award from the Japan National Group for Rock Mechanics.

Acknowledgements

The author sincerely acknowledges Prof. Xia-Ting Feng for inviting the author to contribute to the ISRM Book Series. This book is an outcome of notes and documents prepared by the author for setting the goals and roadmap for the activities of the Rock Dynamics Committee of the Japan Society of Civil Engineers and the studies carried out by the author at Nagoya University, Tokai University and the University of the Ryukyus in Japan for more than three decades and investigations during many reconnaissance visits to regions affected by major earthquakes since the 1992 Erzincan earthquake in Turkey. Discussions with the members of the committee led the author to write this book and sharpen his thoughts. The author would like to thank Emeritus Prof. Dr. Toshikazu Kawamoto of Nagoya University, Emeritus Professor Masanori Hamada of Waseda University, Japan, Emeritus Professor Zeki Hasgür of Istanbul Technical University, Prof. Dr. Reşat Ulusay of Hacettepe University and Prof. Dr. Halil Kumsar of Pamukkale University, Turkey for their guidance, help and suggestions in various stages of his studies quoted in this book. The involvement of the author in earthquake engineering is due to Professor M. Hamada and Professor Z. Hasgür and the reconnaissance visit to Erzincan in April 1992. The author heartfully thanks the late Prof. R. Yarar of Istanbul Technical University for his endless support to the author until he passed away. The author gratefully acknowledges the prize of the Scientific Contribution Award of Matsumae Shigeo of Tokai University, which led the author to acquire the measuring devices used during dynamic testing of rocks and many shaking table experiments described in this book. The author sincerely acknowledges his former students, Mitsuo Daido and Yoshimi Ohta of Tokai University for their help during experiments and computations reported in this study. The author would also like to thank Alistair Bright, acquisitions editor at CRC Press/Balkema for his patience and collaborations during the preparation of this book and Ms. José van der Veer, production editor, for the great efforts to produce this book.

Chapter 1

Introduction

Rock dynamics has become one of the most important topics in the field of rock mechanics and rock engineering (e.g. Aydan *et al.*, 2011; Aydan, 2016; Zhou and Jiao, 2011). The spectrum of rock dynamics is very wide and it includes, the failure of rocks, rock masses and rock engineering structures such as rockbursting, spalling, popping, collapse, toppling, sliding, blasting, non-destructive testing, geophysical explorations, and, impacts (see Figure 1.1).

The fundamental governing equation used in the dynamics of materials and structures is presented in Chapter 2. As rock mass in nature has discontinuities in the form of tiny cracks in small scale to fault zones in large scale, such discontinuities must be

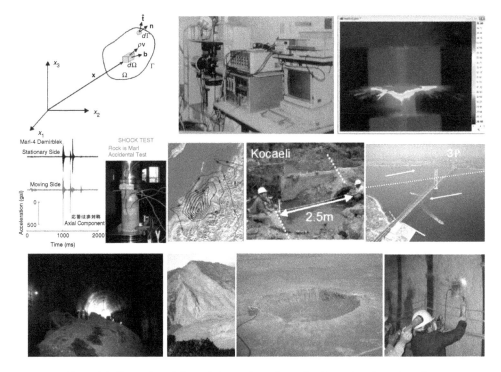

Figure 1.1 Examples of dynamic issues in rock mechanics and rock engineering.

taken into account in any type of analysis involving rock masses. Constitutive models for rocks, discontinuities and rock masses are explained and some fundamental features of the numerical methods used in the rock dynamics problem are outlined.

The experimental techniques and monitoring equipment are quite important for evaluating the dynamic characteristics of rocks, discontinuities and rock masses. Chapter 3 describes current available techniques for measuring the dynamic properties in rock mechanics and points out also some new directions how to deal with actual issues in this field.

The dynamic responses such as acceleration, velocity and displacement of geo-materials during fracturing and slippage have not been studied in the fields of geo-engineering and geo-science as measurement, monitoring and logging technologies were not so advanced in the past. By virtue of the recent advances in measurement, monitoring and logging technologies, it is now possible to carry out such experiments on geo-materials ranging from very soft materials such as clay to hard rocks such as siliceous sandstone by using different loading schemes and loading frames as well as rock discontinuities with different surface roughness characteristics. These experiments and experimental results concerned with the acceleration, velocity and displacement responses of geo-materials during fracturing and slippage under laboratory conditions are given in Chapter 4. The velocity and displacement responses are obtained through the Erratic Pattern Screening (EPS) integration technique proposed by Aydan and Ohta (2011).

Earthquakes are known to be one of the natural disasters resulting in huge losses of human life and properties as experienced in the recent earthquakes. Since there is no way to prevent the occurrence of earthquakes in earthquake-prone countries such as Turkey, Japan, USA and Taiwan, the design of structures and residential and industrial developments must be done according to possible types and magnitude of earthquakes. It is well known that ground motion characteristics, deformation and surface breaks of earthquakes depend upon the causative faults. While many large earthquakes occur along the subduction zones, which are far from the land and their effects appear as severe shaking, the large in-land earthquakes may occur just beneath or nearby urban and industrial zones as observed in the recent great earthquakes. The seismic design of engineering structures is generally carried out by considering the possible shaking characteristics of the ground during earthquakes in a given region. It is a fact that the residual (permanent) relative displacement of the ground is not considered in any seismic code all over the world, except for very long linear structures such as pipelines. This problem is currently considered to be beyond the capability of seismic design concept for structures in earthquake engineering, although it must be dealt with somewhat. The fundamental aspects and features of current methods for estimating strong motions and permanent ground deformations are described in Chapter 5.

Recent earthquakes showed that the foundations of large structures on rock masses may be damaged by permanent ground deformation resulting from faulting or slope failure. Chapter 6 describes some model experiments, case histories on various foundation types of bridges, buildings, highways, railways, dams, pylons, pipelines. Possible methods for evaluating the effects of permanent ground deformation of foundations and associated structures are summarized and several examples of applications are described.

Underground structures are well known as earthquake-resistant structures. However, the recent earthquakes showed that underground structures are also vulnerable to seismic damage. There may be several reasons such as high ground motions and permanent ground movements. Various forms of damage to underground structures such as tunnels, caverns, natural caves and abandoned mines during major earthquakes are presented in Chapter 7. Results of various model tests on underground opening using shaking tables are also presented to show the effect of ground shaking on the response and collapse of underground structures in continuum and discontinuum. Furthermore, some empirical equations are proposed to assess the damage to underground structures, which may be useful for quick assessments of possible damage. Applications of numerical methods on the dynamic responses and stability are also presented in this chapter.

Large inland earthquakes caused many large scale rock slope failures in recent years. The slope failures induced tremendous damage to infrastructures as well as to residential areas, and they involved not only cut slopes but also natural rock slopes. Compared to the scale of soil slope failures, the scale and the impact of rock slope failures are very large and the form of failure differs depending upon the geological structures of rock mass of slopes. Furthermore, the failure of the rock slope failures may involve both active and passive modes. However, the passive modes are generally observed when the ground shaking is quite large. Some model experiments, case histories on slope failures are described in Chapter 8. Possible methods for evaluating the dynamic stability of rock slopes are summarized and several examples of applications are described.

Historical structures are mainly masonry structures, which are composed of blocks made of natural stones, bricks or both, and they are assembled in different patterns with or without mortar. Furthermore, masonry houses presently constitute more than 60 percent of the residential buildings all over the world and they have a very long building history. Despite widespread utilization of masonry structures in building history all over the world, there are a few studies on their seismic response and stability. The observations of damage to actual masonry and historical structures and monuments, the shaking table experiments, available limiting equilibrium and numerical methods for estimating their responses are presented in Chapter 9. Furthermore, a recent example of monitoring multi-parameter response of rock foundations of Nakagusuku Castle during earthquakes and in the long-term is described and its implications are discussed.

It is very rare to see discussion or experimental results on the load-displacement-time or stress-strain-time responses during experiments and loading or excavation of rock engineering structures. The author has found that the responses might be quite different during the transient process from those under static assumptions. It is concluded that the loading of samples and structures as well as excavation of rock engineering structures should be treated as a dynamic phenomenon. Chapter 10 describes some experiments and theoretical and numerical solutions for modeling the loading excavation of rock engineering structures as a dynamic phenomenon.

Blasting is the most commonly used excavation technique in mining and civil engineering applications. Blasting induces strong ground motions and fracturing of rock mass in rock excavations. The characteristics of blasting agents, vibration monitoring in open-pit mines, quarries and underground openings are presented in Chapter 11. Furthermore, negative and positive effects of blasting are also presented and discussed in this chapter. It is also shown how to evaluate and use p-wave explorations to assess the average equivalent mechanical properties of rock masses.

The rockburst phenomenon is one of most dangerous forms of instability in rock engineering. First some available studies on this topic are presented and some effective monitoring and analysis methods for predicting rockburst are explained in Chapter 12. First, the fundamentals of various possible methods such as empirical techniques, analytical approaches and various finite element methods based on conventional elasto-plasticity, energy methods and extension strain method for predicting rockburst are briefly described. Then, some laboratory tests were carried out on the circular openings excavated in sandstone from Tarutoge Tunnel and Third Shizuoka Tunnel intercalated sandstone and shale samples and multi-parameter measurements were done in order to develop some observational and monitoring techniques for predicting rockburst. It is experimentally demonstrated that the combined utilization of monitoring AE, rock temperature, infrared imaging and electric potential may be a quite effective in-situ monitoring tool for predicting rockburst.

Rockbolts and rock anchors are commonly used as principal support members in underground and surface excavations. These support members may be subjected to earthquake loading, vibrations induced by turbines, vehicle traffic and long-term corrosion. Chapter 13 is concerned with rockbolts and rock anchors and some theoretical, numerical and experimental studies on rockbolts and rock anchors under shaking are presented in the first part of this chapter. In the remaining part, the fundamentals of non-destructive techniques for the evaluation of the soundness of rockbolts and rock-anchors are described and several practical applications of non-destructive technique utilizing impact waves are given and discussed.

Chapter 14 is concerned with impact phenomena observed in various fields such as collision of vehicles in transportation engineering, collision of adjacent structures during earthquakes, standard penetration tests in soil mechanics, and impact craters due to meteorites, anchoring of ships or platforms in marine engineering and bullets and missiles destroying targets. The state of art on the impacts of meteorites and their effects are presented. It is shown that the impacts by projectiles such as bullets and missiles and their effects are quite similar to those of the meteorites except their size. Drop tests, which are used in various fields, are explained and several laboratory tests and empirical and analytical formulations for practical applications are presented. In the last part, the impact induced tsunami issue is discussed and some experimental and analytical formulations are presented in relation to water level variations in closed water bodies such as lakes and reservoirs.

The main purpose of the author is to present a treatise on Rock Dynamics. It is hoped that this publication will be a mile-stone in advancing the knowledge in this field and leading to the techniques for experiments, analytical and numerical modelling as well as monitoring in dynamics of rocks and rock engineering structures.

Chapter 2

Fundamental equations, constitutive laws and numerical methods

2.1 FUNDAMENTAL EQUATIONS

Momentum conservation law for rock mass can be given as (e.g. Eringen, 1980) (Figure 2.1):

$$\rho\frac{\partial \mathbf{v}}{\partial t} = -\nabla \cdot \boldsymbol{\sigma} + \mathbf{b} \tag{2.1}$$

where $\rho, \mathbf{v}, \boldsymbol{\sigma}$ and $\mathbf{b}$ are density, velocity, stress tensor and body force, respectively. This governing equation is valid whether rock mass is treated as continuum or discontinuum.

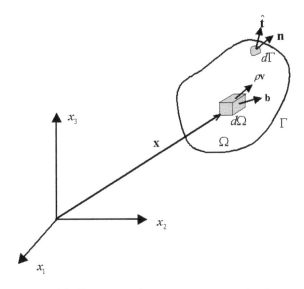

Figure 2.1 Illustration of momentum conservation law.

2.2 CONSTITUTIVE LAWS FOR ROCKS

2.2.1 Linear constitutive laws

When rock or rock mass behaves linearly without any rate dependency, the simplest constitutive law is Hooke's law. This law is written in the following form:

$$\sigma_{ij} = D_{ijkl}\varepsilon_{kl} \tag{2.2}$$

where $\sigma_{ij}, \varepsilon_{kl}$ and D_{ijkl} are stress, strain and elasticity tensors, respectively.

If material is homogenous and isotropic, Eq. (2.2) may be written as

$$\sigma_{ij} = 2\mu\varepsilon_{ij} + \lambda\delta_{ij}\varepsilon_{kk} \tag{2.3}$$

where δ_{ij} is Kronecker delta tensor. λ and μ are Lame coefficients, which are given in terms of elasticity (Young's) modulus (E) and Poisson's ratio (v) as

$$\lambda = \frac{Ev}{(1+v)(1-2v)}; \quad \mu = \frac{E}{2(1+v)} \tag{2.4}$$

When rock or rock mass behaves linearly with rate dependency, which may be called visco-elasticity, one of the simplest constitutive laws is Voigt-Kelvin Law, which may be given in the following form:

$$\sigma_{ij} = D_{ijkl}\varepsilon_{kl} + C_{ijkl}\dot{\varepsilon}_{kl} \tag{2.5}$$

where $\dot{\varepsilon}_{kl}$ and C_{ijkl} are strain rate and viscosity tensors, respectively.

If material is homogenous and isotropic, Eq. (2.5) may be written in analogy to Eq. (2.3) as

$$\sigma_{ij} = 2\mu\varepsilon_{ij} + \lambda\delta_{ij}\varepsilon_{kk} + 2\mu^*\dot{\varepsilon}_{ij} + \lambda^*\delta_{ij}\dot{\varepsilon}_{kk} \tag{2.6}$$

Coefficients λ^* and μ^* may be called viscous Lame coefficients. There are different visco-elasticity models in literature. Since it is difficult to cover all models here, the reader is advised to consult the available literature on the topic of visco-elasticity (e.g. Aydan et al., 2016). Figure 2.2 illustrates one-dimensional versions of linear constitutive laws.

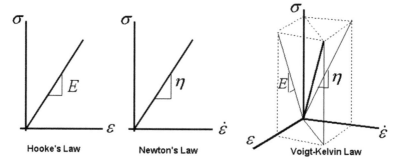

Figure 2.2 Illustrations of linear constitutive laws.

2.2.2 Non-linear behaviour (elasto-plasticity and elasto-visco-plasticity)

Every material in nature starts to yield after a certain stress or strain threshold and rocks or rock masses are no exception. The terms used to describe the material behaviour such as elasticity and visco-elasticity are replaced by the terms of elasto-plasticity or elasto-visco-plasticity as soon as material behaviour deviates from linearity (Figure 2.3). The relation between total stress and strain or strain rate tensors can no longer be used and every relation must be written in incremental form.

The specific derivation of constitutive laws such as elasto-plastic behaviour requires the following (e.g. Aydan and Nawrocki, 1998):

- Existence of a yield function (i.e. Mohr-Coulomb, Drucker-Prager etc.) (Figure 2.4),
- Flow rule (existence of a plastic potential function),
- Prager's consistency condition,

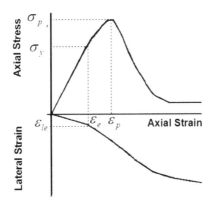

Figure 2.3 Illustration of yielding and failure concepts.

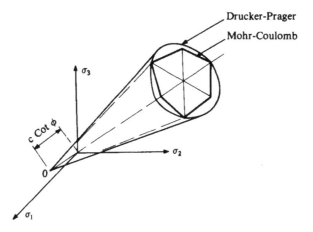

Figure 2.4 Illustration of yield criteria in principal stress space (from Owen and Hinton, 1980).

- Linear decomposition of incremental strain tensor into elastic and plastic components, and
- Existence of Hooke's law between incremental stress and elastic strains.

The incremental elasto-plasticity law may be derived as follows (Owen and Hinton, 1980):

$$\Delta\sigma_{ij} = D^{ep}_{ijkl}\Delta\varepsilon_{kl} \tag{2.7}$$

where

$$D^{ep}_{ijkl} = \left(D_{ijkl} - \dfrac{D_{ijmn}\dfrac{\partial F}{\partial\sigma_{mn}}\dfrac{\partial G}{\partial\sigma_{pr}}D_{prkl}}{h + \dfrac{\partial F}{\partial\sigma_{mn}}D_{mnpr}\dfrac{\partial G}{\partial\sigma_{pr}}} \right) \tag{2.8}$$

$\Delta\sigma_{ij}, \Delta\varepsilon_{kl}, D_{ijkl}, F, G$ and h are incremental stress and strain tensors, elasticity tensor, yield function, plastic potential function and hardening parameter, respectively. When $F = G$, it is called associated flow rule. It is common to introduce the effective stress (σ_e) and effective strain (ε_e) concepts as given below

$$\sigma_e = \sqrt{\dfrac{3}{2}\mathbf{s}\cdot\mathbf{s}} \quad\text{and}\quad \varepsilon_e = \sqrt{\mathbf{e}_p\cdot\mathbf{e}_p} \tag{2.9}$$

where $\mathbf{s}$ and $\mathbf{e}_p$ are deviatoric stress and deviatoric strain tensors respectively, and they are given as follows

$$\mathbf{s} = \mathbf{\sigma} - \dfrac{tr(\mathbf{\sigma})}{3}\mathbf{I}; \quad \mathbf{e}_p = \mathbf{\varepsilon}_p - \dfrac{tr(\mathbf{\varepsilon}_p)}{3}\mathbf{I} \tag{2.10}$$

If $tr(\mathbf{\varepsilon}_p) = 0$, the effective stress and strain would corresponds to those under uniaxial state, that is,

$$\sigma_e = \sigma_1 \quad\text{and}\quad \varepsilon_e = \varepsilon_1 \tag{2.11}$$

This is a very convenient conclusion that the non-linear response can be evaluated under uniaxial state and can be easily extended to multi-dimensional state without any triaxial testing. However, it should be noted that this is only valid when the volumetric component is negligible in the overall mechanical behaviour.

Yielding in theory of plasticity is generally associated with the deviation of lateral strain from linear response and it is a function of plastic strain during hardening, softening and flow stages. On the other hand, failure is defined as the ultimate stress state under a given confining stress and it is a special case of yielding. Nevertheless, yielding and failure functions are used in Rock Mechanics without any discrimination. It is often the case that the functional forms of yield and failure criteria are assumed to have the same form.

Perznya-type elastic-visco-plastic laws are used for representing non-linear rate dependency involving plasticity. In these approaches, the plastic strain is generally assumed to be independent of the volumetric response. Therefore, parameters of such a constitutive model are determined from uniaxial tests.

The constitutive law developed by Aydan and Nawrocki (1998) for elasto-visco-plastic behaviour of rocks is given in the following form:

$$d\sigma = \mathbf{D}^{rp} d\varepsilon + \mathbf{C}^{rp} d\dot{\varepsilon} \tag{2.12}$$

where

$$\mathbf{D}^{rp} = \mathbf{D}^r - \frac{\mathbf{D}^r \dfrac{\partial G}{\partial \sigma} \otimes \dfrac{\partial F}{\partial \sigma} \mathbf{D}^r}{h_{rp} + \dfrac{\partial F}{\partial \sigma} \cdot \left(\mathbf{D}^r \dfrac{\partial G}{\partial \sigma} \right) + \dfrac{\partial F}{\partial \sigma} \cdot \left(\mathbf{C}^r \dfrac{\partial \dot{G}}{\partial \sigma} \right)}$$

$$\mathbf{C}^{rp} = \mathbf{C}^r - \frac{\mathbf{C}^r \dfrac{\partial \dot{G}}{\partial \sigma} \otimes \dfrac{\partial F}{\partial \sigma} \mathbf{C}^r}{h_{rp} + \dfrac{\partial F}{\partial \sigma} \cdot \left(\mathbf{D}^r \dfrac{\partial G}{\partial \sigma} \right) + \dfrac{\partial F}{\partial \sigma} \cdot \left(\mathbf{C}^r \dfrac{\partial \dot{G}}{\partial \sigma} \right)} \tag{2.13}$$

Flow rule is given as

$$d\dot{\varepsilon}^p = \lambda \frac{\partial \dot{G}}{\partial \sigma} \tag{2.14}$$

Flow rule implies that the plastic potential function shrinks in time domain and strain increments consists of time-dependent and time independent parts.

There is no yield (failure) criterion directly incorporating the effect of creep experiments (Aydan and Nawrocki, 1998; Aydan et al., 2012a), although the basic concept has been presented previously (e.g. Ladanyi, 1974) for the time-dependent response of tunnels. Aydan and Nawrocki (1998) discussed how to incorporate the results of creep experiments in yield functions on the basis of results of rare creep triaxial experiments. On the basis of experimental results on various rocks by several researchers (e.g. Ishizuka et al., 1993; Kawakita et al., 1981; Masuda et al., 1987; Aydan et al., 1995), Aydan and Nawrocki (1998) concluded that time-dependency of friction angle is quite negligible and the time-dependency of the cohesive component of yield criteria should be sufficient for incorporation of results of creep experiments. The creep experiments would generally yield the decrease of deviatoric strength in time in view of experimental results, it would correspond to the shrinkage of the yield surface in time as illustrated in Figure 2.6.

Non-linear behaviour requires the existence of yield functions. These yield functions are also called failure functions at the ultimate state when rocks rupture. There are many available yield criteria for rocks. The well-known yield (failure) criteria in geomechanics are Mohr-Coulomb (e.g. Jaeger and Cook, 1979), Drucker-Prager (1952), Mogi (1967), Hoek-Brown (1980, 1997), and Aydan (1995). For two-dimensional cases, it is common to use the Mohr-Coulomb yield criterion given by:

$$\tau = c + \sigma_n \tan\phi \quad \text{or} \quad \sigma_1 = \sigma_c + q\sigma_3 \tag{2.15}$$

where c, ϕ and σ_c are cohesion, friction angle and uniaxial compressive strength. σ_t, σ_c and q are related to cohesion and friction angle in the following form

$$\sigma_t = \frac{2c \cos \phi}{1 + \sin \phi}, \quad \sigma_c = \frac{2c \cos \phi}{1 - \sin \phi}, \quad q = \frac{1 + \sin \phi}{1 - \sin \phi}, \quad \phi = \sin^{-1}\left(\frac{\sigma_c - \sigma_t}{\sigma_c + \sigma_t}\right) \tag{2.16}$$

As the intermediate principal stress is indeterminate in Mohr-Coulomb criterion and there is a corner-effect problem during the determination of incremental elasto-plasticity tensor, the use of Drucker-Prager criterion is common in numerical analyses, which is given by

$$\alpha I_1 + \sqrt{J_2} = k \tag{2.17}$$

where

$$I_1 = \sigma_1 + \sigma_2 + \sigma_3;$$

$$J_2 = \frac{1}{6}\left[(\sigma_1 - \sigma_2)^2 + (\sigma_2 - \sigma_3)^2 + (\sigma_3 - \sigma_1)^2\right] \tag{2.18}$$

and J_2 is the second invariant of deviatoric stress. Nevertheless, it is possible to relate the Drucker-Prager yield criterion with the Mohr-Coulomb yield criterion. On π-plane, if the inner corners of the Mohr-Coulomb yield surface are assumed to coincide with the Drucker-Prager yield criterion, the following relations may be derived:

$$\alpha = \frac{2 \sin \phi}{\sqrt{3}(3 + \sin \phi)}; \quad k = \frac{6c \cos \phi}{\sqrt{3}(3 + \sin \phi)} \tag{2.19}$$

Figure 2.4 compares yield criteria of Mohr-Coulomb and Drucker-Prager in principal stress space.

In Rock Mechanics Rock Engineering, one of the recent yield criteria is Hoek-Brown's criterion (1980), which is written as

$$\sigma_1 = \sigma_3 + \sqrt{m \sigma_{ci} \sigma_3 + s \sigma_{ci}^2} \tag{2.20}$$

where, m and s are some coefficients. While the value of s is 1 for intact rock, the values of m and s change when they are used for rock masses. For the tensile yielding condition, that is, $\sigma_1 = 0$ and $\sigma_3 = -\sigma_{ti}$, coefficient m can be obtained in terms of uniaxial compressive strength (σ_{ci}) and direct tensile strength (σ_{ti}) of intact rock as follows

$$m = \frac{\sigma_{ci}^2 - \sigma_{ti}^2}{\sigma_{ci} \sigma_{ti}} \tag{2.21}$$

Additionally, coefficient m can also be obtained in terms of uniaxial compressive strength (σ_c) and Brazilian tensile strength (σ_{tB}) of intact rock, provided that its stress state is the one derived by Jaeger and Cook (1979), as follows

$$m = \frac{\sigma_{ci}}{\sigma_{tB}} - \frac{16\sigma_{tB}}{\sigma_{ci}} \tag{2.22}$$

Therefore, one must be able to determine parameters of Hoek-Brown criterion for intact rock from uniaxial compression and tensile strength experiments without any triaxial compression testing. Furthermore, this criterion cannot be applied to non-cohesive materials such as granular media, broken rock without tensile strength and discontinuities, as it is clearly seen for the values of parameters such as

$$\sigma_t = 0 \quad \text{and} \quad \sigma_c = 0 \quad \text{or} \quad c = 0 \tag{2.23}$$

It should be also noted that the direct tensile strength should be used if it is available. As such tests are difficult to perform, tensile strength determined from Brazilian tensile tests could be used instead. Although the value of m would be slightly lower than that to be determined for direct tensile strength data. For the known range of ratio of uniaxial compression strength to tensile strength of rocks, the error should not be more than 16%.

Aydan (1995) proposed a yield function for thermo-plastic yielding of rock as given by (Figure 2.5)

$$\sigma_1 = \sigma_3 + \left[\sigma_\infty - (\sigma_\infty - \sigma_c)e^{-b_1\sigma_3} \right]e^{-b_2 T} \tag{2.24}$$

where σ_∞ is the ultimate deviatoric strength and T is temperature while coefficients b_1, b_2 are empirical constants. This is the only yield criterion, which incorporates the effect of temperature besides the confining pressure in Geomechanics. Figure 2.5 shows Aydan's yield (failure) criterion applied to experimental results reported by Hirth and Tulis (1994) as an example.

Under the isothermal condition, the last term of Eq. (2.24) depending upon temperature can be neglected so that it takes the following form

$$\sigma_1 = \sigma_3 + \left[\sigma_\infty - (\sigma_\infty - \sigma_c)e^{-b_1\sigma_3} \right] \tag{2.25}$$

Aydan et al. (2012a) has shown that if all yield (failure) criteria are required representing experimental data from tensile regime to triaxial compressive regime, estimations from Hoek-Brown criterion can be quite different from experimental results and the Hoek-Brown criterion generally fails to represent the whole range of experimental results. Furthermore, parameter m of Hoek-Brown Criterion is purely a function of the uniaxial compressive and tensile strength contrary to the common belief. Therefore, Aydan et al. (2012a) concluded that the applicability of commonly used Hoek-Brown criterion in geo-engineering is quite questionable as the best yield or failure criterion in view of comparisons with experimental results on all rock types.

There is no yield (failure) criterion directly incorporating the effect of creep experiments (Aydan and Nawrocki, 1998; Aydan et al., 2012a), although the basic concept

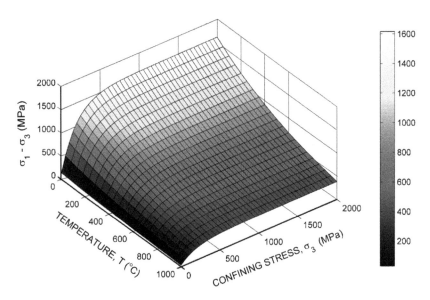

Figure 2.5 A three dimensional representation of Aydan's failure criterion for experimental results of Hirth and Tullis (1994) (from Aydan *et al.*, 2012a).

has been presented previously (e.g. Ladanyi, 1974) for the time-dependent response of tunnels. Aydan and Nawrocki (1998) discussed how to incorporate the results of creep experiments in yield functions on the basis of results of rare creep triaxial experiments. As mentioned in previous sub-section, Aydan and Nawrocki (1998) concluded that time-dependency of friction angle is quite negligible and the time-dependency of the cohesive component of yield criteria should be sufficient for incorporation of results of creep experiments. The creep experiments would generally yield the decrease of deviatoric strength in time in view of experimental results, it would correspond to the shrinkage of the yield surface in time as illustrated in Figure 2.6.

Based on the concept above, the time dependent uniaxial compressive strength ($\sigma_c(t)$) of various rocks is represented in terms of their uniaxial compressive strength (σ_{cs}) and the duration (τ) of a short-term experiment by the following function (e.g. Aydan and Nawrocki, 1998; Aydan and Ulusay, 2013):

$$\frac{\sigma_c(t)}{\sigma_{cs}} = 1.0 - b \ln\left(\frac{t}{\tau}\right) \tag{2.26}$$

where b is an empirical coefficient to be determined from experimental results (e.g. Aydan *et al.*, 2011).

There are almost no proposals to evaluate the triaxial dynamic strength of rocks. Based on various experimental studies summarized in the work of Aydan and Nawrocki

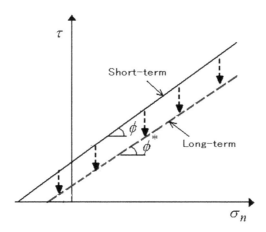

Figure 2.6 An illustration of long-term variation of Mohr-Coulomb yield criterion.

(1998), Mohr-Coulomb relations were modified for strain rate-dependent behaviour of rocks as (Aydan *et al.*, 1998; Ito *et al.*, 1998)

$$\sigma_1 = q(\dot{\varepsilon})\sigma_3 + \sigma_c(\dot{\varepsilon}) \tag{2.27}$$

where σ_1: Maximum principal stress; σ_3: Minimum principal stress; $q(\dot{\varepsilon}) = (1 + \sin\phi)/(1 - \sin\phi)$; $\phi(\dot{\varepsilon})$: Friction angle; $\sigma_c(\dot{\varepsilon})$: strain-rate dependent uniaxial compressive strength. Fortunately, the friction angle of rocks remains almost independent of strain rate (Aydan and Nawrocki, 1998; Matsuda *et al.*, 1987).

Strain-rate uniaxial compressive strength for various rocks and a large range of strain rate is generally represented by the following relation (Aydan, 1997; Aydan and Nawrocki, 1998, Ito *et al.*, 1999):

$$\sigma_c(\dot{\varepsilon}) = \sigma_{co} + b\ln\dot{\varepsilon} \tag{2.28}$$

where σ_{co} is the rate-independent component of uniaxial compressive strength, and b is an empirical constant. The above equation can also be easily adopted when it is expressed in the space of shear and normal stresses as

$$\tau = c_o + b\ln\dot{\gamma} + \sigma_n\tan\phi \tag{2.29}$$

Figure 2.7 shows an example of applications of Eqs. (2.27) and (2.28) to Cappadocia tuffs.

2.3 CONSTITUTIVE MODELING OF DISCONTINUITIES

There are two fundamental theoretical models, namely Hertz's model and Mindlin's model, for contact type modeling of discontinuities (Aydan *et al.*, 1996b). However,

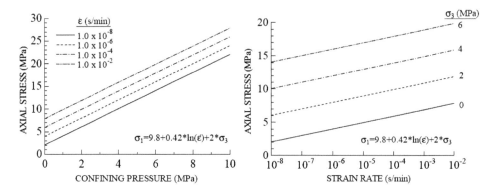

Figure 2.7 Strain-rate dependent of triaxial strength of Cappadocia tuffs.

these models are restricted to a very simple geometry and the elastic behaviour of adjacent materials. Since the configuration of discontinuities (contacts) and the mechanical behaviour of adjacent materials are generally complex, the experimental technique is probably the only way to deal with contact problems. In this respect, the direct shear test technique is widely used to characterize the behaviour of contacts, interfaces and rock discontinuities. There have been mainly three kinds of modeling to interpret and to utilize the responses measured in direct shear tests (Fig. 2.8).

Force-displacement-type modeling: In this case, discontinuities are assumed to have a zero thickness without an explicit definition of contact area A. The responses measured in direct shear tests are directly used in numerical representations.

Stress-displacement-type modeling: This type of modeling is probably the most widely used approach in numerical analyses. The contacts are again assumed to have a zero thickness and the responses measured in direct shear tests are directly used in numerical representations.

Stress-strain type-modeling: Contacts, rock discontinuities and interfaces are considered as bands with a finite thickness. The thickness of the bands is related to the thickness of shear-bands observed in tests or in nature, or the height of asperities along the plane.

A great number of experiments and theoretical studies can be found on the shear strength of rock discontinuities (e.g. Hoek and Brown, 1980; Aydan *et al.*, 1996b). However, we see many yield functions for discontinuities including the effect of asperities explicitly or implicitly (Fig. 2.9). The recent studies on thermo-mechanical behaviour of rock discontinuities indicate that the yield functions of rock discontinuities also depend upon temperature.

As pointed by Aydan and his colleagues (Aydan and Kawamoto, 1990; Aydan and Shimizu, 1995; Aydan *et al.*, 1996a) previously, the surface morphology of rock discontinuities has regularly distributed ridges and troughs with eigen directions (Figure 2.10). Some parameters are available for characterizing the surface morphology of discontinuities. However, they only handle isotropic situations. Aydan and Shimizu (1995) and Aydan *et al.* (1996a) proposed some procedures how to evaluate the anisotropy

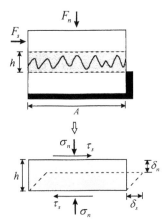

(a) Actual and mechanical models

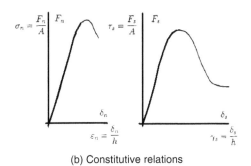

(b) Constitutive relations

Figure 2.8 Modeling of discontinuities (from Aydan et al., 1996a).

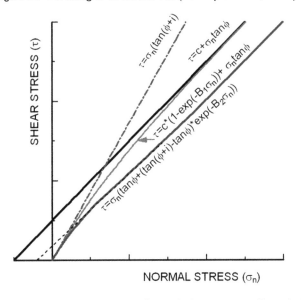

Figure 2.9 Comparison of some yield functions for rock discontinuities (from Aydan et al., 1996a).

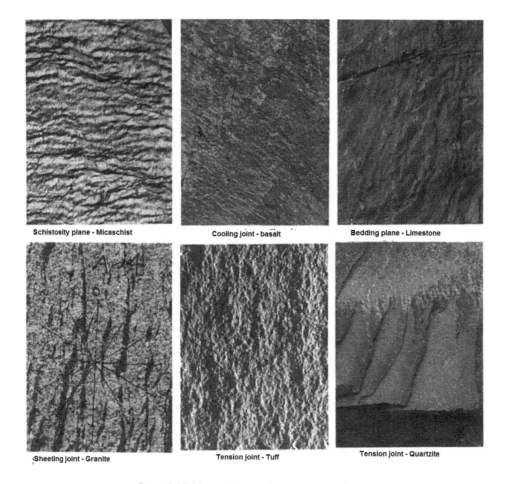

Schistosity plane - Micaschist   Cooling joint - basalt   Bedding plane - Limestone

Sheeting joint - Granite   Tension joint - Tuff   Tension joint - Quartzite

Figure 2.10 Views of some discontinuity surfaces.

of surface morphology parameters, and it is expressed mathematically for any surface morphology $F(\theta)$ parameters as

$$
\begin{aligned}
F(\theta) = & \sum_{i=1}^{n} a_i \cos^i \theta + \sum_{i=1}^{n} b_i \sin^i \theta \\
& + \sum_{i=1}^{n} c_i \cos^i 2\theta + \sum_{i=1}^{n} d_i \sin^i 2\theta + \cdots \\
& + \sum_{i=1}^{n} y_i \cos^i N\theta + \sum_{i=1}^{n} z_i \sin^i N\theta
\end{aligned}
\tag{2.30}
$$

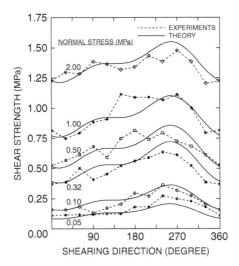

Figure 2.11 Comparison of experimental results with anisotropic strength criterion of Aydan *et al.*
(1996a).

Thus, parameters of thermo-hydro-mechanical constitutive laws of rock discontinuities must take into account the anisotropic nature of surface morphology of rock
discontinuities. For anisotropic shear strength of rock discontinuities, Aydan *et al.*
(1996a) proposed the following yield criterion:

$$\tau = A_1(1 - e^{-B_1\sigma_n}) + \sigma_n\left(\tan\phi_i + A_2 e^{-B_1\sigma_n}\right) \tag{2.31}$$

where $\tau, \sigma_n, A_1, A_2, B_1$ and B_2 are shear stress, normal stress and empirical coefficients
respectively. If the discontinuities are relatively smooth, the cohesive component can
be negligible and the Eq. (2.31) is reduced to the following form:

$$\tau = \sigma_n\left(\tan\phi_i + A_2 e^{-B_1\sigma_n}\right) \tag{2.32}$$

Coefficients A_1, A_2, B_1 and B_2 should be determined from directional surface
morphology measurements and shear strength experiments. Some applications of
Eqs. (2.31) and (2.32) to actual experimental results can be found in Aydan *et al.*,
1996a and one example is shown in Figure 2.11.

2.4 CHARACTERIZATION OF AND CONSTITUTIVE MODELING OF ROCK MASS

There are a number of approaches to characterize rock masses as described by ISRM
(2007). For this purpose there are several rock classifications (e.g. Terzaghi, 1946;
Stini, 1950; Lauffer, 1958) available to practitioners. However, these classifications
are not quantitative and Bieniawski (1974, 1989) and Barton *et al.* (1974) proposed
quantitative rock classifications, which have served the rock-engineering community

well. However, many available rock classification systems have some parameters double counting the effects of discontinuities. In addition, although the effect of water particularly on clay-bearing rocks plays an important role in decrease of their strength, this effect is not adequately considered in the existing rock mass classification systems.

Aydan *et al.* (2013a) proposed a new rock mass quality rating system called Rock Mass Quality Rating (RMQR). This rock mass quality rating system combines relevant parameters of rock masses with the consideration of all available rock mass classifications together with sound mechanical reasoning and provides a quantitative measure for the physical state of rock mass with respect to intact rock. The system consists of six different parameters to describe state of intact rock, discontinuities and groundwater condition and water absorption characteristics. The main parameters specifically are degradation degree of intact rock; discontinuity set number, discontinuity spacing, discontinuity condition, groundwater seepage condition and water absorption condition. The state of rock mass is categorised into six classes.

Rock masses are modeled either through continuum equivalent models or discontinuum models. When the discontinuum models are used, intact rock and discontinuities are modeled separately using discrete element method (DEM), displacement discontinuity analysis (DDA), joint, contact or interface elements in FEM (FEM-J, FEM-I, DFEM) (see Kawamoto and Aydan, 1999 for details). On the other hand, when rock mass is modeled as an equivalent continuum, these models are:

1 Empirical Models
2 Equivalent Elastic Compliance Model
3 Crack Tensor Model
4 Damage Model
5 Micro-structure models and
6 Homogenization Technique.

However, the recent tendency is to estimate the equivalent properties of rock masses using the empirical methods based on some rock mechanics classification parameters. These empirical models are only concerned with elastic modulus and some yield function parameters for feasibility studies and design purposes. It should be, however, noted that most available empirical relations are not experimentally validated.

Based on in-situ experiments on rock masses and information from actual excavations, Aydan *et al.* (2013a) proposed the following relation between the rock mass properties and RMQR with the use of those of intact rock (Figure 2.12):

$$\alpha = \alpha_0 - (\alpha_0 - \alpha_{100}) \frac{\text{RMQR}}{\text{RMQR} + \beta(100 - \text{RMQR})} \tag{2.33}$$

where α_0 and α_{100} are the values of the function at $\text{RMQR} = 0$ and $\text{RMQR} = 100$ of property α, β is a constant to be determined by using a minimization procedure for experimental values of given physical or mechanical properties. The authors proposed some values for these empirical constants with the consideration of in-situ experiments carried out in Japan as given in Table 2.1. When a representative value of RMQR is determined for a given site, geo-mechanical properties of rock mass can be obtained

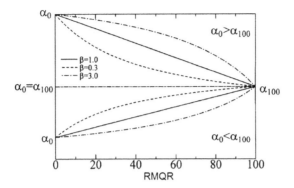

Figure 2.12 Estimation normalized properties of rock mass using Eq. (2.33).

Table 2.1 Values of α_0, α_{100} and β for various properties of rock mass.

Property (α)	α_0	α_{100}	β
Deformation modulus	0.0	1.0	6
Poisson's ratio	2.5	1.0	0.3
Uniaxial compressive strength	0.0	1.0	6
Tensile strength	0.0	1.0	6
Cohesion	0.0	1.0	6
Friction angle	0.3	1.0	1.0

using Eq. (2.33) together with the values of constants given in Table 2.1 and the values of intact rock for a desired property (Figures 2.13 and 2.14).

Jaeger (1960) developed a method to evaluate the effect of single plane of weakness on the anisotropic strength of rocks. Hoek and Brown (1980) extended this method to evaluate rock mass with several discontinuities by simply superimposing yielding with the consideration of orientation and strength properties of discontinuity sets and strength of intact rock in relation to principal stress state. This approach is now known as multiple yield function approach.

2.5 NUMERICAL METHODS

The final forms of discretized form of governing equation (2.1) irrespective of method of solution (FDM, FEM, BEM) and continuum or discontinuum, depending upon the character of governing equation, may be written in the following form:

$$[M]\{\ddot{\phi}\} + [C]\{\dot{\phi}\} + [K]\{\phi\} = \{F\} \qquad (2.34)$$

The specific forms of matrices $[M]$, $[C]$, $[K]$ and vector $\{F\}$ in the equation above will only differ depending upon the method of solution chosen and dimensions of physical space. Viscosity matrix $[C]$ is associated with rate-dependency of geomaterials.

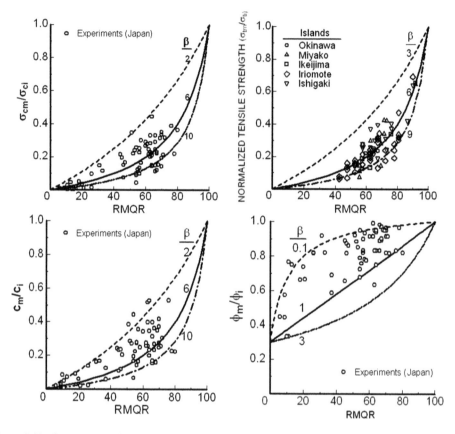

Figure 2.13 Comparison of experimental results with estimations from Eq. (2.33) (from Aydan *et al.*, 2013a).

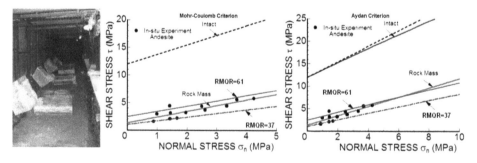

Figure 2.14 Comparison of in-situ rock shear experiments with Mohr-Coulomb and Aydan (1995) yield/failure criteria (from Aydan *et al.*, 2013a).

However, in many dynamic solution schemes, viscosity matrix [C] is expressed in the following form using Rayleigh damping approach as

$$[C] = \alpha[M] + \beta[K] \tag{2.35}$$

where α, β are called proportionality constants. This approach becomes very convenient in large-scale problems if central finite difference technique and mass lumping are used. However, it should be also noted that it is very difficult to determine these parameters from experiments. Again, in non-linear problems, the deformation moduli of rocks are reduced in relation to the straining using an approach commonly used in soil dynamics. In such approaches, the reduction of moduli is determined from cyclic tests. Nevertheless, it should be noted that the validity of such an approach for rock and rock masses is quite questionable.

If material behaviour involves non-linearity, the equation system above must be solved iteratively with the implementation of required conditions associated with the constitutive law chosen. The iteration techniques may be broadly classified as initial, secant or tangential stiffness method (e.g. Owen and Hinton, 1980).

The existence of discontinuities in rock mass have special importance for the stability of rock engineering structures, directional seepage, diffusion or heat transport and its treatment in any analysis requires special attention. Various types of finite element methods with joint or interface elements, discrete element method (DEM), displacement discontinuity analysis (DDA), discrete finite element method (DFEM) and displacement discontinuity method (DDM) are developed so far. Although these methods are mostly concerned with the solution of equation of motion, they can be used for seepage, heat transport or diffusion problems. The fundamental features of the available methods are described in the following by quoting a recent review on these methods by Kawamoto and Aydan (1999).

(a) No-tension finite element method

The no-tension finite element method was proposed by Valliappan in 1969 (Zienkiewicz et al., 1969). The essence of this method lies with the assumption of no tensile strength for rock mass since it contains discontinuities. In the finite element implementation, the tensile strength of media is assumed to be nil. It behaves elastically when all principal stresses are compressive. The excess stress is re-distributed to the elastically behaving media using a similar procedure adopted in the finite element method with the consideration of elastic-perfectly plastic behaviour.

(b) Pseudo discontinuum finite element method

This method was first proposed by Baudendistel et al. in 1970. In this method the effect of discontinuities in the finite element method is considered through the introduction of directional yield criterion in the elasto-plastic behaviour. Its effect on the deformation characteristics of the rock mass is not taken into account. If there is any yielding in a given element, the excess stress is computed and the iteration scheme for elastic-perfectly plastic behaviour is implemented. If there is more than one discontinuity set, the excess stress is computed for the discontinuity set which yields the largest value.

(c) Smeared crack element

Smeared crack element method within the finite element method was initially proposed by Rashid (1968) and adopted by Pietruscak and Mroz (1981) in media having weakness planes or developing fracture planes. This method evaluates equivalent stiffness matrix of element and allows the directional plastic yielding within the element. This approach is adopted by Tang (1997) in the solutions scheme called rock progressive failure analysis (RFPA) with the use of a fine finite element mesh.

(d) Discrete finite element method (DFEM)

Finite element techniques using contact, joint or interface elements, have been developed for representing discontinuities between blocks in rock masses. The simplest approach for representing joints is the contact element, which was originally developed for bond problems between steel bars and concrete. The contact element is a two-noded element having normal and shear stiffnesses. This model was recently used to model block systems by Aydan and Mamaghani (e.g. Mamaghani et al., 1994; Aydan et al., 1996b) by assigning a finite thickness to contact element and employing an up-dated Lagrangian scheme to deal with large block movements. The contact element can easily deal with sliding and separation movements.

(e) Finite element method with joint or interface element (FEM-J)

Goodman et al. (1968) proposed a four-noded joint element for joints. This model is simply a four-noded version of the contact element of Ngo and Scordelis (1967) and it has the following characteristics. In a two-dimensional domain, joints are assumed to be tabular with zero thickness. They have no resistance to the net tensile forces in the normal direction, but they have high resistance to compression. Joint elements may deform under normal pressure, especially if there are crushable asperities. The shear strength is presented by a bi-linear Mohr-Coulomb envelope. The joint elements are designed to be compatible with solid elements. Ghaboussi et al. (1978) proposed a four-noded interface element for joints. This model is a further improvement of the joint element by assigning a finite thickness to joints. Zienkiewicz and Pande (1977) modified the formulation of rectangular elements to model joints and introduced an elastic-visco-plastic-type constitutive law for joints. The thin layer element proposed by Desai et al. (1984) is also similar to that of Ghaboussi (1988). These models are widely used for rock engineering structures in fractured and jointed media.

(g) Displacement discontinuity method (DDM)

This technique is generally used together with the Boundary Element Method (BEM). The discontinuities are modelled as a finite length segment in an elastic medium with a relative displacement. In other words, the discontinuities are treated as internal boundaries with prescribed displacements. As an alternative approach to the technique of Crouch and Starfield (1984) Crotty and Wardle (1985) use interface elements to model discontinuities and the domain is discretized into several sub-domains.

(h) Discontinuous deformation analysis method (DDA)

Shi (1988) proposed a method called Discontinuous Deformation Analysis (DDA). Intact blocks were assumed to be deformable and are subjected to constant strain and stress due to the order of the interpolation functions used for the displacement field of the blocks. In the original model, the inertia term was neglected so that the damping becomes unnecessary. For dynamic problems, although damping is not introduced into the system, the large time steps used in the numerical integration in time-domain results in artificial damping. It should be noted that this type of damping is due to the integration technique for time domain and has nothing to do with the mechanical characteristics of rock masses (i.e. frictional properties). Although the fundamental concept is not very different from Cundall's model, the main difference results from the solution procedure adopted in both methods. In other words, the equation system of blocks and its contacts are assembled into a global equation system in Shi's approach. Recently Ohnishi *et al.* (1995) introduced an elasto-plastic constitutive law for intact blocks and gave an application of this method to rock engineering structures.

(i) Discrete element method (DEM)

Distinct or discrete element method (rigid block models) for jointed rocks was developed by Cundall in 1971. In Cundall's model problems are treated as dynamic ones from the very beginning of formulation. It is assumed that the contact force is produced by the action of springs which are applied whenever a corner penetrates an edge. Normal and shear stiffness were introduced between the respective forces and displacements in his original model. Furthermore, to account for slippage and separation of block contacts, he also introduced the law of plasticity. For the simplicity of calculation of contact forces due to the overlapping of the block, he assumed that the blocks do not change their original configurations. To solve the equations of the whole domain, he never assembled the equilibrium equations of blocks into a large equation system but solved them through a step-by-step procedure, which he called marching scheme. His solution technique has two main merits:

1   Storage memory of computers can be small (note that computer technology was not so advanced during the late 1960s), therefore, it could run on a microcomputer.
2   The separation and slippage of contacts can be easily taken into account since the global matrix representing block connectivity is never assembled. If a large assembled matrix is used, such a matrix will result in zero or very nearly zero diagonals, which subsequently cause singularity or ill conditioning of the matrix system.

As the governing equation is of hyperbolic type, the system could not become stabilized even for static cases unless a damping is introduced into the equation system. In recent years he improved the original model by considering the deformability of intact blocks and their elasto-plastic behaviour. Cundall's model has been actively used in rock engineering structures design by the NGI group in recent years (e.g. Barton *et al.*, 1986, 1987).

Chapter 3

Tests on dynamic responses of rocks and rock masses

3.1 DYNAMIC UNIAXIAL COMPRESSION, BRAZILIAN, TRIAXIAL (HOPKINSON BAR) TEST

The Hopkinson Pressure Bar Testing technique was proposed by Hopkinson in 1914 as a way to measure stress pulse propagation in a metal bar (Figure 3.1). Kolsky (1949) refined Hopkinson's technique by using two Hopkinson bars in series, now known as the Split-Hopkinson bar (SHB), to measure stress and strain, incorporating advancements in the cathode ray oscilloscopes in conjunction with electrical condenser units to record the pressure wave propagation in the pressure bars. This method is initially used to determine dynamic properties of metals initially, and ceramics, polymers, concrete and rocks later. There are several special setups of this technique for uniaxial compression, tensile, torsion, Brazilian and triaxial compression tests of rocks. The most difficult aspect of this technique is the determination of actual straining and stresses in samples of rock, which has to be inferred from strains of incident and transmitter bars. Furthermore, aspect ratio (H/D) of samples is less than that suggested by ISRM SMs (Ulusay and Hudson, 2007) and the control of strain rate beforehand is difficult.

Strains of incident bar and transmitter are measured to infer the strains and stresses of samples (Figures 3.2 and 3.3). Strains of the incident bar consist of incident strain pulse and reflected strain pulse while the strain of transmitted wave is termed transmitted strain pulse. Stresses acting at the interfaces of the sample with incident bar and transmitter bar are given as

$$\sigma_I(t) = \frac{EA}{A_s}(\varepsilon_I(t) + \varepsilon_R(t)) \qquad (3.1a)$$

$$\sigma_T(t) = \frac{EA}{A_s}(\varepsilon_T(t)) \qquad (3.1b)$$

where E, A and A_s are elastic modulus and area of incident bar and area of sample. Thus, the strain rate acting on sample is inferred from the following relation

$$\dot{\varepsilon} = \frac{V_p}{L_s}(\varepsilon_I(t) - \varepsilon_R(t) - \varepsilon_T(t)) \qquad (3.2)$$

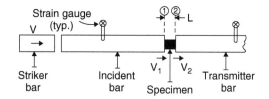

Figure 3.1 A simple illustration of Split-Hopkinson Bar Technique (SHBT).

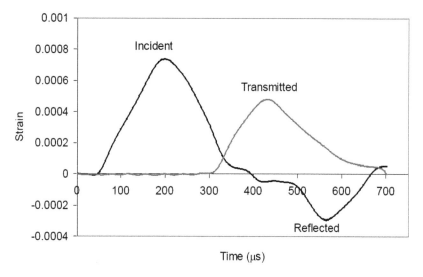

Figure 3.2 Strain pulses measured in a marble sample (from Yavuz et al., 2013).

Average strain and stress of sample in the direction of impact are given in the following form

$$\varepsilon(t) = \frac{V_p}{L_s} \int_0^t (\varepsilon_I(t) - \varepsilon_R(t) - \varepsilon_T(t)) dt \qquad (3.3a)$$

$$\sigma(t) = \frac{EA}{2A_s} (\varepsilon_I(t) + \varepsilon_R(t) + \varepsilon_T(t)) \qquad (3.3b)$$

As the following relation holds among incident, reflected and transmitted strains from the dynamic equilibrium equation

$$\varepsilon_I(t) + \varepsilon_R(t) = \varepsilon_T(t) \qquad (3.4)$$

The stress on sample is directly related to transmitted strain as follows

$$\sigma(t) = \frac{EA}{A_s} \varepsilon_T(t) \qquad (3.5)$$

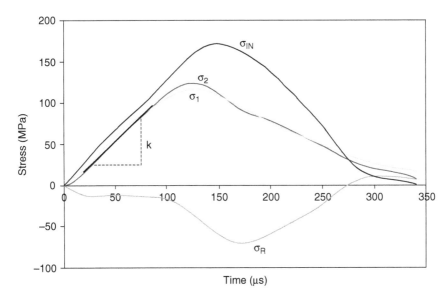

Figure 3.3 Stress pulses measured on a marble sample in a Split-Hopkinson Bar experiment (from Yavuz et al., 2013).

3.1.1 Dynamic uniaxial compression test

The uniaxial compression experiments are generally carried out using cylindrical samples. The height of the sample is generally less than its diameter. In other words, such an aspect ratio of height to diameter introduces shape effect into the experimental results. Figure 3.3 shows dynamic strain-stress relations of various rocks reported by Yavuz *et al.* (2013) with an aspect ratio of 0.6. As noted from the figure, strain-stress relations exhibit elastic-strain hardening response. Nevertheless, it should be noted such behaviour is also caused by the aspect ratio of samples.

Kobayashi (1970) reported many experimental results carried out under different strain rates as shown in Figure 3.5. At a lower strain rate of 10^{-6} to $10^{-1}\,\mathrm{sec}^{-1}$ the increase in the uniaxial compressive strength value was gradual. But at a higher strain rate of 10^{-1} to $10^{3}\,\mathrm{sec}^{-1}$ the uniaxial compressive strength of the rocks increases rapidly. These experimental results are also confirmed in recent studies (e.g. Yavuz *et al.*, 2013). Nevertheless, the inertia resistance resulting from the loading system as well as rock itself may be included may have some effects on the experimental results.

3.1.2 Dynamic tensile strength test (Brazilian, Notch, Slit)

Dynamic tensile strength of rocks is determined from Brazilian Test, Semi-circular Bend tests or Cracked-Chevron Notched Brazilian Test (Figure 3.6). Tensile strength property is determined from well-known relations determined for static cases while using the stresses determined from dynamic impact. Figure 3.7 shows two examples of tensile strength testing of rocks reported by Cadoni *et al.* (2011).

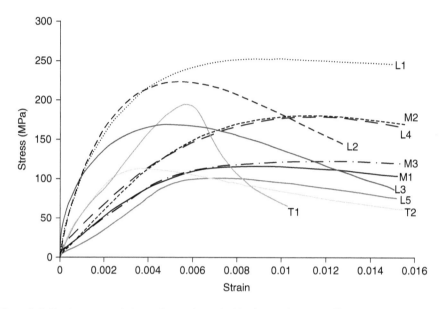

Figure 3.4 Strain-stress relations obtained in split-Hopkinson bar tests (from Yavuz *et al.*, 2013).

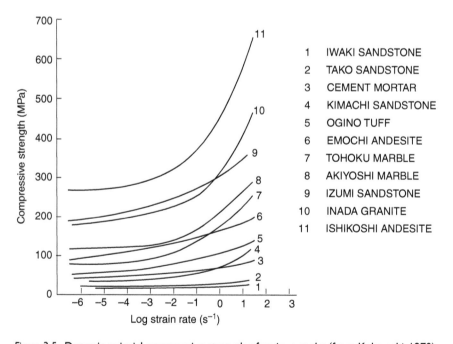

Figure 3.5 Dynamic uniaxial compressive strength of various rocks (from Kobayashi, 1970).

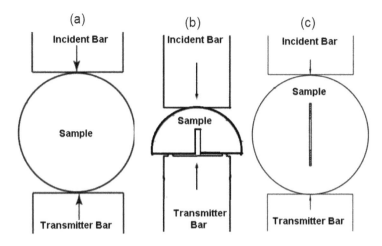

Figure 3.6 Configuration used in determination of tensile strength.

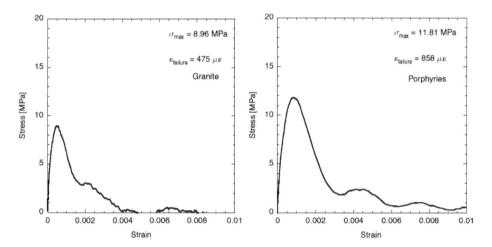

Figure 3.7 Dynamic tensile stress results (from Cadoni et al., 2011).

3.1.3 Dynamic triaxial compression test

There are some attempts to determine the dynamic strength properties under confining pressures. However, such experiments are difficult to perform and some complexities exist in the determination of stress and strains components during the experiments. Christensen (1971) and Christensen *et al.* (1972) reported dynamic triaxial compression experiments on Nugget sandstone under confining pressure up to 30 ksi. The strain rate was in the range of 10^2 to 10^3 per second. They concluded that the failure locus of nugget sandstone at the high strain rate was 15–20 percent higher than that obtained in quasi-static tests for all values of confining pressure, and the main features of the stress-strain response were similar to those seen at low rates (Figure 3.8). In particular, the phenomenon of dilatancy was not significantly affected by the change

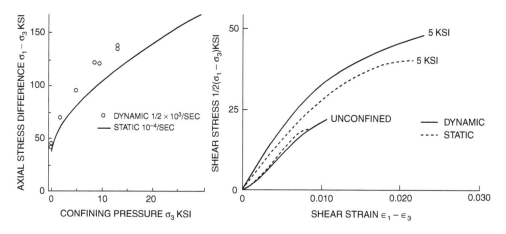

Figure 3.8 Dynamic triaxial test results (from Christensen *et al.*, 1972).

from 10^{-4}/sec to 0.5×10^3/sec axial-strain rate. Furthermore, there was no significant difference on the frictional characteristics under dynamic conditions.

3.1.4   Rate dependency of deformation and strength characteristics of rocks

On the basis of experimental results in the literature of Rock Mechanics and Rock Engineering, the deformation and strength (uniaxial compression, tensile strength, cohesion) characteristics are said to be rate-dependent. For example, the apparent elastic modulus (E^*) of rocks is assumed to be of the following form (i.e. Aydan and Nawrocki, 1968; Ito *et al.*, 1999):

$$E^* = E_o + b \ln(\dot{\varepsilon}) \tag{3.6}$$

where E_o, b and $\dot{\varepsilon}$ are rate-independent elastic modulus, empirical constant and strain rate, respectively.

Similarly, the strength (S^*) (uniaxial compression, tensile and cohesion) may also be given in the following form (e.g. Ito *et al.*, 1999; Aydan, 2016):

$$S^* = S_o + c \ln(\dot{\varepsilon}) \tag{3.7}$$

where S_o, c and $\dot{\varepsilon}$ are rate-independent strength, empirical constant and strain rate, respectively. Figure 3.9 shows data on elastic modulus and uniaxial compressive strength of rocks compiled by Aydan and Nawrocki (1998).

3.2   CYCLIC UNIAXIAL COMPRESSION, TRIAXIAL COMPRESSION AND SHEAR TESTS

The effect of cyclic loading on the elasticity and strengths of geologic materials has long been recognized (Haimson, 1974; Allemandou and Dusseault, 1996; Bagde and

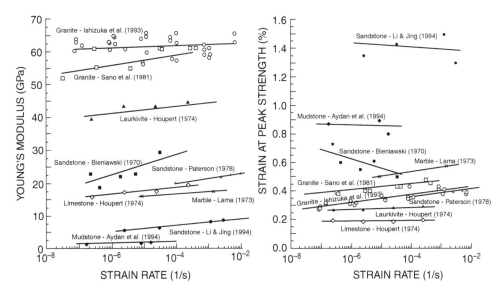

Figure 3.9 Effect of strain rate on dynamic elastic modulus and uniaxial compressive strength of various rocks (from Aydan and Nawrocki, 1998).

Petros, 2005). A common goal of the cyclic testing studies is to determine the fatigue strength of the materials. It has been found that loading cycles can reduce the material strength and elasticity, depending on the loading amplitude and the maximum applied load in each cycle.

Specimens are subjected to sinusoidal loads under a certain level of stress field. In uniaxial and triaxial compression experiments are subjected to fluctuating loads at a mean stress level proportional to the static strength for a given condition. As for cyclic shear testing, the normal stress is kept constant while shear load is applied in a sinusoidal way. The main purpose of this test is to determine the degradation of deformability of strength of rocks and discontinuities as a function of cycle number. It is also known as fatigue tests.

3.2.1   Cyclic uniaxial compression test

Uniaxial cyclic compression tests are generally performed on samples according to the conventional size of specimens in rock mechanics using a servo-controlled universal testing machine. The maximum axial stresses vary among specimens at a ratio of 40% to 100% of the compressive strength while a minimum stress is maintained constant at certain level for all specimens. This small minimum stress is required to ensure that the ends of the specimen remain in contact with the loading platens during the test. The loading frequencies range from A_1 Hz to A_2 Hz. The accumulated axial strain, fatigue stress (S) and time were monitored during loading. Some examples of axial stress-strain curves measured during loading are given in Figure 3.10. Figure 3.11 shows the decrease of the failure (fatigue) stress as the number of loading cycle (N) increases.

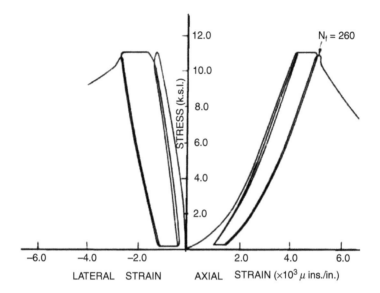

Figure 3.10 Cyclic uniaxial compression tests on Berea sandstone (from Haimson, 1974).

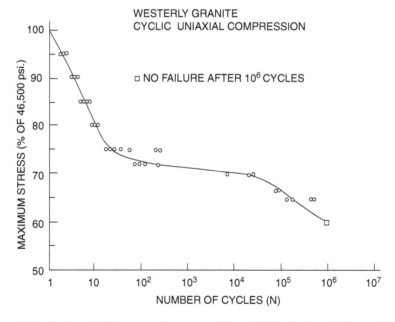

Figure 3.11 Cyclic uniaxial compression tests on Westerly Granite (from Haimson, 1974).

3.2.2 Cyclic tensile (Brazilian) test

Cyclic tensile tests also follow the same procedure. Haimson (1974) reported cyclic direct tension tests on Tennessee marble and the results is show in Figure 3.12. Haimson

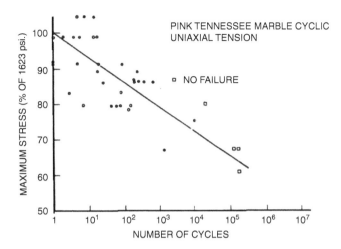

Figure 3.12 The decrease of the failure (fatigue) stress as a function of the number of loading cycle (N) for Tenessee marble in tension.

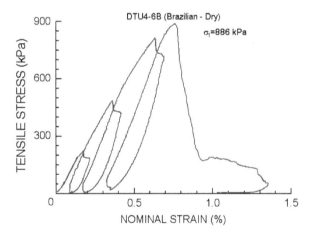

Figure 3.13 An example of cyclic Brazilian test on Derinkuyu tuff.

(1974) pointed out that the most outstanding feature of cyclic loading of rock in tension is the large hysteresis loop in the first cycle compared to that in compression tests.

As direct tensile strength tests are generally difficult to perform, it is quite common to determine tensile strengths from indirect tensile strength (Brazilian) tests. Figure 3.13 shows an example of Brazilian cyclic tests on Derinkuyu tuff.

3.2.3   Cyclic triaxial compression test

Cyclic triaxial compression tests fundamentally follow the same procedure except the application of certain confining pressure. However, it is more cumbersome to perform such tests. Such pioneering experiments were reported by Haimson (1974) and he utilized a triaxial cell for the application of confining pressure to the cyclically loaded

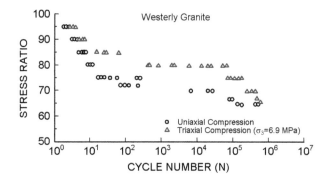

Figure 3.14 The decrease of the failure (fatigue) stress as the number of loading cycle (N) increases in cyclic uniaxial and triaxial compression tests (data from Haimson (1974)).

specimens. The confining pressure was kept approximately constant during the tests while the vertical load was cycled as in the uniaxial case. The specimens used were intact granite. Although the fatigue strength under triaxial compression is slightly higher than that under uniaxial compression, the overall tendency is similar.

3.2.4 Cyclic shearing tests

Cyclic shearing tests on soft rocks, rock discontinuities and interfaces are also important item when structures are constructed on/in rock masses having various kind geological discontinuities. In addition, interfaces may exist within the rock engineering structures (i.e. foundations of superstructures such as dams, nuclear power plants, large span bridges or rock anchors) and they may be critical for superstructures under cyclic loads results from the machinery vibration and/or earthquakes. Figure 3.15 shows one example of shear testing machine used in experiments by Aydan *et al.* (1993, 1994, 1995, 2016). Figures 3.16 to 3.19 show some examples of cyclic shear tests on discontinuities and interfaces and the effect of its cyclic shear strength.

Aydan *et al.* (2016) reported a direct cyclic shear test was carried out on a sample of Oya tuff. The responses measured in the cyclic shear test with a cycle period of 5s are shown in Figs. 3.19 and 3.20. It is interesting to note that the overall behaviour of the Oya tuff is quite similar to samples failing in creep experiments. The shear displacement started to increase drastically after about 130 cycles and failed at the cycle of 142. Furthermore, the displacement started to increase as the behaviour of the specimen entered to the softening stage.

3.2.5 Dynamic shearing tests

There are very few studies on dynamic direct shear tests of rocks and discontinuities in rock mechanics and rock engineering field (Aydan *et al.*, 2014, 2016; Yoshida *et al.*, 2014; Kana *et al.*, 1991). Direct dynamic shear testing is a relatively new field of experimental research and there are no well-established procedures yet. However, the shear loading procedures must be based on the strong motion records. Strong motions

Figure 3.15 A view of shear testing machine used in cyclic shear testing.

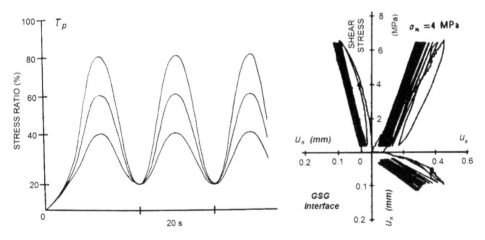

Figure 3.16 An example of cyclic shear loading and measured response (from Aydan et al., 1994, 1995).

can be in the form of acceleration, velocity or displacement (Aydan *et al.*, 2016). Figures 3.21 and 3.22 shows the accelerations together with integrated velocity and displacement responses for 1999 Kocaeli earthquake and Iwate-Miyagi Intra-plate earthquake using the EPS method (Aydan *et al.*, 2011; Aydan and Ohta, 2011; Ohta and Aydan, 2007).

If the shear loading on rock discontinuities due to actual earthquakes is to be considered, the displacement response must be used. As noted from Figures 3.21 and 3.22, the displacements are in the order of meters, no shear testing device can impose such displacement records on the samples. If the acceleration records as illustrated as Pattern 2 in Figure 3.23 are imposed, the asperities would be sheared at first peak and the shear responses of the samples would be reduced to residual state. It seems that the shearing loading based velocity records might be a potential form for shear loading. The shear loading without reversed shearing would be appropriate in view

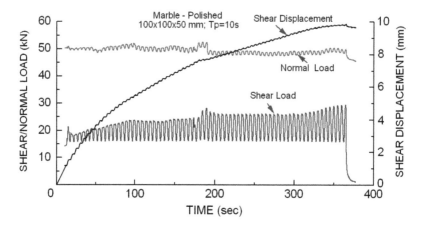

Figure 3.17 An example of cyclic shear loading and measured response of polished marble contact (from Aydan *et al.*, 2016).

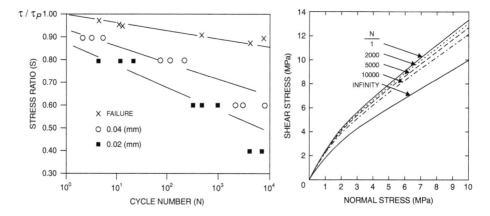

Figure 3.18 Effect of cycle number on shear strength properties of interfaces (from Aydan *et al.*, 1994, 1995).

of the machine capacity as well as the evaluation of linear and non-linear responses during shearing, which is denoted as Pattern 1.

Aydan *et al.* (2016) have recently upgraded the shear testing machine by adding the dynamic shear loading option and the device named as OA-DSTM (Figure 3.15). Furthermore, the operation system of the device was made independent of computer operation systems in order to eliminate many technical problems caused by upgrading the computer operation systems. The set-up of each loading condition can be easily input using the touch-panels. The user can select the testing procedure from the initial MENU panel. The user can see the current load, displacement elapsed time, cycle number and display load, displacement as a function of elapsed time.

The size of direct shear samples can be $100 \times 100 \times 100$ mm or $150 \times 75 \times 75$ mm. The original design size was $150 \times 75 \times 75$ mm with the purpose of eliminating the

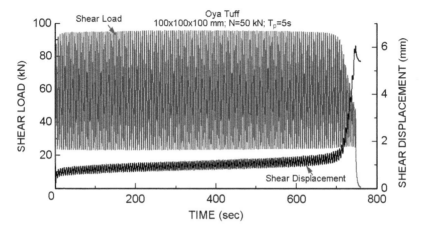

Figure 3.19 Responses of Oya tuff during the cyclic shear test.

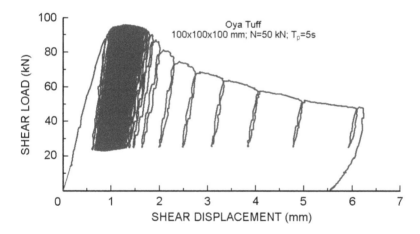

Figure 3.20 Shear displacement-shear load relation of Oya tuff during the cyclic shear test.

rotational effects on the sample. This can be achieved if the ratio of sample length over sample height is greater than 2.

The shear load and displacement, normal load are directly recorded to computers using the outputs from the system, and the outputs are real-time values of shear displacement in mm, and shear and normal loads in kN. The shear loading based on Pattern 1 is adopted by Aydan *et al.* (2016) as illustrated in Figure 3.24.

Yoshida *et al.* (2014) described a dynamic shear testing device based loading Pattern 2. This device can apply shear loads on samples through a normal load actuator and shear load actuator under stress control and displacement control. Test specimens are made from bored-cores and shear boxes have internal dimensions 150 mm × 150 mm square, to house a core of 100 mm in diameter. The lower part of the shear box is movable. The loads can be sinusoidal with a given frequency and constant amplitude or linearly increasing amplitude or loads based on chosen acceleration

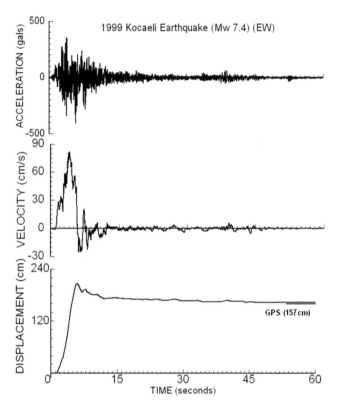

Figure 3.21 EW acceleration record at Sakarya station (DAD-ERD, 1999) by the 1999 Kocaeli earthquake and integrated velocity and displacement responses using the EPS method.

wave forms. Figure 3.25 shows a drawing of the machine and its picture. Yoshida *et al.* (2014) reported that the dynamic shear strength of mudstone joint is higher as compared with the static shear strength using a sinusoidal loading pattern with increasing amplitude. The experimental results indicated that the dynamic cohesion and friction angle were higher than the static ones (Figure 3.26).

Aydan *et al.* (2014) also reported some dynamic shear tests of a schistosity plane in quartzite using the loading Pattern 2 (Figure 3.27). The stroke of shaking table is 20 mm. The shearing motions are monitored using accelerometers and non-contact laser transducers. The specimen consists of lower and upper blocks. The lower block is attached to the movable shaking table while the upper block is attached to unmovable support as illustrated in Figure 3.24. The reason to attach the lower block to the shaking table was to prevent rotational movements during shearing.

Experimental material is blocks of metamorphic quartzite. Geometrical dimensions, mechanical and frictional properties of blocks are given in Tables 3.1–3.3. Schistosity planes are chosen as shearing planes. Schistosity planes also include some micaceous minerals such as muscovite.

The number of cyclic shear tests was 5 and the normal stress was varied. Table 3.4 gives the normal stress levels in a respective cyclic shear test. Although the level of

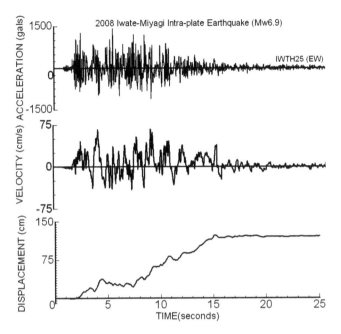

Figure 3.22 EW acceleration record at IWTH25 station (K-NET, 2008) by the 2008 Iwate-Miyagi earthquake and integrated velocity and displacement responses using the EPS method.

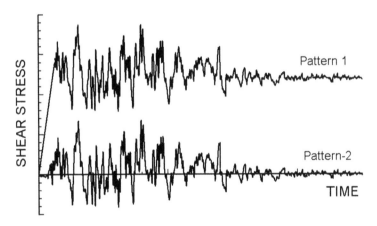

Figure 3.23 Possible shear loading patterns.

normal stress is lower compared with those used in conventional shearing test, these normal stress levels should be appropriate if distinct heating response is achieved. There is no doubt that the heating would be higher if normal stress level and/or shearing velocity becomes higher. Although the normal stress level is comparatively small, the maximum shearing velocity was up to 80 mm/s in all experiments. Furthermore, the duration of shearing is another factor influencing the thermal responses.

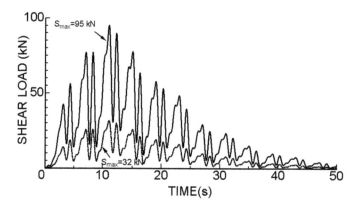

Figure 3.24 Adopted direct shear loading patterns in the dynamic shear testing machine (OA-DSTM).

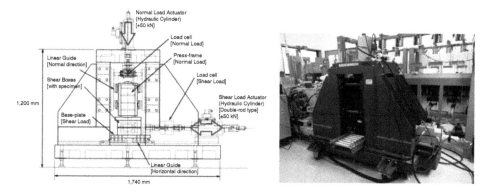

Figure 3.25 Dynamic direct shear testing machine developed by Yoshida et al. (2014).

The responses and results of experiments numbered CST-5 under a normal stress of 11.4 kPa are shown in Figure 3.28. The maximum amplitude of the acceleration was slightly lower than that in other experiments and its maximum amplitude was 1186 gals. The maximum surface temperature rise was 3.6 Celsius with a fluctuation range of ±0.3 Celsius. The initially selected point moves in space during shaking while the selected point in the thermographic image remains the same. The increase in the temperature rise in this experiment compared to that in the previous experiment is directly associated with the increase in normal stress level.

Regarding the temperature rise, it is again worth noticing that temperature rise is much steeper while the amplitude of acceleration is increasing. On the other hand, if the shearing velocity is constant, the heat release or temperature increases at a constant rate. It is also noted that the surface temperature distribution of the specimen was not uniform and the highest temperature rises apparently occurred at contact areas (namely asperities) over the discontinuity plane.

After the experiments, it was noted that a thin powder layer accumulated on the surface of discontinuity surfaces as seen in Figure 3.29. In other words, asperities were

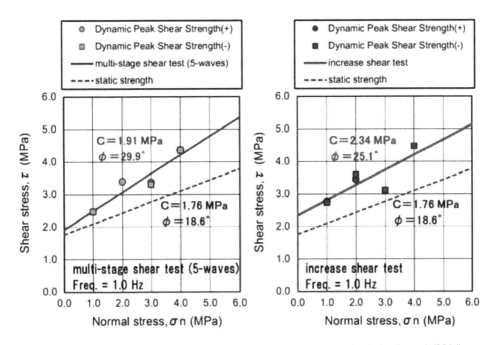

Figure 3.26 Dynamic shear testing result on mudstone joint tested by Yoshiada et al. (2014).

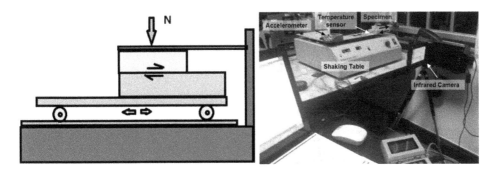

Figure 3.27 A view of a typical experimental set-up and its drawing.

partially worn out and the damage to asperities of the lower mobile block was higher than that of those at the stationary upper block.

The responses of temperature rises for all experiments are plotted in Figure 3.30. As noted from the figure, the temperature rise is highest for the experiment denoted Cyclic-Shear-test-5 while it is lowest for the experiment denoted Cyclic-Shear-test-1. The temperature rise is higher during the increase of shearing rate and temperature increase become linear as the shearing rate becomes constant. The temperature rise is also related to the duration of shearing. The temperature becomes higher as the duration of shearing increases. The responses observed throughout experiments can

Table 3.1 Geometry and weight of blocks.

Block	Length (mm)	Width (mm)	Height (mm)	Weight (gf)
UB	73.1	49.7	14.8	134
LB	90.4	60.8	14.2	197

Table 3.2 Material properties.

Unit Weight (kN/m³)	UCS (MPa)	Elastic Modulus (GPa)	P-wave velocity (km/s)	S-wave velocity (km/s)
24.9–25.2	194	196	4.86	2.47

Table 3.3 Friction properties.

State	Fresh sample	Worn sample
Friction angle	23–26	21–23

Table 3.4 Test names and normal stresses.

Test Name	Normal stress (kPa)
CST-1	0.37
CST-2	2.05
CST-3	3.73
CST-4	5.87
CST-5	11.37

be explained through the consideration of the energy conservation law of the continuum mechanics (e.g. Eringen, 1980; Aydan, 2000, 2009, 2010; Aydan *et al.*, 2011, 2014). The experimental results should have also some implications in the science of earthquakes besides those in geo-engineering. It is often reported that new hotsprings appear soon after the earthquake and some illumination of sky (particularly in nights). The author has personally observed the same phenomena in the 1999 Kocaeli earthquake (Aydan *et al.*, 1999).

3.3 CONCLUSIONS

Well established dynamic experimental techniques are necessary to determine parameters of constitutive laws for rocks and rock masses. The present experimental techniques such as SHBT and cyclic tests are not sufficient enough to determine properties of rocks under different dynamic loading conditions. Furthermore, the dynamic

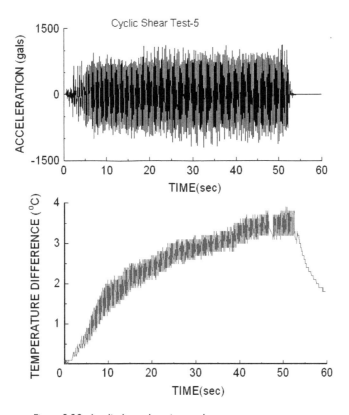

Figure 3.28 Applied acceleration and temperature response.

constitutive laws should be more advanced and those used in soil-dynamics are not appropriate for rocks and rock masses.

The determined properties actually strain rate dependent implying that the strength and deformability properties of rocks, subsequently rock masses are not unique (Figure 3.8). In other words, this may imply that the strength properties should be determined according to anticipated loading conditions. However, it should be noted that inertia resistance of rocks and loading system during rate-dependent test using for example Splitted Hopkinson Bar Technique may cause the apparent increase of the elastic modulus and strength of rocks. For example, Aydan (1997) theoretically showed that the viscous resistance and inertia forces affect the apparent strain-stress response of rocks. This theoretical finding may imply some further discussions on the validity of this testing technique.

The experimental results should have also some implications in the science of earthquakes besides those in geo-engineering. It is often reported that new hot-springs appear soon after the earthquake and some illumination of sky (particularly at night). The author has personally observed the same phenomena in the 1999 Kocaeli earthquake (Aydan *et al.*, 1999) occurring during and after earthquakes. Dynamic shearing induces temperature rises along discontinuities and adjacent rock mass. Temperature

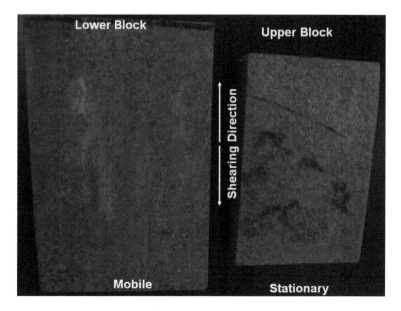

Figure 3.29 Views of sheared surfaces.

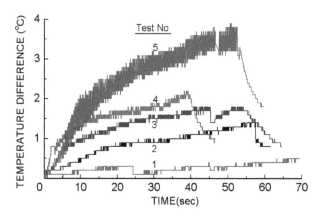

Figure 3.30 Comparison of responses of temperature rises for all experiments.

rise depends on the dynamic shearing rate, normal stress and frictional properties of discontinuities as well as thermal properties of adjacent rocks. The increase of normal stress, dynamic shearing rate and frictional properties proportionally increase the temperature rise. Particularly normal stress and dynamic shearing rate have great influence on the overall rise of temperature. The highest temperature rises occurred at contact areas (namely asperities) over the discontinuity plane. A thin powder layer accumulated on the surface of discontinuity surfaces after completion of the experiment. The experimental results have also some implications in the science of earthquakes such as the illumination phenomenon besides those in geo-engineering.

Chapter 4

Multi-parameter responses and strong motions induced by fracturing of geomaterials and slippage of discontinuities and faulting model tests

4.1 MULTI-PARAMETER RESPONSES AND STRONG MOTIONS INDUCED BY FRACTURING OF ROCKS

Experimental studies for understanding of multi-parameter variations including electric potential, electrical resistivity, magnetic field, and acoustic emissions during deformation and fracturing process of geomaterials, which ranges from crystals, fault gouge-like materials to rocks under different loading regimes and environment have been taken by Aydan and his group (Aydan, 2003, 2013; Aydan *et al.*, 2001, 2003, 2005a, 2005b, 2007, 2010, 2011, 2014) (Figure 4.1). Recently, the experiments have been repeated using an entirely manually operated loading system in order to eliminate the possible electric noise on the system (Aydan *et al.*, 2010). Furthermore, the dynamic responses of geo-materials during fracturing have not received any attention so far. The recent advances in measurement, monitoring and logging technologies enable us to measure and to monitor the dynamic responses of geo-materials during fracturing and slippage.

The applied load and induced displacement were automatically measured and stored on the hard disk of a laptop computer through an electronic logger. Electric potentials induced during the deformation of samples were measured through two electrodes attached to the top and bottom of samples using a voltmeter and logger unit and data were simultaneously stored on the hard-disk of the laptop computer.

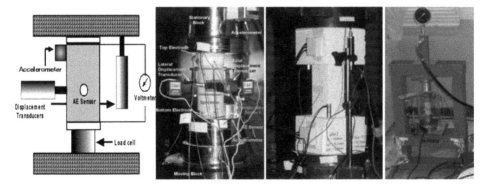

Figure 4.1 Experimental set-up and views of uniaxial compression experiments.

The electrodes were isolated from the loading frame with the use of isolators. Electric potentials were measured either as DC and/or AC. When electrical resistivity is measured, a function generator was used to produce electric current with given amplitude. In some of experiments, magnetic field was also measured. In addition to the above measurement system, acoustic emissions (AE) devices and sensors and temperature sensors were used to measure the acoustic emissions as well as temperature variations during fracturing and sliding of samples. The acceleration responses of the samples during fracturing were measured by using an accelerometer, which can measure three components of accelerations up to 10 g with a frequency range of 0–160 Hz. Various rock samples were tested. Although some of rock samples (e.g. granite, quartzite, sandstone etc.) contain piezo-electric minerals, some rock samples were selected such that they do not contain any piezo-electric substances, such as aragonite crystal, limestone, rocksalt, soapstone and marls.

Various responses measured during some of experiments are shown in Figures 4.2–4.3. The detailed discussions can be found in previous articles (Aydan et al., 2001 2003, 2005a,b, 2010, 2014; Ohta et al., 2008). Nevertheless, one can easily notice the distinct variations of multi-parameters during the deformation and fracturing of rocks. As seen from the experimental results, the deformation and fracturing of rock cause the distinct variations of electric potential, electrical resistivity, magnetic field and acoustic emissions in addition to conventional parameters such as displacement (strain) and force (stress). Despite the possibility of electrical noise from the loading devices in experimental results reported and discussed by Aydan et al. (2001, 2003) and previously, the same statements regarding the electric potential responses can be quoted herein.

Fundamentally, the observed acceleration responses are similar to each other. The acceleration responses start to develop when the applied stress exceeds the peak strength and it attains the largest value just before the residual state is achieved, as seen in Figure 4.4. This pattern was observed in all experiments. Another important aspect is that the acceleration of the mobile part is much larger than that of the stationary part. This is also a common feature in all experiments. The amplitude of accelerations of the mobile part of the loading system is higher than that of the stationary part (Figure 4.5). It is very interesting to note that the maximum acceleration increases as the work done increases. This feature has striking similarities with the strong motion records of nearby earthquake faults observed in the recent large inland earthquakes. Furthermore, the waveforms of the acceleration records of the mobile part are not symmetric with respect to the time axis.

The amplitude of accelerations during the fracturing of hard rocks is much higher than that during the fracturing of soft rocks. This is directly proportional to the energy stored in samples before the fracturing. The chaotic responses in acceleration components perpendicular to the maximum loading direction may be observed. These may have some important implications for the procedures and interpretation of measurements for the short-term forecasting of failure events in geo-engineering and geo-science.

Figure 4.6 shows the frequency characteristics of measured accelerations of a sandstone together with measured accelerations shown in the same figure. The integrated displacement and velocity responses using the EPS Method proposed by Ohta and Aydan (2007a, b) are shown in Figure 4.7. It is important to note that the dominant

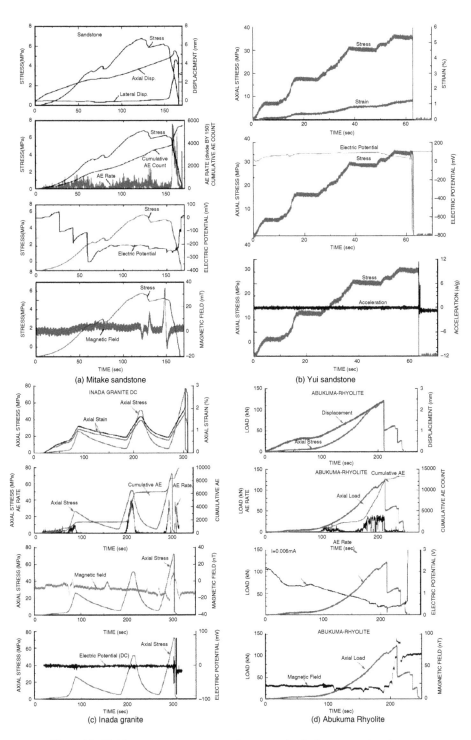

Figure 4.2 Multi-parameter responses of some rocks during conventional tests.

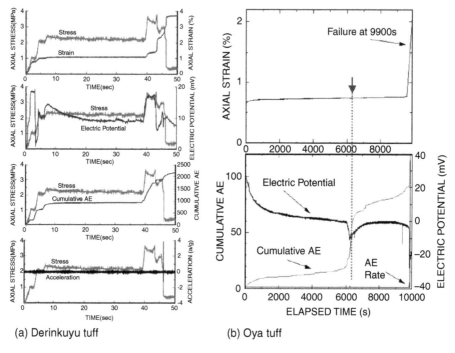

(a) Derinkuyu tuff   (b) Oya tuff

Figure 4.3 Multi-parameter responses of some rocks during compression creep tests.

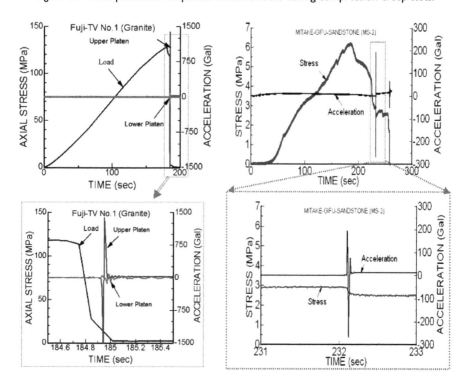

Figure 4.4 Acceleration and axial response of sandstone and granite samples.

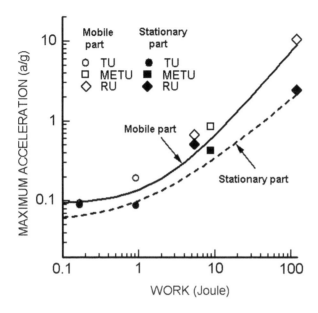

Figure 4.5 The relation between work done on samples and maximum acceleration.

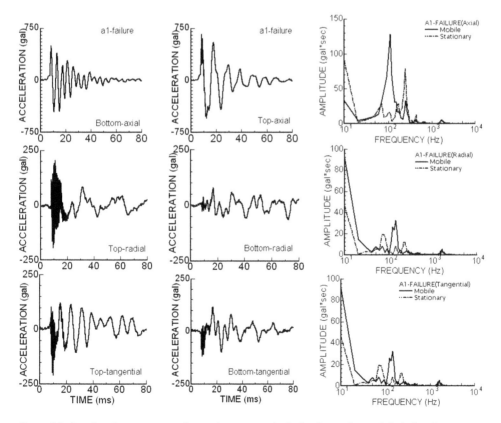

Figure 4.6 Acceleration response of a sandstone sample during fracturing and their Fourier spectra.

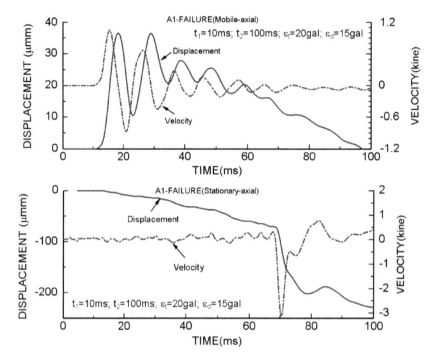

Figure 4.7 Integrated displacement and velocity responses according to EPS method.

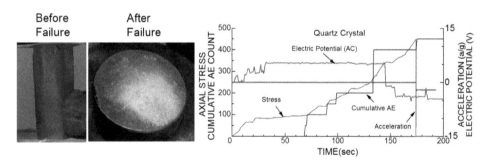

Figure 4.8 Multi-parameter responses of Quartz crystal sample and its view before and after failure.

frequency of the mobile part has a low frequency content compared to that of the stationary part (Ohta and Aydan, 2010).

Aydan *et al.* (2014) have recently performed experiments on minerals and rocks utilizing infrared camera besides the multi-parameters mentioned above. Some experimental results are quoted from this experimental study of Aydan *et al.* (2014).

Responses of several measurable parameters such as stress, strain, electric potential, cumulative AE are shown in Figure 4.8 together with views of quartz crystal before and after the failure. The failure mode of the quartz crystal was like an explosion and

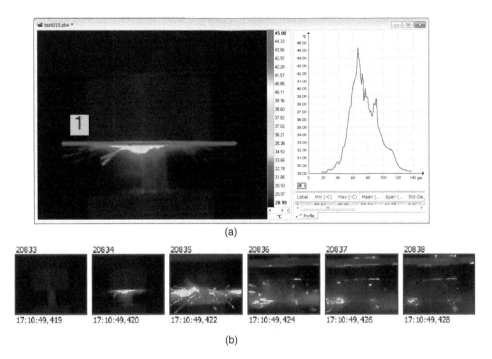

(a)

(b)

Figure 4.9 (a) An infrared thermograph image of the quartz crystal during failure and temperature rise along the selected line, (b) infrared thermograph images during fracture process.

the remains of the sample after the experiment were powderized as seen in the same figure. The maximum acceleration was 13 times the gravitational acceleration. Distinct variations of various measurable parameters such as electric potential, acoustic emission besides load and displacement were observed during deformation and fracturing processes.

An infrared thermograph image and temperature response along the selected line of the quartz crystal sample during the initiation of failure is shown in Figure 4.9. The maximum temperature was observed almost at the center of the top surface of the crystal and temperature difference was about 16 Celsius from the ambient temperature. Furthermore, one can easily notice the ejection trajectories of some fragments from the failing quartz sample. On the other hand, if minerals are ductile like gypsum crystal, the maximum acceleration was very small as compared with that observed during the failure of quartz mineral. Similarly, an infrared thermograph image and temperature response along the selected line in the observed outer surface of gypsum crystal sample during the initiation of failure was very small and it was very low compared with that observed during the experiment on the quartz crystal.

Experimental results on mudstone sample from Seyitömer open-pit lignite mine of Turkey are presented herein. Figure 4.10 shows an example of infrared thermographic image of Seyitömer mudstone and temperature distribution along the line indicated

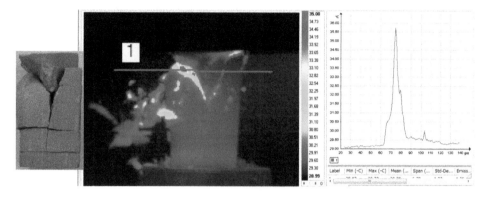

Figure 4.10 Actual view and infrared thermograph image of Seyitömer mudstone together with temperature distribution along the chosen line.

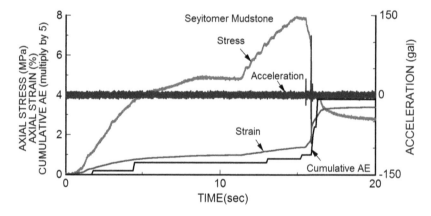

Figure 4.11 Multi-parameter responses of Seyitömer mudstone.

in the image. It is also interesting to note that a thermal band, which coincides with fracture zone, is observed. This may be of great value for identifying the possibility of rockbursts and their locations in rock engineering and earthquake faults in geoscience.

Figure 4.11 shows the multi-parameter response of the mudstone sample from Seyitömer. It is of great interest that acceleration waves with lower amplitude occurred when the macroscopic crack initiation starts to occur at the peak strength while the maximum acceleration observed during the failure state as noted in previous experiments reported by Aydan *et al.* (2010) and Ohta and Aydan (2010). Furthermore, the maximum acceleration is about 120 gals.

Inada granite from Ibaraki Prefecture of Japan was selected as another example of hard rocks. Figure 4.12 shows a view of the failed sample and an infrared thermograph image together with temperature distribution along the selected line shown in the figure. The temperature difference was more than 47 Celsius at the time of initiation measured by the Infrared Thermal Camera. The temperatures were higher at the

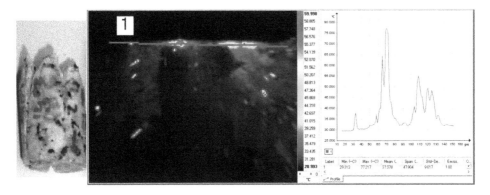

Figure 4.12 View after the experiment and infrared thermograph image of Inada granite together with temperature distribution along the chosen line.

corners where crack initiation started. The sample failed in a violent manner and the trajectories of fragments of the rock sample together with a powder-cloud are easily noticed from the image. From the images it is also possible to obtain ejection velocities, which may be of great importance for assessing the possible ejection distance as well as destruction potential of the rockbursts in actual constructions.

4.2   STRONG MOTIONS INDUCED IN STICK-SLIP TESTS

An experimental device consisted of an endless conveyor belt and a fixed frame (see Ohta and Aydan, 2010b for details). The inclination of the conveyor belt can be varied so that the tangential and normal forces can be easily imposed on the sample as desired. The belt itself was made of fiber reinforced rubber. In order to study the actual frictional resistance of interfaces of rock blocks, the lower block was stuck to rubber belt while the upper block was attached to the fixed frame through a spring as illustrated in Figure 4.13. During experiments, displacement, load, acceleration and acoustic emissions were measured simultaneously.

When the upper block moves together on the base block at a constant velocity (stick phase), the spring is stretched at a constant velocity. The shear force increases to some critical value and then a sudden slip occurs with an associated spring force drop. During slip phase, the upper block slips violently over the base block.

To measure the shear force acting on upper block, a load cell is installed between spring and fixed frame. During experiments, the displacement of the block is measured through a laser displacement transducer produced by KEYENCE and a contact type displacement transducer with a measuring range of 70 mm, while the acceleration responses parallel and perpendicular to the belt movement are measured by a three component accelerometer (TOKYO SOKKI) attached to the upper block. The measured displacement, acceleration and force are recorded onto laptop computers.

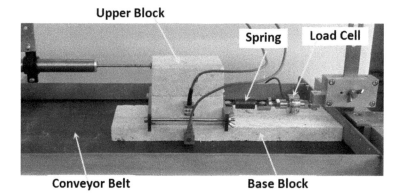

Figure 4.13 A view of the stick-slip apparatus and a typical experimental set-up.

During the stick-slip experiments, the following conditions were investigated:

1 Variation of stiffness by either using different springs or weight of upper block,
2 Variation of the velocity of the base plate and
3 Variation of interface friction properties of blocks.

Stick-slip experiments were carried on various natural rock blocks as well as other types of blocks made of foam, plastic, wood and aluminium. Responses of displacement, associated acceleration and acoustic emissions for stick-slip experiment on granite sample are shown in Figure 4.14 for rough to rough combination discontinuity together with an expanded view of responses of displacement and acceleration of a typical slip event, in which rise time, relative slip and stress drop can be clearly seen. There are many interesting observations in these responses, which are relevant to earthquake prediction as well as strong motions during earthquakes. As noted from the figure, the velocity of the upper block starts to change before the slippage. Similar responses can be also noticed in Figure 4.14.

Figure 4.15 shows the relation between the amount of relative slip and maximum accelerations. It is noted that a linear relation holds and it should be also noted from Figure 4.14 that the induced acceleration waves are not generally symmetric with respect to time axis. Stick-slip experiments on the saw-cut surface of Ryukyu limestone (coral) reported by Ohta and Aydan are shown in Figures 4.16 and 4.17. Despite difference of rock types, results are quite similar.

Figure 4.18 shows a mechanical model for stick-slip experiments. Stick-slip response consists of stick phase, during which mechanical energy is accumulated and slip phase when the accumulated energy is released. The formulation presented herein is based on the formulation of Bowden and Leben (1939).

During stick phase, the following relations hold:

$$\dot{x} = v_s, \quad F_s = k \cdot x \tag{4.1}$$

where v_s is belt velocity, k spring stiffness.

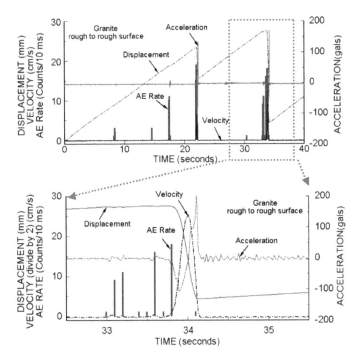

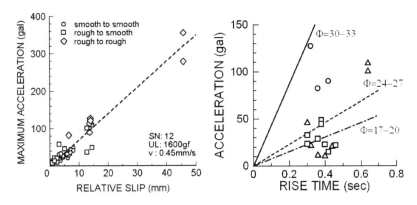

Figure 4.14 Multi-parameter responses of a discontinuity in granite during a stick-slip experiment.

Figure 4.15 The relation between the maximum acceleration and relative slip and rise-time.

The maximum frictional resistance is given by

$$F_y = \mu_s N \tag{4.2}$$

where μ_s is static friction angle and N is normal force. Normal force N is equal to block weight (W) shown in Figure 4.18 and it is equal to mg (m is mass and g is gravitational acceleration).

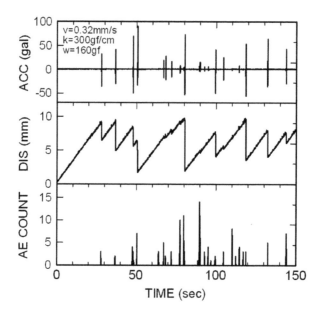

Figure 4.16 Stick-slip test of Ryukyu Limestone sample with saw-cut surface (lower base velocity and lower load of the upper part).

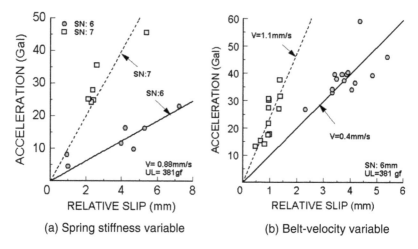

(a) Spring stiffness variable

(b) Belt-velocity variable

Figure 4.17 The relation between the maximum acceleration and relative slip while spring stiffness or belt velocity is varied.

During slip phase, the following relation holds:

$$-kx + \mu_k W = m \frac{d^2 x}{dt^2} \qquad (4.3)$$

where μ_k is kinetic friction angle.

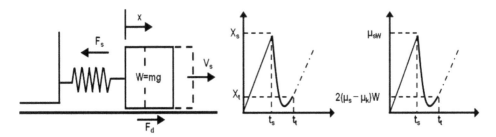

Figure 4.18 Mechanical model for stick-slip phenomenon.

The solution of the differential equation above can be obtained as

$$x = A_1 \cos \Omega t + A_2 \sin \Omega t + \mu_k \frac{W}{k} \qquad (4.4)$$

The integral constants in Eq. 4.4 can be obtained from the initial conditions, that are, $\dot{x} = v_s$ and $x = x_s$ at $t = t_s$ as

$$
\begin{aligned}
x &= \frac{W}{k}(\mu_s - \mu_k) \cos \Omega(t - t_s) + \frac{v_s}{\Omega} \sin \Omega(t - t_s) + \mu_k \frac{W}{k} \\
\dot{x} &= -\frac{W}{k}(\mu_s - \mu_k)\Omega \sin \Omega(t - t_s) + v_s \cos \Omega(t - t_s) \qquad (4.5) \\
\ddot{x} &= -\frac{W}{k}(\mu_s - \mu_k)\Omega^2 \cos \Omega(t - t_s) - v_s \Omega \sin \Omega(t - t_s)
\end{aligned}
$$

where $\Omega = \sqrt{k/m}$ and $x_s = \mu_s \frac{W}{k}$.

As the velocity of the block would be equal to the belt velocity, that is, $\dot{x} = v_s$ at termination of the slip $t = t_t$, we have the following relation:

$$t_t = \frac{2}{\Omega}\left(\pi - \tan^{-1}\left(\frac{(\mu_s - \mu_k)W\Omega}{k \cdot v_s}\right)\right) + t_s \qquad (4.6)$$

where $x_s = v_s \cdot t_s$.

The slip duration, which is called "rise time" is given by the following relation:

$$t_r = t_t - t_s \qquad (4.7)$$

Thus, the rise time is obtained from Eqs. (4.6) and (4.7) as

$$t_r = \frac{2}{\Omega}\left(\pi - \tan^{-1}\left(\frac{(\mu_s - \mu_k)W\Omega}{k \cdot v_s}\right)\right) \qquad (4.8)$$

If $v_s \approx 0$, rise time takes the following form

$$t_r = \pi\sqrt{\frac{m}{k}} \qquad (4.9)$$

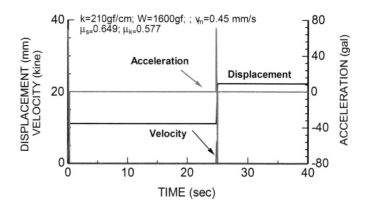

Figure 4.19 Computed stick-slip response for a discontinuity in granite with a rough surface.

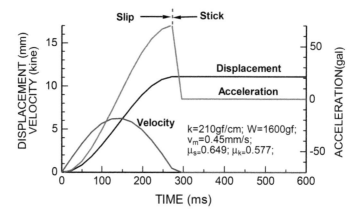

Figure 4.20 Computed responses during slip phase for a discontinuity in granite with a rough surface.

Accordingly, the absolute slip can be obtained as

$$x_r = |x_t - x_s| = 2\frac{W}{k}(\mu_s - \mu_k) \tag{4.10}$$

Furthermore, force drop during the slip event takes the following form

$$F_d = 2(\mu_s - \mu_k)W \tag{4.11}$$

The theory of stick-slip behaviour presented above is applied to experiments carried out on granite discontinuity with a friction angle ranging between 30 and 33 degrees. The computational results are shown in Figure 4.19 and 4.20. The values of the parameters used in computations are also shown in respective figures. As noted from the figure, the rise time is about 280 ms and the acceleration can be up to 70 gals.

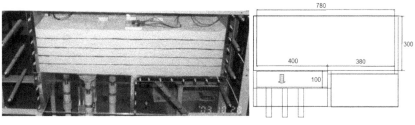

<table>
<tr><td>(a) View of an experimental set-up</td><td>(b) Drawing of experimental device</td></tr>
</table>

Figure 4.21 A view of an experimental set-up and a drawing of the experimental device.

When the results are compared with those shown in Figure 4.15, they are remark-ably similar to each other. Furthermore, the theory reveals the effects of fundamental parameters involved with the stick-slip phenomenon of rock discontinuities, which may have direct implications on the actual slip behaviour of earthquake faults.

4.3   STRONG MOTIONS INDUCED IN MODEL FAULTING EXPERIMENTS

The author and his group (Aydan, 2003; Aydan *et al.*, 2011; Ohta, 2011; Ohta *et al.*, 2014) developed a device to study the motions during faulting events under constant acceleration environment and performed a series of laboratory model experiments on various rock engineering structures subjected to thrust and normal faulting as well as strike-slip faulting movements. One side of the device is moveable in a chosen direction to induce base movements similar to normal or thrust faulting with a different inclination. The device can simulate from 45 degrees normal faulting movement to 135 degrees thrust faulting. The box is 780 mm long, 300 mm high and 300 mm wide. The length of the moveable side is 400 mm. The motion of the moveable side of the device is achieved through its own weight by removing a stopper. The amount of the vertical movement of the moving base can be up to 200 mm and it can be set to a certain level as desired. The device is equipped with non-contact laser displacement transducers and contact type accelerometers with three components.

The maximum displacement of faulting of the moving side of the faulting experi-ments was varied between 25 and 100 mm. The vertical displacements of the fault was 25, 50, 75 and 100 mm. Due to the nature of the physical model, the vertical compo-nent of accelerations becomes maximum among other components. Figure 4.22 shows the vertical accelerations and displacement measured simultaneously for 200 mm thick soil deposit for 90 degrees normal faulting and its motion at several time steps. As seen from Figure 4.22a, the maximum acceleration of the movable side is greater than that of the stationary side. Furthermore, the maximum acceleration is observed when the movement of the movable side is restrained and the acceleration response is entirely unsymmetric while the acceleration response of the stationary side is almost symmetric. The responses measured during these experiments are quite similar to the

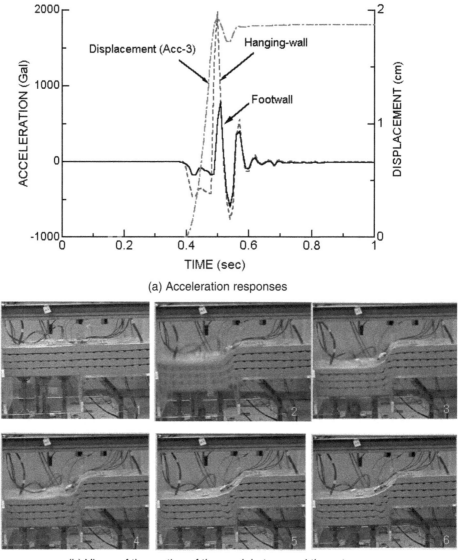

(a) Acceleration responses

(b) Views of the motion of the model at several time steps

Figure 4.22 Motions of 200 mm thick soil deposit for 90° normal faulting.

observations in actual earthquakes as well as rock fracture experiments shown previously and described in detail in Ohta (2011), Ohta and Aydan (2010) and Ohta et al. (2014). The variation of soil deposit thickness has a certain effect on the resulting acceleration. However, if the displacement of the fault is the same, its effect would be small as compared with that of the variation of the fault displacement.

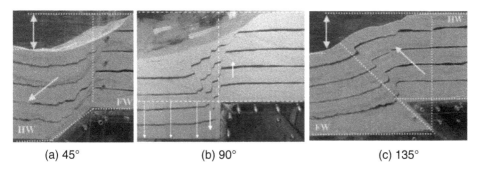

(a) 45°   (b) 90°   (c) 135°

Figure 4.23 Views of faulting experiments.

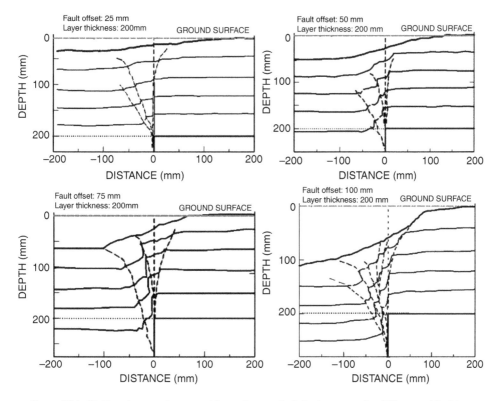

Figure 4.24 Sliplines in experiments with varying vertical displacement for 90° normal faulting.

The observations on deformation and slip-lines in experiments carried out for inclinations of 45, 90 and 135 degrees are shown in Figure 4.23. Comparisons are done for three different inclinations of faulting for the same amount of vertical displacements. The most extensive studies are carried out on experiments with the faulting inclination of 90 degrees and in which the effects of allowable vertical displacements and soil thickness were investigated. Figure 4.24 shows the slip lines fault offsets of 25, 50,

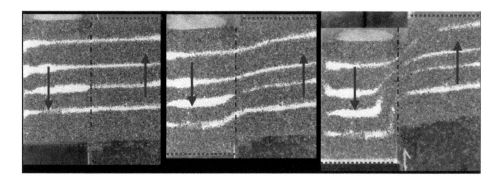

Figure 4.25 Negative views of several stages in a strike-slip faulting experiment in granular medium.

75 and 100 mm. As pointed out previously, the top soil deposit on the hanging wall (mobile) side is highly deformed while the soil deposit on the footwall (stationary) side is much less. Furthermore, the number of slip-lines on the hanging-wall (mobile part) of the soil layer is greater than that in the footwall side. This is probably associated with the amount of displacement of the mobilized soil in the hanging-wall side. It is also interesting to note that thrust type slip-lines occur at the hanging-wall side while normal type slip-lines develop in the footwall side. Such slip-lines may be of great significance when ground deformation and slip-lines are interpreted for faults in-situ.

Some experiments were carried out using an experimental faulting device of constant velocity type. Although the basic pattern of ground deformation and slip-line formation is similar to that of the experimental results such that the volume of the deformed wedge-like body is smaller and the inclination of slip-lines are less and slightly steeper. Furthermore, the ground surface profiles are not sharp as they are in the experiments in this section. Similar conclusions are also valid for normal faulting experiments.

In addition, some strike-slip experiments are carried out under constant velocity condition. Figure 4.25 shows negative images of several stages of a strike-slip faulting experiment on a granular media. Although slip-lines were not apparent from the figure, a wide deformation band with a thickness equivalent to the amount of relative displacement developed on both sides of the projected fault-line.

Chapter 5

Ground motions due to earthquakes and estimation procedures

It is well known that the earth's crust is ruptured and contains numerous faults and various kinds of discontinuities and it is almost impossible to find a piece of land without faults. During the construction of structures such as tunnels, dams, power plants, roadways, railways, power transmission lines, bridges, elevated expressways etc. it is almost impossible not to cross a fault or faults. Therefore, one of the most important items is how to identify which fault segments observed on ground surface will move or rupture during an earthquake. Ground motion characteristics, deformation and surface breaks of earthquakes depend upon the characteristics of causative faults. Their effects on the seismic design of engineering structures are not considered in the present codes of design although there are attempts to include them in some countries (i.e. USA, Japan, Taiwan).

5.1 CHARACTERISTICS OF EARTHQUAKE FAULTS

The fault is geologically defined as a discontinuity in geological medium along which a relative displacement took place. Faults are broadly classified into three big groups, namely, normal faults, thrust faults and strike-slip faults as seen in Figures 5.1 and 5.2. A fault is geologically defined as active if a relative movement took place in a period of less than 2 million years.

It is well known that a fault zone may involve various kinds of fractures as illustrated in Figure 5.3 and it is a zone having a finite volume (Aydan *et al.*, 1997; Ulusay *et al.*, 2002). In other words, it is not a single plane. Furthermore, the faults may

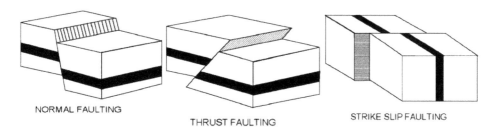

NORMAL FAULTING

THRUST FAULTING

STRIKE SLIP FAULTING

Figure 5.1 Fault types (from Aydan, 2003, 2012).

(a) Normal faulting   (b) Strike-slip faulting   (c) Thrust faulting

Figure 5.2 Some examples of faulting.

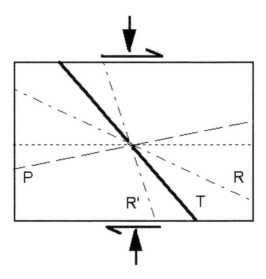

Figure 5.3 Fractures in a shear zone or fault (from Aydan *et al.*, 1997).

have a negative or positive flower structure as a result of their trans-tensional or trans-pressional nature and the reduction of vertical stress near the earth surface (Aydan *et al.*, 1999). For example, even a fault having a narrow thickness at depth may cause a broad rupture zones and numerous fractures on the ground surface during earth-quakes (Figure 5.4). Furthermore, the movements of a fault zone may be diluted if a thick alluvial deposit is found on the top of the fault (e.g. 1992 Erzincan earthquake (Hamada and Aydan, 1993)). The appearance of ground breaks is closely related

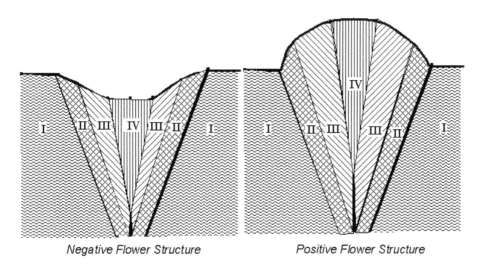

Negative Flower Structure   Positive Flower Structure

Figure 5.4 Negative and positive flower structures due to trans-tension or trans-pression faulting and zoning (modified from Aydan *et al.*, 1997, 1999; Aydan, 2003).

to geological structure, characteristics of sedimentary deposits, their geometry, the magnitude of earthquakes and fault movements.

5.2 OBSERVATIONS ON STRONG MOTIONS AND PERMANENT DEFORMATIONS

5.2.1 Observations on maximum ground accelerations

It is observationally known that the ground motions induced by earthquakes could be much higher in the hanging-wall block or mobile side of the causative fault as observed in the recent earthquakes such as the 1999 Kocaeli earthquake (strike-slip faulting), the 1999 Chi-chi earthquake (thrust faulting), the 2004 Chuetsu earthquake (blind thrust faulting) and the 2000 Shizuoka earthquake and l'Aquila earthquake (normal faulting) (Ohta, 2011; Chang *et al.*, 2004; Somerville *et al.*, 1997; Tsai and Huang, 2000; Aydan *et al.*, 2009; Abrahamson and Somerville, 1996; Aydan, 2003; Aydan *et al.*, 2007) as seen in Figures 5.5–5.7.

Figure 5.8 illustrates the effect of hanging-wall effect on the attenuation of maximum ground accelerations observed in the 1999 Chi-chi earthquake (Taiwan), 1999 Düzce earthquake (Turkey) and 2001 Geiyo earthquake (Japan) with different faulting mechanisms (Ohta and Aydan, 2010).

Figure 5.9 shows the records of accelerations at the ground surface and at bedrock 260 m below at Ichinoseki strong motion station (IWTH25) of KiK-NET (2008) strong motion network of Japan measured during the 2008 Iwate-Miyagi earthquake. The strong motion station was located on the hanging-wall side of the fault and it was very close to the surface rupture. As noted from the figure, the ground acceleration of the

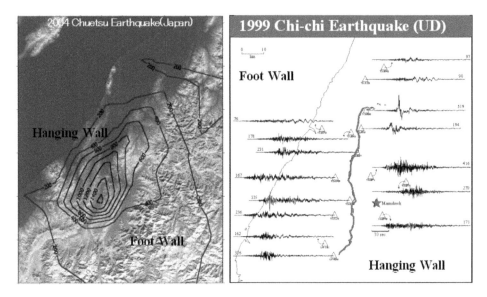

Figure 5.5 The footwall and hanging-wall effects on the maximum ground accelerations (thrust faulting).

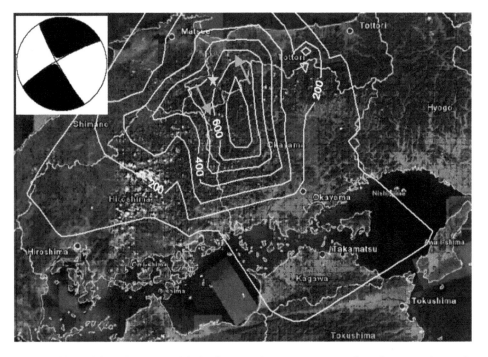

Figure 5.6 The mobile and stationary block effects on the maximum ground accelerations observed in 2000 Tottori Seibu earthquake (strike-slip faulting).

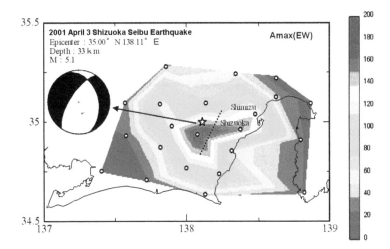

Figure 5.7 The footwall and hanging-wall effects on the maximum ground accelerations (normal faulting).

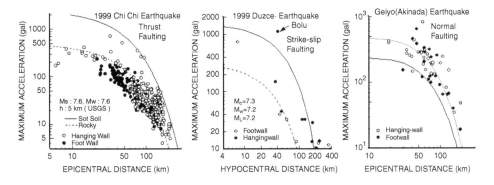

Figure 5.8 Attenuation of maximum ground accelerations for some earthquakes.

UD component was amplified by 5.67 times that at the bedrock and the acceleration records are not symmetric with respect to time axis. This record is also the highest strong motion recorded in the world so far.

5.2.2 Permanent ground deformation

The recent global positioning system (GPS) also showed that permanent deformations of the ground surface occur after each earthquake (Figures 5.10 and 5.11). The permanent ground deformation may result from different causes such as faulting, slope failure, liquefaction and plastic deformation induced by ground shaking (Aydan *et al.*, 2010). These types of ground deformation have limited effect on small structures as long as the surface breaks do not pass beneath those structures. However, such deformations may cause tremendous forces on long and/or large structures such as rock

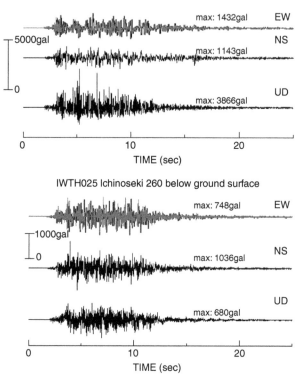

Figure 5.9 Acceleration records at ground surface and bedrock at Ichinoseki strong motion station IWTH25 of KIK-NET in Iwate-Miyagi earthquake.

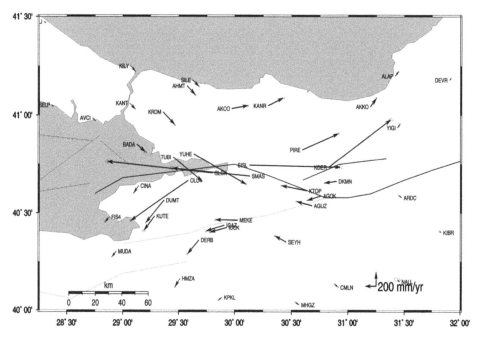

Figure 5.10 Permanent ground deformations and associated straining induced by the 1999 Kocaeli earthquake (Reilinger *et al.*, 2000).

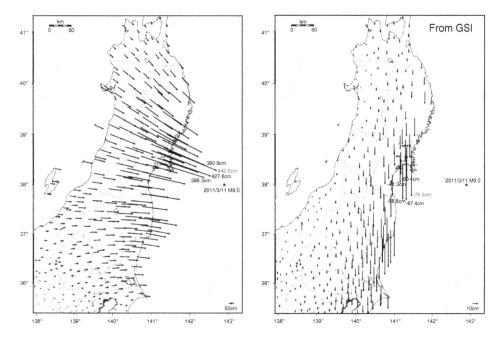

Figure 5.11 Ground deformation induced by the great East Japan earthquake (measured using GPS by GSI, 2011).

engineering structures. The ground deformation may induce large tensile or compression forces as well as bending stresses in structures depending upon the character of permanent ground deformations. Blind faults and folding processes may also induce some peculiar ground deformations and associated folding of soft overlaying sedimentary layers. Such deformations caused tremendous damage to tunnels during the 2004 Chuetsu earthquake although no distinct rupturing took place.

5.3 STRONG MOTION ESTIMATIONS

5.3.1 Empirical approach

There are many empirical attenuation relations for estimating ground motions in literature (i.e. Joyner and Boore, 1981; Campbell, 1981; Ambraseys, 1988; Aydan *et al.*, 1996). Including so-called next (?) generation attenuation (NGA) relations, all these equations are essentially spherical or cylindrical attenuation relations and they cannot take into account the directivity effects. As is shown in the beginning of this section, ground motions such as maximum ground acceleration (A_{MAX}) and maximum ground velocity (V_{MAX}) have strong directivity effects in relation to fault orientation. Furthermore, these relations are generally far below the maximum ground acceleration and they are incapable of obtaining the maximum ground acceleration (A_{MAX}) or the preferred term "peak ground acceleration (PGA)".

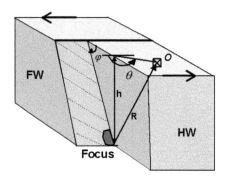

Figure 5.12 Illustration of geometrical fault parameters (R, θ, φ).

Aydan (2012, 2015) proposed an attenuation relation by combining their previous proposals (Aydan *et al.*, 1997; Aydan, 1997, 2001, 2007; Aydan and Ohta, 2011) together with the consideration of the inclination and length of earthquake fault using the following functional form (Figure 5.12):

$$\alpha_{\max} = F_1(V_s) * F_2(R, \theta, \varphi, L^*) * F_3(M) \tag{5.1}$$

where V_s, θ, φ, L^* and M are the shear velocity of ground and the angle of the location from the strike and dip of the fault (measured anti-clockwise with the consideration of the mobile side of the fault) and earthquake magnitude. The following specific forms of functions in Equation (5.1) were put forward as

$$F_1(V_s) = Ae^{-V_s/B} \tag{5.2a}$$

$$F_2(R, \theta, \varphi, L^*) = e^{-R(1-D\sin\theta + E\sin^2\theta)(1+F\cos\varphi)/L^*} \tag{5.2b}$$

$$F_3(M) = e^{M/G} - 1 \tag{5.2c}$$

The same form is also used for estimating the maximum ground velocity ($V_{\max}$). L^* (in km) is a parameter related to the half of the fault length. And it is related to the moment magnitude in the following form

$$L^* = a + be^{cM_w} \tag{5.3}$$

The specific values of constants of Eqs. (5.2a,b,c)–(5.3) for this earthquake are given in Tables 5.1, 5.2 and 5.3.

The most important parameter in this approach is the estimation of magnitude of the potential earthquake. If a very reliable data base exists for a given region, one may estimate the magnitude of the most-likely earthquake from such a data base. Another approach may be the estimation from the characteristics (length, area, maximum relative slip) of active faults. Matsuda (1975), Sato (1989), Wells and Coppersmith (1994), and Aydan (1997) proposed empirical relations. Aydan (2007, 2012, 2015) recently

Table 5.1 Values of constants in Equation (5.1) for inter-plate earthquakes.

	A	B (m/s)	D	E	F	G (Mw)
A_{max}	2.8	1000	0.5	1.5	0.5	1.05
V_{max}	0.4	1000	0.5	1.5	0.5	1.05

Table 5.2 Values of constants in Equation (5.1) for intra-plate earthquakes.

	A	B (m/s)	D	E	F	G (Mw)
A_{max}	2.8	1000	0.5	1.5	0.5	1.16
V_{max}	0.4	1000	0.5	1.5	0.5	1.16

Table 5.3 Values of constants in Equation (5.3) for earthquakes.

Faulting type	a	b	c
Normal faulting	30	0.002	1.35
Strike-slip faulting	20	0.002	1.40
Thrust faulting	20	0.002	1.27

Table 5.4 Values of constants for Eq. (5.1) for each fault parameter.

Fault type	Parameter	L (km)	S (km²)	U_{max} (cm)
Normal	A	0.0014525	0.003	0.0003
Faulting	B	1.21	1.50	1.6
Strike-slip	A	0.0014525	0.001	0.00035
Faulting	B	1.19	1.70	1.6
Thrust	A	0.0014525	0.0032	0.0014
Faulting	B	1.25	1.50	1.6

L: rupture length; S: rupture area; U_{max}: maximum displacement.

established several relations between moment magnitude and rupture length (L), rupture area (A) and net slip (U_{max}) of fault given below and checked their validity with available data as well as the data from the most recent event of the 2011 Great East Japan earthquake:

$$L, S \text{ or } U_{max} = A \cdot M_w e^{M_w/B} \tag{5.4}$$

The functional form of the empirical relations is the same while their constants A and B differ depending upon the faulting sense, which are given in Table 5.4. If striation or sense of deformation of the potential active fault is known, it is also possible to infer its focal mechanism. Such a method is proposed by Aydan (2000) and compared with

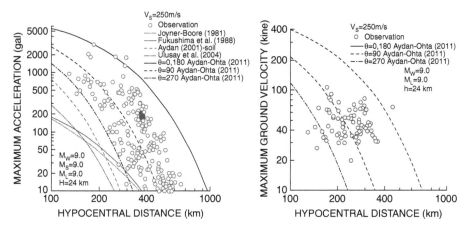

Figure 5.13 Comparison of estimated attenuation of maximum ground acceleration and ground velocity with observations for the 2011 Great East Japan earthquake.

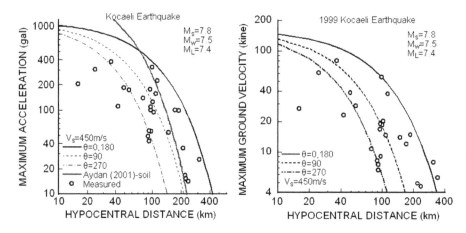

Figure 5.14 Comparison of estimated attenuation of maximum ground acceleration and ground velocity with observations for the 1999 Kocaeli earthquake.

the focal mechanism solutions inferred from fault striations or sense of deformation with those from telemetric wave solutions.

The attenuation relation given by (Eq. 5.1) were used to evaluate the maximum ground acceleration and ground velocity of the 2011 Great East Japan Earthquake (GEJE) and the 1999 Kocaeli earthquake and compared with actual observation data in Figures 5.13 and 5.14. The same equation is used to evaluate the areal distribution of maximum ground acceleration and velocity for the Kocaeli earthquake and compared with observational data in Figure 5.15. For large earthquakes, the application of Eq. (5.1), the estimations based on the segmentation of faults may be more appropriate. Figure 5.16 shows the single and double source models for the 2008 Wenchuan earthquake.

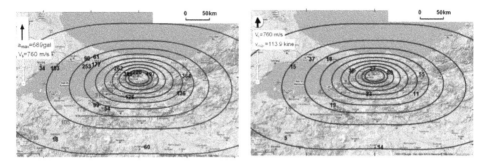

Figure 5.15 Comparison of estimated contours of maximum ground acceleration and ground velocity with observations for the 1999 Kocaeli earthquake.

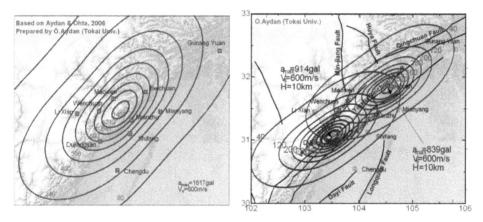

Figure 5.16 Comparison of single and double source models for maximum ground acceleration for the 2008 Wenchuan earthquake.

5.3.2 Green-function based empirical wave form estimation

The Empirical Green's function method was initially introduced by Hartzell (1978). Follow-up methods proposed by Hadley and Helmberger (1980) and Irikura (1983) are modifications of Hartzell's method of summing empirical Green's functions. In empirical Green's function approach rupture propagation and radiation pattern were specified deterministically and the source propagation and radiation effects were included empirically by assuming that the motions observed from aftershocks contained this information (Somerville et al., 1991). A semi-empirical Green's function summation technique has been used by Wald et al. (1998), Cohee et al. (1991) and Somerville et al. (1991) which allows gross aspects of the source rupture process to be treated deterministically using a kinematic model based on first motion studies, teleseismic modelling and distribution of aftershocks. Gross aspects of wave propagation are modeled using theoretical Green's functions calculated with generalized rays (Figure 5.17). Empirical Green's function method can be used only for a region where small events (i.e., aftershocks or foreshocks) of the target event are available.

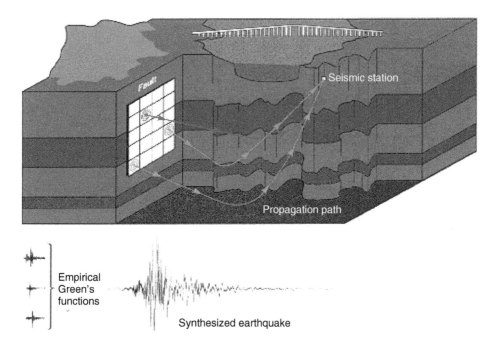

Figure 5.17 An illustration of the fundamental concept of the empirical Green function method (from Hutchings and Viegas, 2012).

Ikeda *et al.* (2016) recently performed an analysis of strong motion induced by the 2014 Nagano-ken Hokubu earthquake using the empirical Green's function method. The earthquake fault was assumed to be 9.6 km long and 7.2 km wide with a dip angle of 50 degrees. The stress drop was about 12.6 MPa and the rupture time was 2.7 km/s with a rise-time of 0.6 s. Figure 5.18 shows the observed and simulated responses of acceleration, velocity and displacement for NS direction for K-NET Hakuba strong motion station. As seen from the figure the simulated strong motions are close to the observations.

5.3.3 Numerical approaches

As explained in the Chapter 2, there are several numerical techniques, which are known to be Finite Difference Method (FDM), Finite Element Method (FEM), Boundary Element Method (BEM). The FDM is the earliest numerical model while FEM and BEM have become available after the 1960s and 1970s, respectively. Therefore, the first application of the numerical methods for strong motion estimation is related to the FDM. When this method is applied for strong motion estimation, one needs to solve Eq. (2.1) together with appropriate constitutive laws for the medium and the assumption of a rupture plane. Particularly the geometrical definition of the rupture plane and its rupture velocity would be also the key parameters of the simulations. Furthermore,

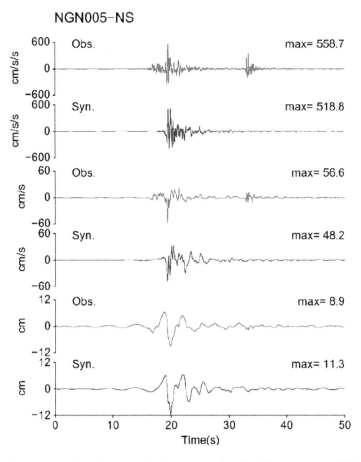

Figure 5.18 Comparison of simulations with observations for the NS component of Hakuba record.

both the FDM and FEM consider finite size domain, the prevention of reflections of waves from the boundaries would be necessary. This issue is generally dealt with by the introduction of the Lysmer-type viscous boundaries into the numerical model. Both the FDM and FEM would evaluate the wave propagation without any assumption on how waves generated at the source are transferred to any point of particular interest, which is a major issue in Green function based strong motion simulations. While the FEM can easily handle the irregular boundaries such as the surface of the model with its topography as free boundary, the FDM has a severe restriction dealing with such boundaries with irregular geometry. Nevertheless, some procedures dealing with irregular surface topography have been proposed (i.e. Hestholm, 1999; Gravers *et al.*, 1996). For irregular surface topography, FDM and FEM are also combined (i.e. Ducellier and Aochi, 2012). In addition, there are also some proposals to combine FDM or FEM with BEM in order to deal with newly developing ruptures.

(a) Finite Difference Method (FDM)

Constitutive law for the medium adjacent to rupture plane is generally assumed to be visco-elastic (i.e. Graves *et al.*, 1996). The most difficult aspect is the simulation of the fault plane associated with rupture process in the FDM schemes. The most conventional technique is to assume that fault plane to coincide with grid planes. Forced displacement field is introduced at domain where two points occupies the same space initially and can move relative to each other after the rupture. Many schemes also explore the incorporation of finite element method or boundary integral model to simulate the fault plane and the rest of the domain is discretized using the FDM. Figure 5.19 shows a simulation of strong motion induced by the 1995 Kobe earthquake using the FDM by Pitarka *et al.* (1998).

(b) Finite Element Method (FEM)

Toki and Miura (1985) and Tsuboi and Muira (2000) utilized Goodman-type joint elements in 2D-FEM to simulate both rupture process and ground motions (Figure 5.20). Fukushima *et al.* (2010) utilized this method to simulate ground motions caused by the 2000 Tottori earthquake (Figure 5.21). Later Mizumoto *et al.* (2004) extended the same method to 3D.

Iwata *et al.* (2016) recently investigated the strong motions induced by the 2014 Nagano-Hokubu earthquake. The model is based on 3D FEM version. Figure 5.22(a) shows the fault parameters and Figure 5.22(b) shows the 3D mesh of the earthquake fault and its vicinity. Figure 5.23(a) shows the time histories of surface acceleration at distances of 1 km and 2 km from the surface rupture in 3D-FEM model. Rupture time is about 7-8 seconds. The maximum acceleration is higher in the east-side hanging-wall than that in west-side (footwall), which is close to the general trend observed in strong motion records. Nevertheless, the computed acceleration was less than the measured accelerations. Figure 5.23(b) shows the time histories of surface displacement at distances of 1 km and 2 km from the surface rupture. The east side of the fault moves upward with respect to the footwall together with movement to the north direction and the vertical displacement of the east-side is larger than that of the footwall and the computed results are close to the observations. However, it is necessary to utilize finer meshes for better simulations of ground accelerations, which requires use of the super-computers.

Recently, Romano *et al.* (2014) simulated the rupture process of the 2011 Great East Japan earthquake using 3D-FEM together with joint elements to simulate the rupture process as well as ground motions (Figure 5.24).

5.4 ESTIMATION OF PERMANENT SURFACE DEFORMATION

These surface ruptures and permanent ground deformations may cause the failure of foundations of super-structures such as bridges, dams, viaducts and pylons. The recent large earthquakes caused severe damage to pylons and the foundations of viaducts and bridges. Therefore, it is an urgent issue how to assess the effects of possible surface ruptures in potential earthquakes and how to minimize their effect on structures.

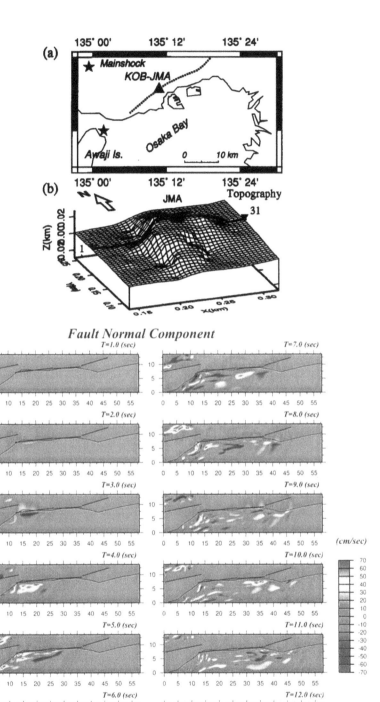

Figure 5.19 Fault normal ground velocity propagation induced by the 1995 Kobe earthquake (from Pitarka *et al.*, 1998).

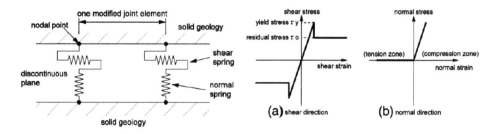

Figure 5.20 Representation of joint elements for faults and its constitutive law (from Fukushima *et al.*, 2010).

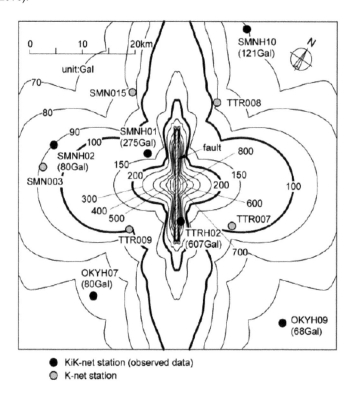

- ● KiK-net station (observed data)
- ◎ K-net station

Figure 5.21 Comparison of computed and observed maximum ground acceleration for 2000 Tottori earthquake (from Fukushima *et al.*, 2010).

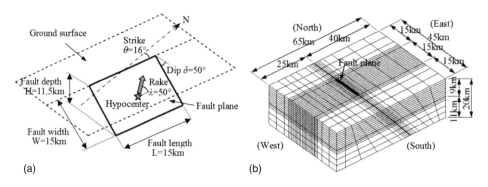

Figure 5.22 Fault model and 3D FEM mesh.

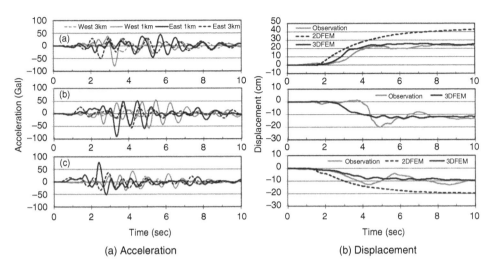

(a) Acceleration   (b) Displacement

Figure 5.23 Computed acceleration (a) and displacement (b) responses.

5.4.1 Observational surface deformation by GPS and ground surveying

The recent global positioning system (GPS) also showed that permanent deformations of the ground surface occur after each earthquake (Figure 5.10, 5.11). The permanent ground deformation may result from different causes such as faulting, slope failure, liquefaction and plastic deformation induced by ground shaking (Aydan *et al.*, 2010). These types of ground deformation have limited effect on small structures as long as the surface breaks do not pass beneath those structures. However, such deformations may cause tremendous forces on long and/or large structures. The ground deformation may induce large tensile or compression forces as well as bending stresses in structures depending upon the character of permanent ground deformations. As an example, the ground deformations reported by Reilinger *et al.* (1998) are shown in Figure 5.25, which were caused by a strike-slip fault during the 1999 Kocaeli earthquake in Turkey. Blind faults and folding processes may also induce some peculiar ground deformations and associated folding of soft overlaying sedimentary layers. Such deformations caused tremendous damage to tunnels during the 2004 Chuetsu earthquake although no distinct rupturing took place.

5.4.2 InSAR method

Interferometric synthetic aperture radar, abbreviated **InSAR** or **IfSAR**, is a radar technique used in geodesy and remote sensing. This geodetic method uses two or more synthetic aperture radar (SAR) images to generate maps of surface deformation or digital elevation, using differences in the phase of the waves returning to the satellite or aircraft. The technique can potentially measure centimetre-scale changes in deformation over spans of days to years. It has applications for geophysical monitoring of natural hazards, for example earthquakes, volcanoes and landslides, and in structural engineering, in particular monitoring of subsidence and structural stability. Figure 5.26

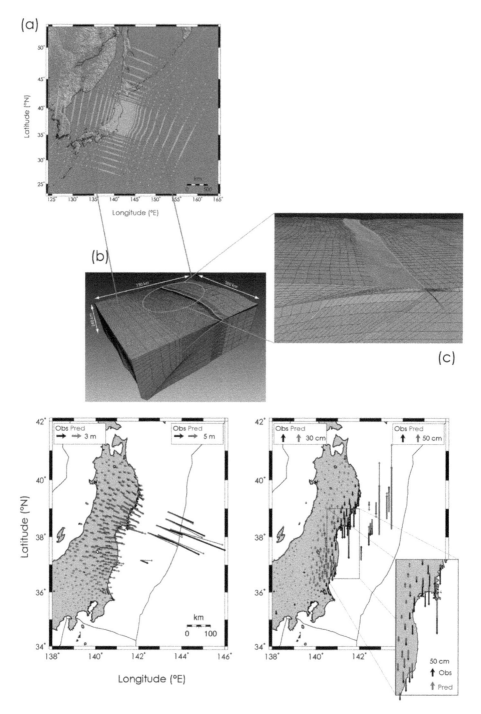

Figure 5.24 3D-FEM simulations of ground motions associated with the 2011 Great East Japan earthquake (from Romano *et al.*, 2014).

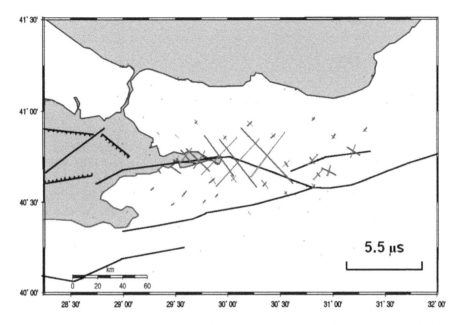

Figure 5.25 Computed ground straining from GPS measurements (from Aydan et al., 2010).

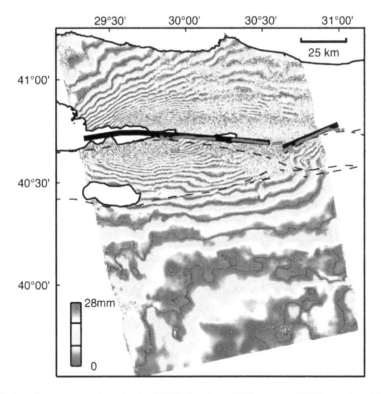

Figure 5.26 Interferogram produced using ERS-2 data from 13 August and 17 September 1999 for the 1999 Kocaeli earthquake (NASA/JPL-Caltech).

shows an application of the InSAR to estimate ground deformations induced by the 1999 Kocaeli earthquake.

5.4.3 EPS method

Ohta and Aydan (2007a,b) have recently shown that the permanent ground deformations may be obtained from the integration of acceleration records. The erratic pattern screening (EPS) method proposed by Ohta and Aydan (2007a,b) and Aydan and Ohta (2011) can be used to obtain the permanent ground displacement with the consideration of features associated with strong motion recording. The duration of shaking should be naturally related to the rupture time t_r. Depending upon the arrival time difference of S-wave and P-wave, the shaking duration would be a sum of rupture duration and S-P arrival time difference Δt_{sp}. If ground exhibits plastic response due to yielding or ground liquefaction, the duration of shaking would be elongated. If co-seismic crustal deformations are to be obtained, the integration duration should be restricted to the rupture duration with the consideration of S-P arrival time difference. The existence of plastic deformation can be assessed by comparing the effective shaking duration and the sum of rupture duration t_r and S-P arrival time difference Δt_{sp}. If the effective shaking duration is longer than the sum of rupture duration t_r and S-P waves arrival time difference Δt_{sp}, the integration can be carried out for both durations and the difference can be interpreted as the plastic ground deformation. The most critical issue is the information of rupture duration. The data on rupture duration are generally available for earthquakes with a moment magnitude greater than 5.6 worldwide. The effective shaking duration may be obtained from the acceleration records using the procedure proposed by Housner (1961). When such data are not available, the empirical relation proposed by Aydan (2007) may be used.

$$t_r = 0.0005 M_w \exp(1.25 M_w) \tag{5.5}$$

S-P arrival time difference Δt_{sp} can also be easily evaluated from the acceleration record.

They divided an acceleration record into three sections and apply filters in Section 1 and Section 3 and the integration is directly carried in Section 2 without any filtering. The times to differentiate sections are t_1 and t_2. Time t_1 is associated with the arrival of P-wave while time t_2 is related to the arrival time of P-wave, rupture duration and S-P waves arrival time difference for the crustal deformations and it is given below

$$t_2 = t_1 + t_r + \Delta t_{sp} \tag{5.6}$$

Any deformation after time t_2 must be associated with deformations related to local plastic behaviour of ground at the instrument location. We show one example for defining times t_1 and t_2 on a record taken at HDKH07 strong motion station in the 2003 Tokachi-oki earthquake, Japan (Figure 5.27). The estimated rupture time for this earthquake is about 40 seconds (Kikuchi, 2003) with about 18 seconds S-P waves arrival time difference (Δt_{sp}).

Another important issue is how to select filter values in Sections 1 and 2. This is a somewhat subjective issue and it depends upon the sensitivity of the accelerometers. The filter value ε_1 is generally small and this stage is associated with pre-trigger value of instruments. The experience with the selection of ε_1 for K-NET and KIK-NET

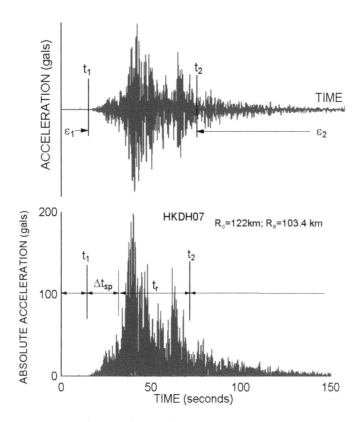

Figure 5.27 Definition of sections in the EPS method.

accelerometer records implies that its value be less than ±2 gals. As for the value of ε_2, higher values must be assigned. Again the experiences with the records of K-NET and KIK-NET accelerometers imply that its value should be ±6 gals. As for the experiences for the records of Turkey and Italy, these values are much less than those high-sensitive accelerometers of networks of Japan.

This method is applied to results of laboratory faulting and shaking table tests, in which shaking was recorded using both accelerometers and laser displacement transducers, simultaneously. Furthermore, the method was applied to strong motion records of several large earthquakes with measurements of ground movements by GPS as seen in Figure 5.28. The comparison of computed responses with actual recordings was almost the same, implying that the proposed method can be used to obtain actual recoverable as well as permanent ground motions from acceleration recordings. Figure 5.29 shows the application of the EPS method to the strong motions records of 2009 L'Aquila earthquake to estimate the co-seismic permanent ground displacements. These results are very consistent with the GPS observations. However, it should be noted that permanent ground deformations recorded by the GPS do not necessarily correspond to those of the crustal deformation. Surface deformations may involve crustal deformation as well as those resulting from the plastic deformation of ground due to ground shaking. The records at ground surface and 260 m below the ground

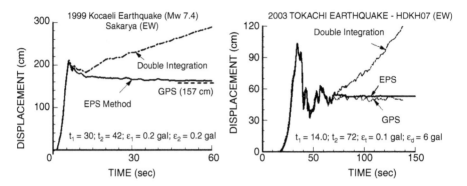

Figure 5.28 Comparison of the permanent ground deformation by the EPS method with measured GPS recordings (from Ohta and Aydan, 2007b).

Figure 5.29 Estimated permanent ground displacements by EPS method (from Aydan et al., 2009e).

surface taken at IWTH25 during the 2008 Iwate-Miyagi earthquake clearly indicated the importance of this fact in the evaluation of GPS measurements (KIKNET, 2008).

5.4.4 Okada's method

Okada (1992) proposed closed form solutions for dislocation in a half-space isotropic medium (Figure 5.30 and 5.31). Closed form analytical solutions are presented in a

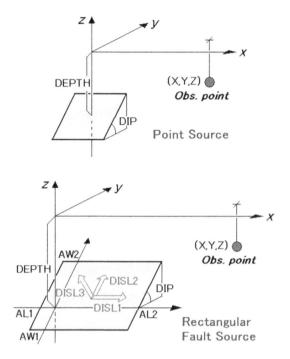

Figure 5.30 Geometrical illustration of assumed fault and relative displacements.

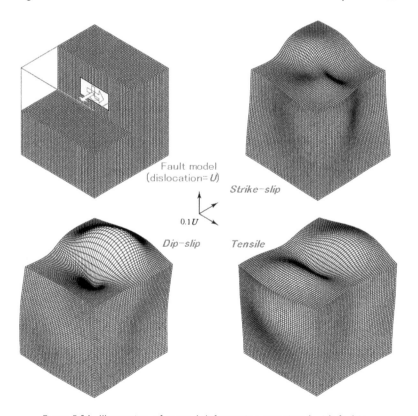

Figure 5.31 Illustration of ground deformation associated with faulting.

Figure 5.32 Relations between static stress changes and seismicity (from Toda *et al.*, 2002).

unified manner for the internal displacements and strains due to shear and tensile faults in a half-space for both point and finite rectangular sources. These expressions evaluate deformations in an infinite medium and a term related to surface deformation is obtained through multiplying by the depth of observation point. Stein (2003) utilized the solutions of Okada's method in his software to compute permanent ground deformation and associated stress changes. This method is also used to forecast earthquakes with the introduction of superposing displacement field and associated stress changes together with the use of Mohr-Coulomb criterion (King *et al.*, 1994; His method is upgraded by his research group and applied to various earthquakes in recent years). Figure 5.32 shows an example of computations by Toda *et al.* (2002) for earthquake activity in Izu Islands.

5.4.5 Numerical methods

FDM, FEM, BEM or combined FDM, FEM and BEM produce permanent ground deformation as a natural output of computations provided that the rupture process is well simulated. Figures 5.23 and 5.24 show an example of such a computation for the 2014 Nagano-Hokubu earthquake and 2011 Great East Japan earthquake (or Tohoku earthquake). Although the displacement response could be more easily simulated, acceleration responses simulation requires fine meshes, which undoubtedly require the use of super-computers. It should be also noted that FEM models could easily simulate both permanent displacements in addition to strong ground motions.

Chapter 6

Dynamic responses and stability of rock foundations

6.1 MODEL EXPERIMENTS ON FOUNDATIONS

Several experiments were carried out to investigate the effect of faulting on a bridge and its foundations by Ohta (2011) and Aydan *et al.* (2011). The bridge was a truss bridge just over the projected fault line. Figure 6.1 shows truss bridge models above the jointed rock mass foundation. Figure 6.1(a) shows views of the bridge model before and after the experiments subjected to the forced displacement field of vertical normal faulting mode. Bridge foundations were pulled apart and tilted. The vertical offset was 0.37 times the bridge span. Similarly Figure 6.1(b) shows the bridge model before and after the experiments subjected to the forced displacement field of 45° thrust faulting mode. Bridge foundations were also pulled apart at the top and compressed at the bottom and tilted as seen in Figure 6.1(b).

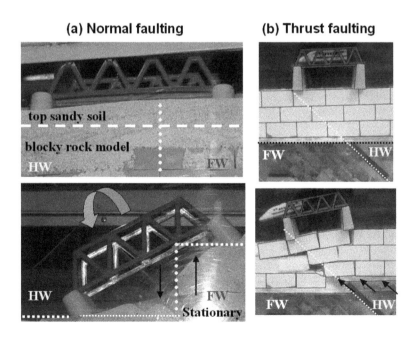

Figure 6.1 Views of the model truss bridge before and after experiments.

Table 6.1 A list of major earthquakes, which caused damage to bridge foundations.

Earthquake	Year	Mw	Number of damage and comments
Nobibeya	1891	8.0	Collapse of bridges over Nagara and Kiso Rivers
Alaska (USA)	1964	9.2	Million Dollar Railway bridge collapsed
Niigata	1964	7.3	Collapse of Showa bridge piers on Shinano River
Manjil	1990	7.6	Sefid Rud gravity buttress dam damaged by nearby faulting
Kobe	1995	6.9	Akashi Suspension Bridge Awajima side pier is off set by 100 cm dextral slip
Kocaeli (Turkey)	1999	7.5	Settlement and lateral movement of piers of bridges over Sakarya River
Düzce (Turkey)	1999	7.2	Rotation and 2–3 m dextral offset of 30–40 m high expressway piers at Kaynaşli
Chi-Chi	1999	7.6	Mingju, Wufeng, Peyfong Bridges by thrust faulting
Denali	2002	7.9	Tanana roadway bridge damaged
Kashmir	2005	7.5	Offset of girders of a RC bridge at Balakot
Iwate-Miyagi	2008	6.9	Coumnar toppling of Akita-side abutment at Matsurube bridge, failure of abutment at Shosenkyo Bridge
Wenchuan	2008	7.5	Bridges in Yingxiu, Beichuan towns, Baihua Bridge, Gaoyuan, Bailu
Kumamoto	2016	7.0	Choyo Bridge failed due to columnar joints, Aso Bridge foundation failed due to faulting and columnar jointing

Table 6.2 A list of major earthquakes, which caused damage to dam foundations.

Earthquake	Year	Mw	Number of damage and comments
San Francisco (USA)	1906	7.9	Crystal springs and San Andreas dams 1.8–2.4 m dextral offset however no leakage due to clay core
Alaska (USA)	1964	9.2	Eklutna dam due to shaking
Kern (USA)	1952	7.5	Damage to gravity dams along Kern River
Chi-chi	1999	7.6	Shikang gravity dam ruptured by 8 m thrust faulting
Wenchuan	2008	7.6	Zipungpu Rockfill dam is damaged by settlement of relative upward movements
Iwate-Miyagi	2008	6.9	Isawa Rockfill dams
Kumamoto	2016	7.0	Ohkirihata earthfill dam, fault passed through right abutment

6.2 OBSERVATIONS OF DAMAGE TO FOUNDATIONS BY EARTHQUAKES

In this section, observations of damage to foundations due to earthquakes are described. Tables 6.1 to 6.4 list some major earthquakes, which caused damage to various types of foundation of superstructures. The details of damage can be found in the reconnaissance reports of earthquakes published by several organizations such as JSCE, JGS, ASCE, PEER, TDV and numerous papers in the literature of rock engineering, earthquake engineering and architecture.

Table 6.3 A list of major earthquakes, which caused damage to pipelines.

Earthquake	Year	Mw	Number of damage and comments
San Francisco (USA)	1906	7.9	Rupture of water pipelines by dextral faulting
Koaceli	1999	7.5	Oil pipeline rupture at Tepertarla by 2 m dextral offset, water pipeline rupture at Kullar by 4 m dextral offset, pipelines of SEKA in Sapanca Lake
Düzce (Turkey)	1999	7.2	Rupture of pipes at Kaynaşlı, Findikli
Denali	2002	7.9	Dextral offset by 3 m dextral slip of Denali fault but no damage due to earthquake-proof design
Iwate-Miyagi	2008	6.9	Rockfill dams

Table 6.4 A list of major earthquakes, which caused damage to pylons.

Earthquake	Year	Mw	Number of damage and comments
Kobe (Japan)	1995	6.9	Dextral slip along Nojima fault
Koaceli	1999	7.5	Secondary normal faulting at Kavakli
Chi-Chi	1999	7.6	Thrust faulting
Kashmir	2005	7.5	Damage to pylons by thrust faulting near Balakor
Wenchuan	2008	7.6	Damage by rock slope failures and rock falls at Yingxiu, and thrust faulting
Kumamoto	2016	7.0	Damage to pylons in Tateno and Minami Aso Village due to slope failure or faulting

6.2.1   Roadways and railways

The Trans-European Motorway (TEM) of Turkey was damaged at three different locations by the earthquake fault caused by the 1999 Kocaeli earthquake of Turkey (Aydan et al., 1999; Ulusay et al., 2001). The motorway with east and west bounds having 3 lanes each was slightly elevated through embankments in the earthquake affected region. The surfacing of the motorway was damaged by rupturing and buckling as seen in Figure 6.2. The railways were also built on the existing ground surface. The railways were buckled near Tepetarla station where the earthquake fault crossed the railways at an angle of 50–55° with well known 'S' shape (Figure 6.3).

6.2.2   Bridges and viaducts

The earthquake fault of the 1995 Kobe earthquake passed through the piers of Akashi Suspension Bridge and the piers were displaced by more than 100 cm horizontally (Figure 6.4). Nevertheless, the earthquake faulting did not cause any major damage to the bridge as the girders of the bridge was not constructed yet.

Along the damaged section of the TEM motorway mentioned above, there were several overpass bridges. Among them, a four span overpass bridge at Arifiye junction collapsed as a result of faulting (Figure 6.5). The fault rupture passed between the northern abutment and the adjacent pier. The overpass was designed as a simply supported structure according to the modified AASHTO standards and girders had elastrometric bearings. However, the girders were connected to each other through

Figure 6.2 Buckling of roadway surfacing.

Figure 6.3 Buckled railways.

prestressed cables. The angle between the motorway and the strike of the earthquake fault was approximately 15° while the angle between the axis of the overpass bridge and the strike of the fault was 65°. The measurements of the relative displacement in the vicinity of the fault ranges between 330 and 450 cm. Therefore an average value of 390 cm could be assumed for the relative displacement between the pier and the abutment of the bridge. The Pefong bridge collapsed due to thrust faulting in the 1999 Chi-chi earthquake, which passed between the piers near its southern abutment as seen in Figure 6.6.

During the 1999 Düzce earthquake, a 50 m high pier of the expressway was rotated by faulting just below the pier. As the skew angle between the expressway viaducts and earthquake fault was quite narrow, the faulting did not cause the fall of girders as seen in Figure 6.7 (Aydan *et al.*, 2000).

The foundation failure of Matsurube Bridge by the 2008 Iwate-Miyagi earthquake was caused due to columnar toppling (Figure 6.8). The abutment rock mass is tuff and fractured by columnar cooling joints. The column rock mass at Akita side toppled towards the valley so that the upper deck of the bridge buckled and piers were displaced and rotated.

Figure 6.4 A view of Akashi bridge after construction with seabed fault trace.

Figure 6.5 The collapse of the overpass bridge at Arifiye.

The largest bridge damaged by the 2005 Kashmir earthquake was located in Balakot town, where the NW tip of the causative fault terminated (Figure 6.9). The girder of the bridge was displaced by about 1 m in the southward direction. While the west side of the bridge sits over the pier beneath, the eastern side of the bridge is offset from the pier by about 35–50 cm. In other words, it seems that the distance between

Figure 6.6 Collapse of Pefong Bridge (note the uplifted ground on RHS).

Figure 6.7 Damage to the piers of the TEM expressway at Kaynaşlı.

the piers decreased. The ground on the eastern side of the bridge is likely to experience the permanent displacement of ground resulting from the slope failure as well as from thrust faulting. There is no doubt that the inferred ground motion is quite high at Balakot and this may also cause the permanent displacement of the bridge deck as the girders were not fixed against horizontal ground motions.

One of the heavly damaged bridges was the four span reinforced arch Xiaoyudong bridge in Longmenshan town (Lin and Chai, 2008), after which the Longmenshan fault zone was named. The arch sections of the bridge were sheared as shown in Figure 6.10. Besides high ground motions in the vicinity of the bridge, the shearing of arched section in the direction of the longitudinal axis of the bridge was caused by permanent ground movement resulting from thrust faulting.

The Kumamoto earthquake with a moment magnitude of 7.0 on April 16, 2016 caused some damage to bridge foundations (Aydan *et al.*, 2017) (Figure 6.11). The foundations of bridges failed due to discontinuous structure of rock masses. The failure at the abutment on Tateno side of the Aso-Choyo Bridge took place due to the planar

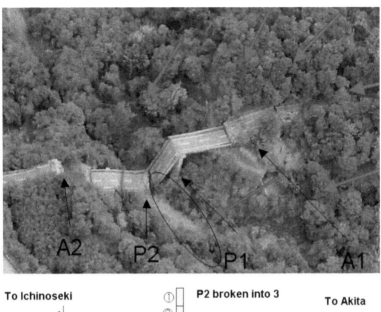

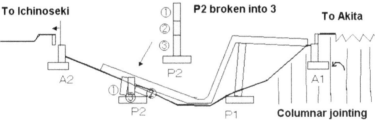

Figure 6.8 An aerial view of damage to Matsurube bridge by the 2008 Iwate-Miyagi intraplate earthquake.

sliding of rock blocks beneath. The rock mass beneath the footing was intensely jointed in the form of inclined columnar joints.

The rock mass beneath the foundation of the Great Aso Bridge (Aso-Ohashi), which was an arched truss structure on the Tateno side failed due to probably a combined form of sliding and toppling of rock blocks. The heavy ground shaking, permanent movements of foundation rockmass and the close proximity of the bridge to the trace of the earthquake fault were probably the main causes of the total failure of the bridge.

The south abutment of the Minami-Aso bridge also suffered the effect of permanent ground deformations induced by partial slope movements and dextral type faulting movements. The stoppers of this bridge were ruptured and the bridge deck was rotated.

An interesting failure of the Toshita Bridge was caused by the toppled rock slope falling over the bridge deck. The piers of the bridge were of T-type. One of the piers of this bridge was broken due to bending resulting from the shock load of the failed slope, and the pier together with the deck of the bridge fell to the river bottom as seen in Figure 6.12.

Figure 6.9 Displaced girder of the Balakot Bridge due to permanent displacement of piers and heavy shaking.

Figure 6.10 A view of the four span reinforced arch Xiaoyudong Bridge in Longmenshan town.

Figure 6.11 Damage to foundations of bridges.

Figure 6.12 Views broken pier and deck of the Toshita Bridge at the bottom of the river.

Figure 6.13 Failure of Shihkang dam due to thrust faulting.

6.2.3   Dams

There are many dams damaged to some extent in various earthquakes in the past. Among them, the most catastrophic damage occurred at Shihkang dam during the 1999 Chi-Chi earthquake in Taiwan. The Shihkang dam, which is a concrete gravity dam with a height of 25 m, was ruptured by the thrust type faulting during the 1999 Chi-Chi earthquake (Figure 6.13). The relative displacement between the uplifted part of the dam was more than 980 cm. Liyutan rockfill dam with a height of 90 m and a crest width of 210 m, which was on the overhanging block of Chelongpu fault was not damaged even though the acceleration records at this dam showed that the acceleration was amplified 4.5 times of that at the base of the dam (105 Gal).

The deformation zone of faulting during the 2008 Wenchuan earthquake in Tibet (China) caused some damage at Zipingpu Dam with concrete facing (JSCE, 2008; Aydan *et al.*, 2008). The Zipingpu Dam is about 9 kilometers upstream from the city of Dujiangyan. The surveying indicated that the abutments of the dam moved inward and the maximum displacement was 101.6 mm. The settlement of the crest of the dam was 734.6 mm with a 179.9 mm displacement towards the downstream side (Figure 6.14). The seepage of the dam changed from 10.38 l/s to 15.07 l/s. The Zipingpu dam was designed for the base acceleration of 0.26 g but the acceleration at the crest during the earthquake was greater than 2 g. The gates and power generating units were damaged by the earthquake.

Ohkirihata Dam, which is a earthfill dam with a height of 23 m for irrigation purposes, was damaged by faulting induced by the 2016 Kumamoto earthquake and the leakage of reservoir water occurred. The earthquake fault passed through the dam embankment near the right-abutment and the relative displacement of the fault was about 140 cm in the vicinity of the dam (Figure 6.15).

The wall of the water reservoir of the Kurokawa Hydroelectric Power Plant failed due to slope failure, which induced a debris flow causing damage and casualties in the Shinjo area of Tateno Village.

6.2.4   Power transmission lines

Power transmission lines generally consist of pylons and power transmission cables and they are quite resistant to earthquakes. The design of pylons and cables is generally

Figure 6.14 Views of Zipingpu dam after the 2008 Wenchuan earthquake.

Figure 6.15 Views of damage and faulting at Okirihata Dam.

(a) 1999 Kocaeli earthquake   (b) 1995 Kobe Earthquake   (c) 1999 Chi-chi Earthquake

Figure 6.16 Damage to pylons due to faulting.

based on the wind loads resulting from typhoons or hurricanes. The cables do not fail during earthquakes unless the pylons are toppled due to faulting, shaking or slope failure.

During the 1999 Kocaeli earthquake, only one pylon was damaged nearby Ford-Otosan automobile factory at Kavakli district of Gölcük town. At this site a normal fault, which is a secondary fault to the main lateral strike-slip faulting event, crossed through the foundations of the pylon and its vertical throw was about 240 cm. One of the foundations of the pylon was pulled out of the ground and was exposed as seen in Figure 6.16(a). Some of its truss elements were slightly buckled. Similar type damage to pylons straddling the Nojima fault break in the 1995 Kobe earthquake (Figure 6.16(b)) and the Chelungpu fault during the 1999 Chi-Chi earthquake were observed as shown in Figure 6.16(c).

The site investigations and additional data and information other sources clearly indicated that the damage to pylons were either caused by slope failures (rock falls) or faulting. Although the effect of ground shaking on the damage to pylons seems to be small in comparison, the selection of foundation ground for high pylons and their structural design and construction should have been very carefully implemented.

Many pylons were damaged by the Kumamoto earthquake on April 2016 due to permanent deformation of their foundations as a result of ground movements induced by faulting and slope failures. Figure 6.17 shows several examples of damage to pylons. Although the earthquake did not cause the toppling of the pylons, they were either tilted or deformed, and some segments became loose due to rupture of the bolts.

6.2.5 Tubular structures

Tubular structures may be specifically designated as petrol and gas pipe-lines, water pipes and sewage systems as seen in Figure 6.18. They can be also classified as line-like structures. These structures may fail due to either buckling or separation during a faulting event. Five such incidents were observed during the 1999 Kocaeli earthquake

Figure 6.17 Views of damage to pylons.

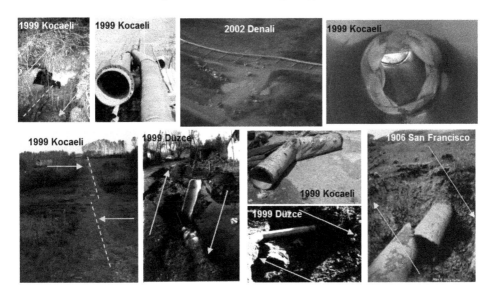

Figure 6.18 Faulting induced damage to pipelines.

(Figure 6.18). One of the incidents involved the separation of a ductile iron pipe as a result of faulting near the collapsed overpass bridge. The second incident took place at the pumping facility of the Seka papermill plant at Sapanca Lake. The third incident occurred near Tepetarla village where the railways were buckled. The fourth and fifth incident took place at Arifiye and nearby Basiskele. The fifth incident was quite important since the fault caused a heavy damage to the main water-pipe having a diameter of 2 m. Similar types of failure took place in the sewage pipe networks whenever faulting breaks were observed. The natural gas pipelines crossing the Izmit Gulf between Yalova and Pendik were undamaged. Some brittle asbestos water pipes were also damaged in

Kaynasli and Findikli due the fault rupture of the 1999 Düzce earthquake. Although it is difficult to prevent damage to tubular structures, the use of flexible joints may be effective in such active fault zones when they are embedded. Similar observations were done in various recent large earthquakes.

6.3 ANALYTICAL AND NUMERICAL STUDIES ON ROCK FOUNDATIONS

The interaction of foundations of superstructures under the imposed displacement field and shaking conditions due to faulting and earthquake shaking is quite a complex phenomenon. The most simple technique is to use a forced displacement field on the structure. Pseudo-dynamic techniques simulate ground deformation during faulting while eliminating the inertia effects. The most proper evaluation is to do analyses purely under dynamic conditions including the forced displacement and shaking conditions although such analyses are quite rare.

6.3.1 Simplified techniques

(a) Analyses of a pipeline structure under forced displacement field

Let us consider a simple beam-like structure with a length L as shown in Figure 6.19. Shear and bending moment acting at a given section under a forced displacement y can be obtained as follows:

$$S(x) = \frac{12EI}{L^3}y \tag{6.1}$$

$$M(x) = \frac{6EI}{L^2}y\left(\frac{2x}{L} - 1\right) \tag{6.2}$$

where E and I are elastic modulus and areal inertia moment. The displacement along the beam takes the following form:

$$v(x) = y\left(\frac{x}{L}\right)^2\left(3 - \frac{2x}{L}\right) \tag{6.3}$$

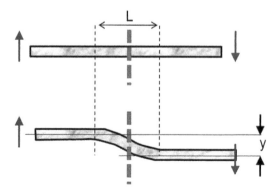

Figure 6.19 Modeling of a beam-like structure under forced displacement field.

This type model is quite often used to design tubular structures under forced displacement field due to earthquake faulting. Figure 6.20 shows an example of computation for equations presented above.

The same methodology is followed for structures under more complex boundary and geometrical conditions using truss or frame-type numerical analyses (Ohta, 2011). Figure 6.21 shows the deformation response of a tubular structure under normal and

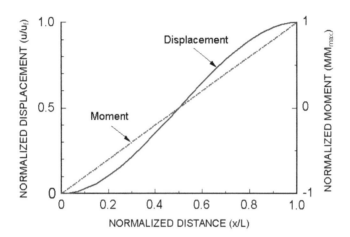

Figure 6.20 Distribution of displacement and moment along the fault zone (from Ohta, 2011).

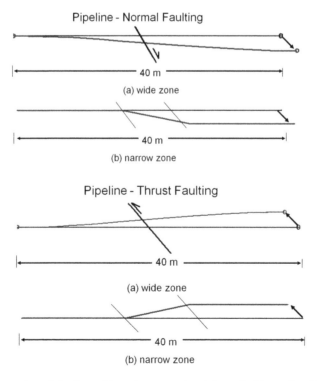

Figure 6.21 Deformation of pipeline subjected to normal and thrust faulting action (from Ohta, 2011).

Table 6.5 Material and geometrical properties of pipeline.

Material	Elastic Modulus (GPa)	Thickness (mm)	Diameter (m)
Pipe	210	10	2

283 cm relative normal or thrust faulting.

Table 6.6 Material and geometrical properties of truss structures.

Material	Elastic Modulus (GPa)	Area (m²)
Bridge	210	0.025
Pylon	210	0.025

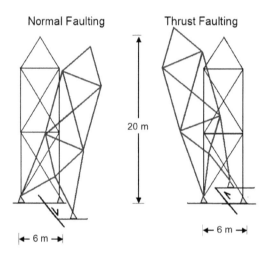

Figure 6.22 Deformation response of 20 m high pylon subjected to normal and thrust faulting action.

thrust faulting conditions with a net throw of 283 cm along the fault plane with the use of frame-type finite element approach. Material properties used in the analysis are given in Table 6.5. In the simulations the width of the fault zone was varied and its effect of displacement response and resulting shear force and moment distributions were also investigated.

A two-dimensional truss-type finite element method was used to simulate responses of pylons and bridges under forced displacement conditions of normal and thrust type faulting with a relative net-slip of 283 cm using the material properties given in Table 6.6. Figures 6.22 and 6.23 shows the computational results for the assumed conditions.

Figure 6.24 shows the displacement response of a bridge modeled as a frame-structure subjected to normal and thrust faulting conditions with relative slip of 283 cm.

The final example is associated with the response of a building subjected to normal and thrust faulting action with a relative slip of 283 cm with material properties given

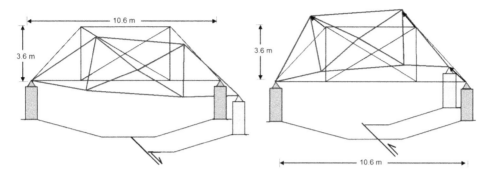

Figure 6.23 Deformation response of a truss bridge subjected to normal and thrust faulting action.

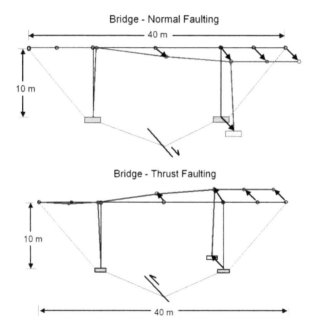

Figure 6.24 Deformation response of a moment resisting bridge subjected to normal and thrust faulting action (from Ohta, 2011).

Table 6.7 Material and geometrical properties of bridge.

Material	Elastic Modulus (GPa)	Area (m²)	Inertia (m⁴)
Girder	21	$4(4 \times 1)$	$1.333(4 \times 1^3/12)$
Piers	21	$4(2 \times 2)$	$0.333(2 \times 2^3/12)$

Table 6.8 Material and geometrical properties of building.

Material	Elastic Modulus (GPa)	Thickness (mm)	Width (mm)
Column/beam	21	300	300

283 cm relative normal or thrust faulting

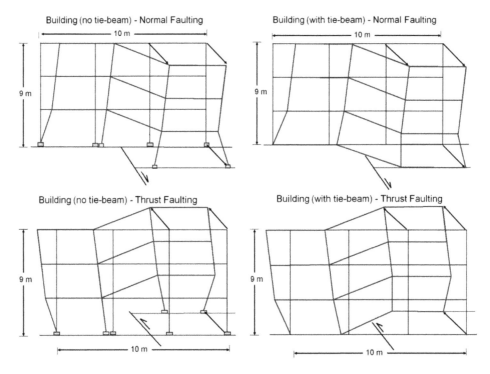

Figure 6.25 Deformation response of a moment resisting building subjected to normal and thrust faulting action (from Ohta, 2011).

in Table 6.8. The effect of tie-beams is investigated for two faulting conditions. The computational results are shown in Figure 6.25. Although the overall deformation is not affected by the existence of tie-beams, they result in the reduction of member forces and moments, which are generally critical for the overall stability of the structures.

6.3.2 Pseudo-dynamic techniques

The most important aspect in earthquake engineering is the interaction between structures and fault breaks. For this purpose, a truss structure straddling over the projected fault trace on the ground surface was considered, and normal faulting and thrust faulting conditions are imposed through prescribed displacement at selected points as in the previous computations using a pseudo-dynamic version of the discrete finite element

Table 6.9 Materials properties used in discrete finite element method simulations.

Material	λ (MPa)	μ (MPa)	γ (kN/m³)	c (MPa)	ϕ (°)	σ_t (MPa)
Solid	2000	2000	26	–	–	–
Fault	50	50	–	0.0	40	0.0

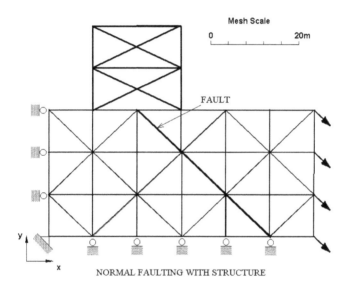

NORMAL FAULTING WITH STRUCTURE

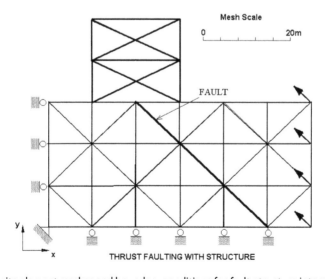

THRUST FAULTING WITH STRUCTURE

Figure 6.26 Finite element meshes and boundary conditions for fault-structure interaction simulations.

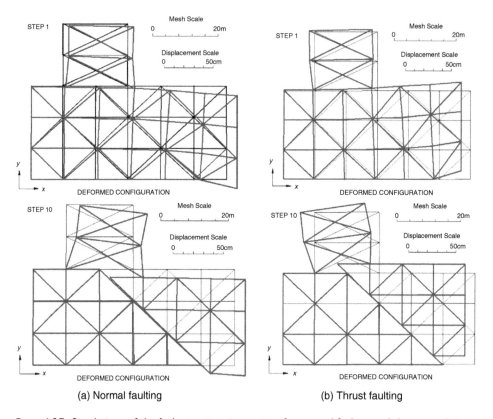

Figure 6.27 Simulations of the fault-structure interaction for normal faulting and thrust conditions.

method (Aydan *et al.*, 1996; Aydan, 2003). The material properties for solid and fault are given in Table 6.9. The elastic modulus and the cross-section area of a typical truss were chosen as 90 GPa and 0.1 m^2 and their behaviour was assumed to be elastic. Figure 6.26 shows the finite element meshes and boundary conditions used in simulations. Figure 6.27 shows the deformed configurations at computation steps 1 and 10 for normal and thrust type faulting modes. In both cases, the truss structure tilts. While the thrust type faulting causes the contraction of trusses, the normal faulting condition results in the extension of trusses and separation of the supporting members fixed to the ground. Although trusses were assumed to be behaving elastically in these simulations, it is quite easy to implement their elasto-plastic behaviour within the DFEM.

6.3.3   Pure dynamic techniques

Nakano and Ohta (2008) proposed a elasto-plastic finite element method to simulate the dynamic response of frame-structures under forced time-series displacement condition applied to one of the abutments (Figure 6.28). They evaluated the damage

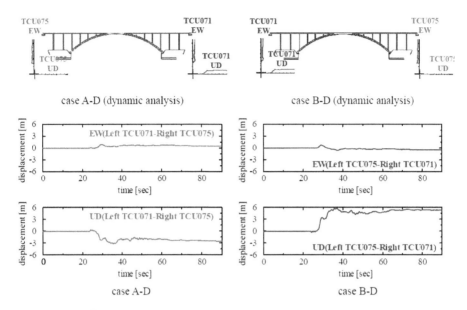

Figure 6.28 Numerical model and imposed dynamic responses on the models.

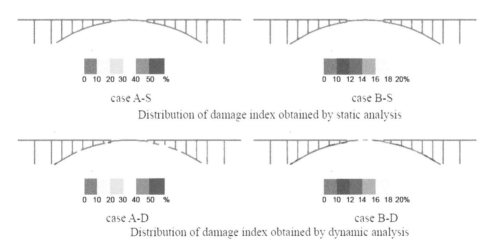

Figure 6.29 Comparison of damage distribution under static and dynamic type simulations.

situation in the structure and compared the results with static type analyses similar to those described in Sections 6.3.1 and 6.3.2. They concluded that the consideration of the inertial effects is quite important if the structure response involves plastic behaviour (Figure 6.29).

Chapter 7

Dynamic responses and stability of underground excavations in rock

7.1 GROUND MOTIONS IN UNDERGROUND STRUCTURES

It is well known that the ground motions are generally smaller than those at ground surface. Nasu (1931) carried out the first instrumental studies on tunnels during the aftershock activity following the 1924 Izu earthquake with a 2.4 m offset. Kanai and Tanaka (1951) measured ground acceleration in underground caverns and at the ground surface. These measurements indicated that the surface acceleration was generally twice or greater than twice of that at depth as expected theoretically.

Komada and Hayashi (1980) reported the results of an extensive monitoring program on ground motions during earthquakes on underground caverns and adjacent tunnels. They also investigated the frequency content and amplification in relation to magnitude and distance of earthquakes. However, these caverns were not in the epicentral area.

Figure 7.1 shows the acceleration records measured on the ground surface (GSA) and underground (GSG) during the 2009 Mw 6.3 L'Aquila earthquake. The GSA station is at Assergi and the GSG station is located in an underground gallery of Gran Sasso Underground Physics Laboratory of Italy. Both stations are founded on Eocene limestone with a shear wave velocity of about 1 km/s. Although the epicentral distances and ground conditions are almost the same, the acceleration at ground surface is amplified almost 6.4 times that measured in the underground gallery (Aydan *et al.*, 2010a,b).

The recent global positioning system (GPS) also showed that permanent deformations of the ground surface occur after each earthquake (Figure 5.10, 5.11) as pointed out previously. The permanent ground deformation may result from different causes such as faulting, slope failure, liquefaction and plastic deformation induced by ground shaking. These types of ground deformation have limited effect on small structures as long as the surface breaks do not pass beneath those structures.

Blind faults and folding processes may also induce some peculiar ground deformations and associated folding of soft overlaying sedimentary layers. Such deformations caused tremendous damage to tunnels during the 2004 Chuetsu earthquake although no distinct rupturing took place.

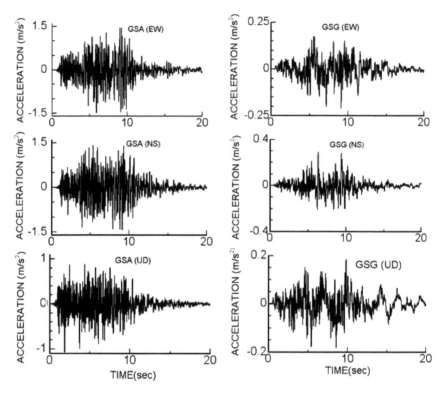

Figure 7.1 Acceleration records at GSA and GSG strong motion stations (Aydan *et al.*, 2009, 2010a,b, 2011).

7.2   MODEL EXPERIMENTS ON SHALLOW UNDERGROUND OPENINGS

Many model experiments on underground openings have been performed by Aydan and his group (e.g. Aydan *et al.*, 1994, 2011; Genis and Aydan, 2002). The initial series of experiments on shallow underground openings in discontinuous rock mass using non-breakable blocks were reported by Aydan *et al.* (1994), in which a limit equilibrium method was developed for assessing their stability. These experiments have been recently repeated using the breakable material following the observations of damage to tunnels caused by the 2008 Wenchuan earthquake. The inclination of continuous discontinuity plane varied between 0° and 180°. Figure 7.2 shows views of some of the experiments. Unless blocks fail themselves, the failure modes were very similar to those of the model experiments using hard blocks. In some of the experiments with discontinuities dipping into the mountain side, flexural toppling of the rock mass model occurred. The comparison of the preliminary experimental results with the theoretical estimations based on Aydan's method (Aydan, 1986; Aydan *et al.*, 1994) are remarkably close to each other (Ohta and Aydan, 2010).

Aydan *et al.* (2011) have also performed some model experiments on the effect of faulting on the stability and failure modes of shallow underground openings. Figure 7.3

Figure 7.2 Failure modes of shallow tunnels adjacent to slopes with breakable material (from Aydan *et al.*, 2011).

shows views of some model experiments on shallow underground openings subjected to the thrust faulting action with an inclination of 45°. Underground openings were assumed to be located on the projected line of the fault. In some experiments three adjacent tunnels were excavated. While one of the tunnel was situated on the projected line of faulting, the other two tunnels were located in the footwall and hanging wall side of the fault. As seen in Figure 7.3, the tunnel completely collapsed or was heavily damaged when it was located on the projected line of the faulting. When the tunnel was located on the hanging wall side, the damage was almost nil in spite of the close proximity of the model tunnel to the projected fault line. However, the tunnel in the footwall side of the fault was subjected to some damage due to relative slip of layers pushed towards the slope. This simple example clearly shows the damage state may differ depending upon the location of tunnels with respect to fault movement as well as the slope geometry.

7.3 TUNNELS

Aydan *et al.* (2011) compiled a number of case histories such as tunnel damage, caves and underground power houses from various reports, papers and their own observations (Kuno, 1935; Okamoto, 1973; Rozen, 1976; Tsuneishi *et al.*, 1978; Kawakami, 1984; Asakura *et al.*, 1996; Prentice and Ponti, 1997; Asakura and Sato, 1998; Berberian *et al.*, 1999; Hashimoto *et al.*, 1999; Sakurai, 1999; Aydan *et al.*, 2000; Ueta *et al.*, 2001; Wang *et al.*, 2001; Ulusay *et al.*, 2002; Aydan, 2003; Esghi and

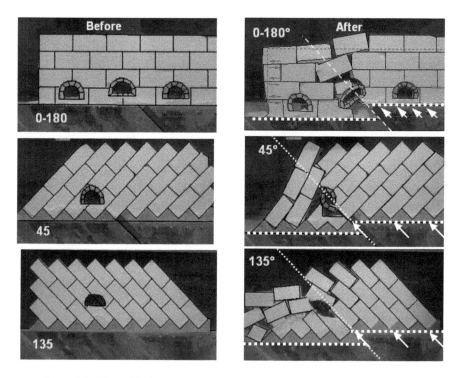

Figure 7.3 Effect of faulting on underground openings (from Aydan *et al.*, 2011).

Zare, 2003; Aydan and Kawamoto, 2004; TEC-JSCE, 2005; Yashiro *et al.*, 2007; Aydan *et al.*, 2009a,b,c). The earthquakes, which caused the damage to tunnels, are listed in Table 7.1. The damage to underground structures may be classified as:

a)   portal damage (Figure 7.4) and
b)   shaking induced damage (Figure 7.5)
c)   permanent ground deformation induced damage (Figure 7.6).

Permanent ground deformation induced damage is generally caused either by faulting or slope movements.

The past experience on the performance of tunnels through active fault zone during earthquakes indicates that the damage is restricted to certain locations (Aydan *et al.*, 2010; Hashimoto *et al.*, 1999, Yashiro *et al.*, 2007; Asakura *et al.*, 1996). Portals and the locations where the tunnel crosses the fault may be damaged as occurred in the 2004 Chuetsu, 2005 Kashmir and 2008 Wenchuan earthquakes (Figure 7.4). A section nearby the Elmalık portal of Bolu Tunnel collapsed (Figure 7.5). This section of the tunnel was excavated under very heavy squeezing conditions (Aydan *et al.*, 2000). The well known examples of damage to tunnels at locations, where the fault rupture crossed the tunnel, are mainly observed in Japan. The Tanna fault that ruptured during the 1930 Kita-Izu earthquake caused damage to a railway tunnel and

(a) 2004 Chuetsu   (b) 2005 Kashmir   (c) 2008 Wenchuan

Figure 7.4 Examples of damaged portals of tunnels.

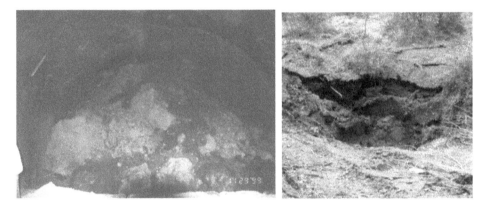

Figure 7.5 Collapse of Bolu tunnel during the 1999 Düzce earthquake.

the relative displacement was about 100 cm (Kuno, 1935; Sakurai, 1999). The 1978 Izu-Oshima Kinkai earthquake induced damage to Inatori railway tunnel (Kawakami, 1984; Tsuneishi *et al.*, 1978). Similar type damage with a small amount of relative displacements due to motions of Rokko, Egeyama and Koyo faults to the tunnels of Shinkansen and subway lines through the Rokko mountains were also observed (Aydan, 1996). During the 1999 Chi-Chi earthquake, the portal of the water intake tunnels was ruptured over a length of 10 m as a result of thrust faulting. Except this section, the tunnel was undamaged over its entire length.

Jiujiaya Tunnel is a 2282 m long double lane tunnel. It is 226.6 km away from the earthquake epicenter and it is about 3–5 km away from the earthquake fault of Wenchuan earthquake. The tunnel face was 983 m from the south portal at the time of the earthquake. The concrete lining follows the tunnel face at a distance of approximately 30 m. 30 workers were working at the tunnel face and one worker was killed by the flying pieces of rockbolts, shotcrete and bearing plates caused by intense deformation of the tunnel face during the earthquake (Aydan *et al.*, 2010a,b).

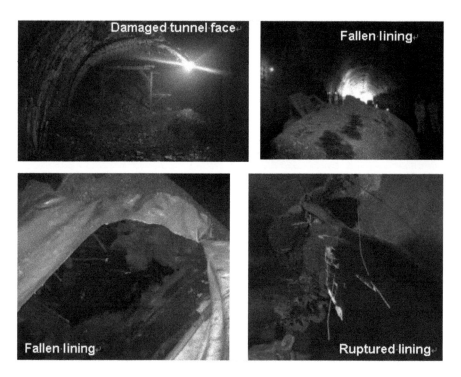

Figure 7.6 Earthquake damage at Jiujiaya Tunnel due to permanent deformations.

The concrete lining was ruptured and collapsed at several section (Figure 7.6). However, the effect of the unreinforced lining rupturing was quite large and intense in the vicinity of the tunnel face. The rupturing of the concrete lining generally occurred at the crown sections although there was rupturing along the shoulders of the tunnel at several places. Furthermore, the invert was uplifted due to buckling at the middle sections.

The earthquake on April 16, 2016 caused heavy damage to several tunnels in the vicinity of Tateno and Minami-Aso villages. Damage to Tawarayama Roadway Tunnel and Aso Railway Tunnel and Minami-Aso Tunnel were publicized (i.e. Manabe, 2016; Ikeda *et al.*, 2016; Kumamoto Prefecture, 2016; MLIT, 2016; Aydan *et al.*, 2017). The damage level index (DLI) of Tawarayama tunnel was 6 at two locations according to the classification proposed by Aydan *et al.* (2010) (Figures 7.7, 7.8, 7.9). The first damage occurred approximately 50–60 m from the west portal of the tunnel and the concrete lining was displaced by about 30 cm almost perpendicular to tunnel axis. The heaviest damage occurred over a length of 10 m about 1600 m away from the west portal and about 460 m from the east portal. The angle between the relative movement and tunnel axis was about 20–30 degrees. At this location, the non-reinforced concrete lining collapsed over a length of about 5 m. Although the tunnel is located about 2 km away from the main fault, the tunnel was damaged by secondary faults associated with the trans-tension nature of the earthquake fault.

Figure 7.7 Views of damage and their locations at Tawarayama Tunnel (from Aydan *et al.*, 2017).

Permanent ground deformation induced damage is generally caused either by faulting or slope movements. Most tunnels have non-reinforced concrete linings. Since the lining is brittle, the permanent ground movements may induce the rupture of the linings and falling debris may cause disasters with tremendous consequences to vehicles passing through. Therefore, this current issue needs to be urgently addressed. It should be also noted that the same issue is valid for the long-term stability of high-level nuclear waste disposal sites.

7.4 OBSERVATIONS ON ABANDONED MINES AND QUARRIES

Nishida *et al.* (1984) reported that damage to abandoned lignite mines occurred in Ohira and Esashi lignite field in Tohoku region during the 1978 Off Miyagi earthquake (M7.4) and the number of sinkholes and subsidence damage was 212. They also mentioned that damage occurred at Esashi lignite field, which was 30 m deep, even though the estimated ground acceleration was only 50 gals.

The earthquake caused sinkholes at 5 locations in Iwaki City in Fukushima Prefecture, 11 locations in Kurihara City, 7 locations in Osaki, 11 locations in Higashi

Figure 7.8 Views of damage and their locations at Aso railway tunnel (from Aydan *et al.*, 2017).

Figure 7.9 Views of damage at Minami-Aso Roadway Tunnel (from Aydan *et al.*, 2017).

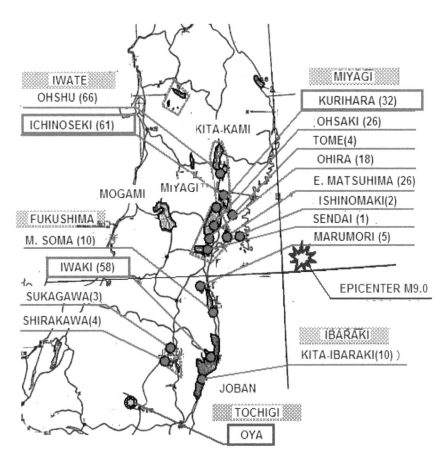

Figure 7.10   Coal and lignite fields in Tohoku Region together of sinkhole event number (Aydan & Tano, 2012b).

Matsushima and 3 locations in Kurogawa in Miyagi Prefecture, 11 locations in Ichinoseki City and 10 locations in Oshu City in Iwate Prefecture according to the first preliminary report released by the Ministry of Economy, Trade and Industry (METI) on March 30, 2011 (Aydan and Tano, 2012a). However, the press release on July 25, 2011 by the METI revealed that the number of events was more than 316 (Aydan and Tano, 2012b). Figure 7.10 also shows the number of events at each major abandoned mine areas. The drastic increase of the event number may be related to the attention of the ministry, which was initially concentrated on the Fukushima nuclear power plant incident. As the effects of the Fukushima nuclear power plant incident become gradually controlled, the ministry started to pay attention to damage caused by the abandoned mines. It was reported that the number of sinkhole occurrences increased almost 20 times the annual number of sinkholes in Iwate and Miyagi prefectures. In the following, the direct observations of the authors on abandoned lignite and coalmines are presented.

Figure 7.11 Views of sinkhole locations LI–L3 (Aydan & Tano, 2012b).

The most recent numbers of events related abandoned mines in Iwaki are 58. The maximum ground acceleration at Iwaki strong motions station of KiK-net was 481 gals at the ground surface. There were 5 sinkholes at the abandoned Yoshima mine of Jobando coal field as shown in Figure 7.11. The collapses were associated with inclined shafts. The sloshing of the underground water in inclined shafts caused the failure of roof rock resulting in sinkholes as seen in Figures 7.11 and 7.12. The ground water spouted from the sinkholes and ejected fragments of lignite, surrounding rock and sandy material. Similar types of event were also observed in previous earthquakes such as in the 2003 Miyagi Hokubu earthquake (Aydan and Kawamoto, 2004).

The Ministry of Economy, Trade and Industry of Japan (2011) reported that collapses occurred at 114 (32 initially) locations in Miyagi Prefecture and 127 (21 initially) locations in Iwate Prefecture. One of the reasons for such a difference may be that the attention of the Ministry was concentrated on the Fukushima Nuclear Power incident as mentioned before. The additional reason may be the effects of large aftershocks causing further collapses or sinkholes of already damaged zones in abandoned mines by the main shock. The section responsible for abandoned mines incidents in Miyagi Prefecture did not release the names of locations of sinkholes to the authors and the sites were located by the efforts of the authors together with the generous help of the local people who suffered from sinkholes caused by this earthquake.

The ground motions records differed from place to place. The largest ground motions were recorded at the Tsukidate strong motion station of K-NET. Except

Figure 7.12 Views of sinkholes locations L4–L5 (Aydan & Tano, 2012b).

top 1 m thick clay layer, this station is based on sedimentary rocks, which are commonly observed in lignite mine areas. Figure 7.13 shows the acceleration record at the Tsukidate strong motion station of K-NET. The maximum ground acceleration at Tsukidate strong motion station of K-NET was 2933 gals at the ground surface. However, the maximum ground acceleration measured at the strong motion station in Wakayanagi operated by the JMA was only 630 gals.

In Wakayanagi town of Kurihara City, 5 sinkholes occurred in residential area as seen in Figure 7.14. Luckily, these sinkholes did not cause the collapse of the residential houses on the ground surface. However, the fracturing of the base concrete occurred and one corner of the building was overhanging with a slight tilting. The overburden of the sinkholes ranged between 1 to 3 m at this locality.

In a large rice field near Ohira Village of Kurihara City, more than 10 sinkholes with a diameter ranging from 1 to 5 m were observed as seen in Figure 7.15. Besides ground shaking, the sloshing phenomenon of ground water in the underground openings abandoned lignite mine areas contributed to the collapse of the roof layers. As a result of this phenomenon, sandy material together with fragments of rocks and lignite appeared on the ground surface and this ejected sandy material is sometimes mis-interpreted as ground liquefaction by soil-mechanics researchers and engineers. The overburden of the sinkholes ranged between 2 to 4 m at this locality.

Besides sinkholes in the rice field, some sinkholes were observed along the roadway and in the parking lot next to residential buildings as seen in Figure 7.16.

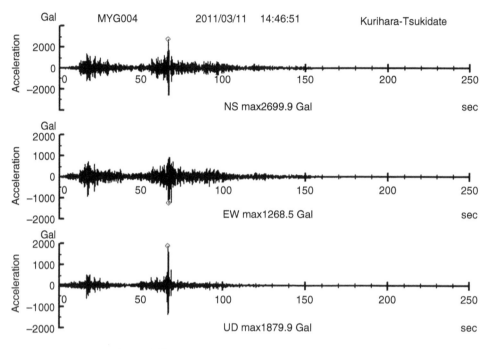

Figure 7.13 Acceleration records at Tsukidate station.

Figure 7.14 Some views of sinkholes in Wakayanagi town (Aydan & Tano, 2012b).

Figure 7.15 Views of sinkholes in a rice field near Ohira Village (Aydan & Tano, 2012b).

The overburden of the sinkholes with a diameter of 2–3 m ranged between 1 to 2 m at this locality. The ground shaking caused the collapse, as there was no trace of ground water spouted from the sinkhole. Two sinkholes in residential areas and one sinkhole in rice field occurred in the Tsukidate town of Kurihara City. Although there were no casualties in these events, they caused some anxiety for the people living in such areas.

The 2011 East Japan Mega Earthquake caused a number of sinkholes in both residential areas and fields in Ichinoseki. The sinkholes increased in size after each large aftershock. Figures 7.17 and 7.18 shows the sinkholes beneath or in close proximity of houses in Hanaizumi town of Ichinoseki City. Figure 7.18 shows a sinkhole along a roadway in Hanaizumi town of Ichinoseki City. In addition to these there were sinkholes in rice fields.

In addition to the sinkholes detailed in this chapter, there were also some sinkholes formations in abandoned lignite mine areas in Oshu City (Iwate Prefecture) and Matsushima City (Miyagi Prefecture). Sinkholes were observed in both residential areas and fields. However, there was no casualty in any incident.

Damage to abandoned underground quarries occurred in Oya Town (Tochigi Prefecture). Oya tuff or Oya stone, has been quarried in the Oya region, Utsunomiya, Japan. Over 200 underground quarries have been exploited for more than 100 years and some of those are below residential areas. Oya town is about 295 km away from the earthquake epicenter. Although ground motions were recorded in the vicinity of abandoned underground quarries in Oya, the instruments are mainly velocity based and

Figure 7.16 Views of sinkholes along a roadway and parking lot near Ohira Village.

Figure 7.17 A sinkhole beneath a house in Hanaizumi town of Ichinoseki. The size of the sinkhole enlarged after the April 7, 2011 aftershock.

(a) A large sinkhole extending below the storage room and the main building

(b) Cracking of the concrete base and walls in the main building.

Figure 7.18 Sinkholes beneath a house in Hanaizumi town of Ichinoseki.

Figure 7.19 Sinkhole beneath the roadway in Hanaizumi town of Ichinoseki.

Figure 7.20 Totally collapsed semi-underground opening.

they do not measure three components of the ground motions. The maximum ground acceleration at Utsunomiya strong motions station of K-NET was 358 gals at the ground surface. Total or partial collapse of semi-underground quarries occurred as seen in Figures 7.20 and 7.21. The toppling of pillar (P1) caused the collapse of the thin roof of the abandoned quarry as seen in Figure 7.21. In some cases, partial collapses of the overhanging parts occurred as seen in Figure 7.21.

The subsidence of back-filling material of three shafts of abandoned quarries occurred. In the example shown in Figure 7.22, the subsidence of back-filling material was more than 5 m and the diameter of the shaft was about 7 m. The back-filling material was non-cohesive. The back-filling material of the abandoned shaft settled more than 3 m as seen in Figure 7.23. There was a hut on top of this shaft. The hut

Figure 7.21 Partially collapsed semi-underground opening (same as above).

Figure 7.22 The subsidence of back-filling material of the shaft (Aydan & Tano, 2012b).

and the car parked next to the hut had fallen into the settled shaft. These observations are very similar to those observed in Yamoto town during the 2003 Miyagi-Hokubu earthquake and model experiments on shaft with backfilling described briefly in the previous section (Aydan, 2004; Aydan and Kawamoto, 2004; Kawamoto *et al.*, 2004).

Settlement is
more than 3m.

Figure 7.23 The subsidence of back-filling material of the shaft and the fall of the hut and car into the shaft (Aydan & Tano, 2012b).

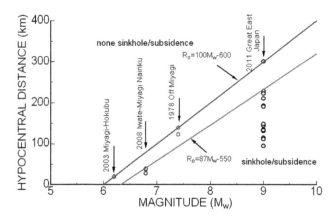

Figure 7.24 Comparison of case history data with an empirical relation for assessing the possibility of sinkhole/subsidence occurrence as a function of earthquake magnitude.

The authors plotted the case history data described in the previous section together with those from other earthquakes occurring in Tohoku region in the space of earthquake moment magnitude versus hypocentral distance of the locality where sinkhole or large subsidence occurred as shown in Figure 7.24. The data was fitted to a linear function whose coefficients are shown in the same figure. The line in the figure should be interpreted as a limiting line between surface damage and non-damage on the ground surface. This figure may serve as a guideline to assess the risk of sinkhole or large subsidence due to abandoned mines and quarries exploited using the room and pillar method as a function of earthquake magnitude. However, there is still a great necessity to compile more data to confirm the validity of the proposed relation.

Besides the possibility of caving of abandoned mines during earthquakes due to either pillar failure and/or roof failure, the ground water in mines may present additional effects on the immersed abandoned mines. These effects may be observed as sloshing, which may further weaken the surrounding rockmass and cause additional collapses.

7.5 UNDERGROUND POWERHOUSES

It is well known that the underground structures such as tunnels and powerhouses are generally resistant against earthquake-induced motions. However, they may be damaged when permanent ground movements occur in/along the underground structures. The reports on damage to underground powerhouses so far are almost none (Aydan, 2015). The damage by the 1999 Chi-chi earthquake in Taiwan to powerhouses in the hanging wall side of the wall was almost none. A slight damage to the Singkarak powerhouse in Sumatra Island of Indonesia by the M6.4 Singkarak earthquake was reported.

7.6 EMPIRICAL APPROACHES

Permanent ground deformations may result from relative movement along the earthquake faults, earthquake induced slope failures involving tunnels and plastic deformation of ground due to high ground accelerations. The author compiled cases histories listed in Table 7.1 and developed databases for three different categories of damages, namely, faulting induced (18 cases histories), shaking induced (98 case history) and slope failure induced (47 cases histories). The parameters of these databases are the name of tunnel, earthquake parameters (magnitude, hypocenter depth, relative slip), distances from epicenter and the earthquake fault surface trace (extrapolated surface trace when earthquake fault does not appear on the ground surface), geometry of tunnel, overburden, lining thickness, rockbolt density, rock unit, damage level index (DLI) defined in Table 7.2. Case histories are plotted in Figure 7.25 with different symbols depending upon the damage mode, the moment magnitude (Mw) of the earthquake and hypocentral distance (R) of the underground structure. As expected, the hypocentral distance of the damaged underground structures increases as the magnitude of the earthquake becomes larger. Furthermore, the limiting relations for fault-induced and ground shaking induced damage on tunnels would be different. The available case history data also imply that there is no damage to underground openings by earthquakes when the magnitude is less than 6.

The limiting distance of damage for portals of underground openings would be more far-distant and the relation proposed by Aydan (Aydan, 2007; Aydan et al., 2009a) for slope failure given below may be used for this purpose.

$$R = A^*(3 + 0.5 \sin\theta - 1.5 \sin^2\theta)^* e^{B \cdot M_w} \tag{7.1}$$

where θ is the angle of the location from the strike of the fault. Constants A and B of Eq. (7.1) are given in Table 7.2 according to ground conditions. Since ground

Table 7.1 A list of major earthquakes, which caused damage to underground structures.

Earthquake	Year	Mw	Number of damage and comments
Meiwa (Japan)	1771	7.4	Natural caves in Ishigaki and Miyakojima Islands were damaged and some roof collapses occurred.
San Francisco (USA)	1906	7.9	Wright tunnels damaged by relative offset displacement of 170–180 cm. Tunnels abandoned.
Great Kanto (Japan)	1923	7.9	93 tunnels were affected and 25 tunnels had to be repaired. The tunnel damage was heavy on the hanging-wall side of the earthquake fault.
North-Izu (Japan)	1930	7.2	The heading of Tanna tunnel under construction was offset by more than 100 cm.
Erzincan (Turkey)	1939	7.9	Roadway tunnel adjacent to the slope was heavily damaged and abandoned.
Kern (USA)	1952	7.5	Tunnels were damaged by thrust faulting with a maximum offset of 122 cm.
Alaska (USA)	1964	9.2	Whittier railway tunnel was damaged.
Izu-Oshima (Japan)	1970	7.0	Inatori tunnel was crossed by the secondary fault and portal damage occurred at several tunnels.
San Fernando (USA)	1971	6.6	Balboa inlet tunnel was damaged for 300 m length on either side of the earthquake fault.
Irpinia (Italy)	1980	6.8	Pavoncelli roadway tunnel was damaged at several fault zones along the tunnel.
Nagano-seibu (Japan)	1983	6.8	Otaki tunnel was damaged by faulting.
Erzincan (Turkey)	1992	6.7	Portal damage at three railway tunnels.
Noto-oki (Japan)	1993	6.6	Roof collapse at Kinoura Tunnel occurred.
Kobe (Japan)	1995	6.9	Out of 107 tunnels, damage was observed in 24 tunnels. 12 tunnels had been repaired. Some tunnels were crossed by faults. Rokko tunnel had to be repaired at 12 locations where fracture-fault zones were encountered during excavation.
Zirkuh (Iran)	1997	7.2	250 unlined qanats, which is an ancient system of underground irrigation canals, and 20 deep wells either collapsed or were heavily damaged.
Golbaf (Iran)	1998	6.6	25 out of 64 qanats were damaged.
Iwate (Japan)	1998	6.1	Outlet tunnel was crossed by a thrust fault with an offset of 10–20 cm and the concrete lining collapsed.
Düzce (Turkey)	1999	7.2	Bolu tunnel was damaged near Asarsuyu portal by faulting and collapses occurring in squeezing rock sections at Elmalik side due to shaking.
Chi-chi (Taiwan)	1999	7.6	Out of 57 tunnels, 49 tunnels experienced various degrees of damage. Most damaged tunnels were on the hanging wall side of the fault and 16% of the tunnels were heavily damaged. Several tunnels were damaged by faulting and some of tunnels were chopped away due to slope failures.
Tottori (Japan)	2000	7.3	Headrace tunnel 200 below the ground surface was crossed by the fault with an offset of 10–20 cm and the concrete lining was ruptured.
Miyagi-Hokubu (Japan)	2003	6.2	There were 29 events of sinkholes due to collapse of shallow room and pillar abandoned lignite mines in Yamato town.

(continued)

Table 7.1 Continued.

Earthquake	Year	Mw	Number of damage and comments
Bam (Iran)	2003	6.5	40% of 126 qanats and deep wells either collapsed or suffered heavy damage.
Niigata-Chuetsu (Japan)	2004	6.8	24 Railway tunnels were damaged and 5 of them had to be repaired. 10 roadway tunnels were damaged.
Kashmir (Pakistan)	2005	7.6	The portal of a roadway tunnel 4 km south of Muzaffarabad collapsed. A railway tunnel in India was also damaged by the earthquake.
Nias (Indonesia)	2005	8.6	Partial roof collapses at Tögindrawa cave.
Singkarak (Indonesia)	2006	6.4	Cracking of concrete of turbine housing structure. The underground cavern is not damaged.
Antofagasta (Chile)	2007	7.7	Galleguillos highway tunnel was damaged.
Wenchuan (Tibet-China)	2008	7.9	110 tunnels were damaged. Jijiuya (226 km from the epicenter) and Longqi tunnels experienced fault offset and concrete linings were ruptured and collapsed.
L'Aquila (Italy)	2009	6.4	Three karstic caves caused sinkholes on the ground surface.
Nagano-Hokubu	2014	6.3	Portal of an abandoned tunnel, portal of a new tunnel.
Kumamoto	2016	7.0	Tawarayama, Tateno, Minami-Aso Tunnels damaged with various degree of damage due to secondary faulting, collapses of rock slopes at portals.

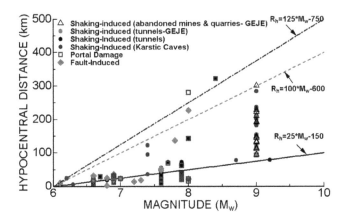

Figure 7.25 Empirical relations between magnitude and limiting damage distance.

accelerations differ according to the location with respect to fault geometry, the empirical bounds proposed herein can provide some basis for the scattering range of observations.

The definitions of damage to underground structures are generally too broad and a more refined classification of damage is felt to be necessary. The author proposes a classification for this purpose as given in Table 7.3.

Table 7.2 Parameters of Eq. (7.1).

Condition	A	B
Disrupted ground	0.10	0.9
Coherent ground	0.08	0.9

Table 7.3 Earthquake-induced damage level index (DLI) for underground structures with the consideration of support members.

Damage Level Index (DLI)	Remarks
1	No cracking of concrete lining and shotcrete, no plastic deformation of rockbolts and steel ribs, no invert heaving
2	Hair cracking of concrete lining and shotcrete, non-noticeable deformation of rockbolt platens and steel ribs, no invert heaving
3	Visible cracking of concrete lining, shotcrete, noticeable plastic deformation of rockbolt platens and steel ribs, slight invert heaving
4	Exfoliation of concrete lining and shotcrete, noticeable bending deformation of rockbolt platens and steel ribs, invert heaving. However, it is structurally stable
5	Spalling of concrete lining and shotcrete, and considerable plastic deformation of rockbolt platens and bending of steel ribs, invert heaving. It is structurally problematic and requires repairs and reinforcement
6	Collapse of concrete lining, shotcrete, and extreme deformation of rockbolt platens and rupturing rockbolts and buckling of steel ribs, buckling and rupturing of invert. Collapse of blocks of ground from roof and shoulders. It is structurally unstable and requires immediate repairs and reinforcement
7	Complete closure of the section by failed surrounding ground. Crushing of concrete lining and shotcrete, rupturing of rockbolts and twisted steel ribs and extreme heaving of invert. Underground openings are either to be abandoned or re-excavated with extreme precautions

Figure 7.26 shows the re-plotted data shown in Figure 7.25 as a function of distance (R_f) from the surface trace or extrapolated surface trace of the earthquake fault. The vertical axis is the damage level index, whose minimum and maximum values are 1 and 7, respectively. When the tunnel response is purely elastic, the damage level is assigned as 1.

The functional form for the damage level index of underground openings subjected to earthquakes may be given in the following form

$$DLI = Q(V_s, R_f, \theta, M, \delta_{max}) \tag{7.2}$$

where V_s, R_f, θ, M and δ_{max} are the shear velocity of ground and the distance from the actual or extrapolated surface fault surface and the angle of the location from the strike of the fault (measured anti-clockwise with the consideration of the mobile side of the fault), earthquake magnitude and maximum relative slip of the earthquake fault. It is an extremely difficult task how to select the specific functional form of Eq. (7.2). Aydan (2007) (see also Aydan et al., 2009a) proposed several empirical relations between the various characteristics and moment magnitude of earthquakes.

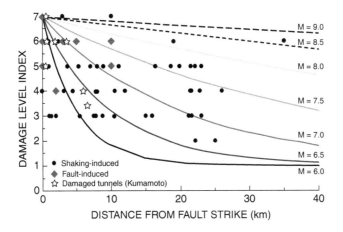

Figure 7.26 Relation between distance (Rf) from surface trace of the fault and damage level index (DLI).

For example, the attenuation relation for maximum ground motion parameters (maximum ground acceleration, or velocity) are given in the following functional form

$$A_{max} \text{ or } V_{max} = F(V_s)^* G(R, \theta)^* H(M) \tag{7.3a}$$

or more specifically

$$A_{max} \text{ or } V_{max} = Ae^{-V_s/B} e^{-R(1-D\sin\theta+E\sin^2\theta)/C}(e^{M_w/F} - 1) \tag{7.3b}$$

where parameters A and C depend upon the nature of earthquake (i.e. interplate or intraplate). Taking into account the empirical relation (Eq. 7.3) by Aydan (2007) with a slight change and the maximum and minimum values of DLI, we propose the following functional form for the damage level index (DLI) and plot it for different magnitudes in Figure 7.26:

$$DLI = Ae^{-V_s/B} e^{-R_f(1-D\sin\theta+E\sin^2\theta)/C^*} + 1 \tag{7.4}$$

where C^* is assumed to be a function of moment magnitude as given below

$$C^* = 10 \cdot 2^{2(M_w-6)} \tag{7.5}$$

It should be noted that the value of $Ae^{-V_s/B}$ must not be greater than 6 in view of the maximum value of the DLI. The authors plotted Eq. (7.4) in Figure 7.22 for different magnitudes by assuming that $\theta = 90°$ and fixing the value of $Ae^{-V_s/B}$ to 6 with $D = 0.5$ and $E = 2.5$. As noted from the Figure 7.26, the chosen function can closely estimate the observed damage level index of underground openings subjected to earthquakes. Nevertheless, the authors also feel the necessity of including a function related to the relative slip of the fault in Eq. (7.4).

7.7 LIMITING EQUILIBRIUM METHODS

7.7.1 Shallow underground openings in discontinuous media

It is well known that shallow underground openings are more vulnerable to stability problems as compared to deep underground openings. The shallow underground openings may completely fail during earthquakes as seen from Figures 7.3 to 7.4. There are also many reports of such case histories in literature (i.e. Wang *et al.*, 2001; Aydan and Kawamoto, 2004; Aydan *et al.*, 2009b, 2009c; Karaca *et al.*, 1995). The author and his group have been studying the stability of shallow underground openings under both static and dynamic conditions (Aydan *et al.*, 1994; Genis and Aydan, 2002). The model tests (see also Figure 7.3) revealed that there are two or three regions potentially unstable (denoted Regions I, II and III) in the close vicinity of shallow underground openings as illustrated in Figure 7.27. Aydan (Aydan *et al.*, 1994) derived the following condition for the horizontal seismic coefficient (α_H) to initiate the sliding of a region along a discontinuity set emanating from the opening for a shallow underground opening as illustrated in Figure 7.2 in view of his experimental studies.

$$\alpha_H > \frac{\sin(\theta_B + \phi_B)\sin 2\phi_A - \dfrac{W_{II}}{W_I}\sin(\theta_A - \phi_A)\sin\theta^*}{\sin(\theta_B + \phi_B - \beta)\sin 2\phi_A + \dfrac{W_{II}}{W_I}\sin(\theta_A + \beta - \phi_A)\sin\theta^*} \tag{7.6}$$

where $\theta^* = \theta_A + \theta_B + \phi_A + \phi_B$. W_I, W_{II}, θ_A, θ_B, ϕ_A, ϕ_B and β are weights of region I and II, inclinations and friction angles of discontinuity sets A and B and angle of seismic force with respect to horizontal, respectively. The same method can be used for the stability of region II. Aydan *et al.* (1994) have checked experimentally the validity of the limit equilibrium conditions for layered media and media with two joint sets.

7.7.2 Shallow room and pillar mines and shallow karstic caves

Many collapses were observed in the abandoned mine area of Yamoto town due to the 2003 Miyagi-Hokubu earthquake (Aydan and Kawamoto, 2004). However, there is no available approach to consider the effect of earthquakes on the stability of abandoned room and pillar mines. Although the appropriate consideration would require the solution of the equation of motion, Aydan *et al.* (2006) proposed a simple approach based on the seismic coefficient approach used in the seismic design of structures in earthquake engineering (Figure 7.28). When the pillar is in full contact with overburden layers, the stress on a pillar may be given as follows

$$\sigma_p = \rho g H \frac{A_t}{A_p}\left[1 + 6\alpha_H \frac{H}{w}\cdot\frac{y}{w}\right] \tag{7.7}$$

where $\rho, g, H, A_t, A_p, w, \alpha_H$ and y are density, gravity, overburden height, area supported by a pillar, pillar area, pillar width, horizontal seismic coefficient, distance from center of the pillar, respectively. The maximum compressive and tensile stress would occur at both sides of the pillar. If the tensile strength between the roof layer

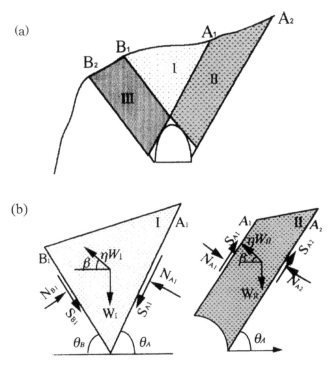

Figure 7.27 Illustration of mechanical model for stability analysis of shallow underground openings (from Aydan et al., 1994).

and pillar is assumed to be nil and the yielding takes to a certain depth from the side of pillar, the maximum compressive stress on the pillar side can be obtained as follows

$$\sigma_p = \sigma_{po} \frac{1}{(1-\xi)^2} \left[1 + \frac{6}{(1-\xi)} \left(\alpha_H \frac{H}{2w} - \xi \right) \right] \tag{7.8}$$

where $\xi = \frac{e}{w}$, $e = w - w^*$, $A_p^* = w^* \cdot w^*$, $\sigma_{po} = \rho g H \frac{A_t}{A_p}$ and w^* is effective pillar width. The relation between the normalized yield distance and seismic coefficient α for no-tension condition between roof layer and pillar is obtained after some manipulations as follows,

$$\alpha_H = \frac{w}{3H} [5\xi + 1] \tag{7.9}$$

The horizontal seismic coefficient at the collapse of the pillar may be obtained for the condition that the maximum compressive stress obtained from equations (7.7) and (7.8) is the same. In Figure 7.28, H and h stand for overburden height and pillar height, respectively. Model tests on room and pillar mines on shaking table indicated that the roof layers also fail by bending. Therefore, the load condition under horizontal shaking is assumed to consist of gravitational load inducing the bending stresses and

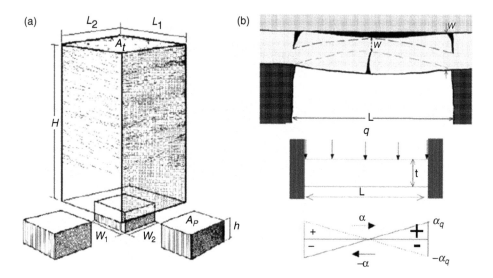

Figure 7.28 Mechanical model for roof stability under earthquake load for roof.

the linearly varying axial stress along the roof axis from tension to compression due to horizontal shaking. On the basis of this assumption, the horizontal seismic coefficient at the time of roof layer failure by bending is obtained as follows:

$$\alpha_H = \frac{\sigma_t}{\rho g t (A_t/A_p - 1)} - \frac{t}{2H}(A_t/A_p - 1) \tag{7.10}$$

where t are σ_t are thickness and tensile strength of roof layer, respectively. Figure 7.29 shows a computed diagram for the relations between overburden ratio and seismic coefficient for various failure modes for the chosen parameters shown in the same figure. As noted from the figure, there are 4 different regions in the diagram, namely, Region 1: roof layer and pillar are stable; Region 2: roof layer is stable, pillar is unstable; Region 3: roof layer is unstable and pillar is stable; Region 4: roof layer and pillar are both unstable. These results indicate that the shallow mines are prone to roof failure while the deeper mines are prone to pillar failure for actual strong ground motions during earthquakes.

The stability of the roof of the shallow karstic caves subjected to earthquake loads may be also analyzed using a similar mechanical model presented above. The seismic stability of shallow sinkholes is evaluated as a built-in beam subjected to horizontal and vertical seismic loads. On the basis of experimental observations the horizontal seismic load will cause tension on one side while it will become compression on the other side (Figure 7.30)

$$\alpha_H = \frac{\dfrac{\sigma_t}{\gamma t} - \left(1 + \dfrac{t^*}{t}\right)\dfrac{1}{2}\left(\dfrac{L}{t}\right)^2}{\left(\dfrac{L}{t}\right)\left(1 + \dfrac{t^*}{t}\right)\left[1 + \dfrac{\beta}{2}\left(\dfrac{L}{t}\right)\right]} \tag{7.11}$$

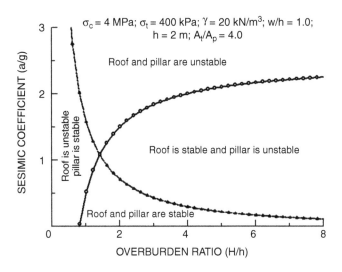

$\sigma_c = 4$ MPa; $\sigma_t = 400$ kPa; $\gamma = 20$ kN/m³; w/h = 1.0;
h = 2 m; $A_t/A_p = 4.0$

Roof and pillar are unstable

Roof is stable and pillar is unstable

Roof and pillar are stable

Roof is unstable / pillar is stable

SESIMIC COEFFICIENT (a/g)

OVERBURDEN RATIO (H/h)

Figure 7.29 Seismic stability charts for abandoned mines.

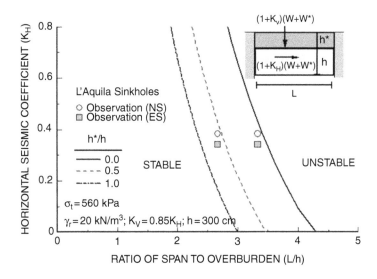

HORIZONTAL SEISMIC COEFFICIENT (K_H)

L'Aquila Sinkholes
○ Observation (NS)
□ Observation (ES)

h^*/h

——— 0.0
- - - - 0.5
-·-·- 1.0

STABLE

UNSTABLE

$(1+K_v)(W+W^*)$

h^*

$(1+K_H)(W+W^*)$ h

L

$\sigma_t = 560$ kPa
$\gamma_r = 20$ kN/m³; $K_V = 0.85 K_H$; h = 300 cm

RATIO OF SPAN TO OVERBURDEN (L/h)

Figure 7.30 Comparison of computational results with observational results.

where
γ: unit weight of roof material
t: roof thickness
t^*: surcharge load height
L: opening width
β: ratio of vertical seismic coefficient (α_V) to horizontal seismic coefficient (α_H)

This approach was applied to the sinkholes caused by 2009 L'Aquila earthquake (Aydan *et al.*, 2009c). The computational results are shown together with the

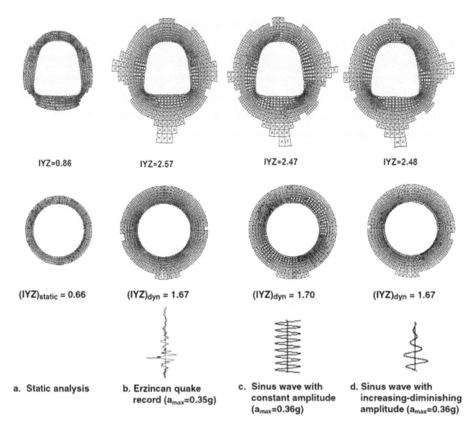

IYZ=0.86   IYZ=2.57   IYZ=2.47   IYZ=2.48

$(IYZ)_{static} = 0.66$ $(IYZ)_{dyn} = 1.67$ $(IYZ)_{dyn} = 1.70$ $(IYZ)_{dyn} = 1.67$

a. Static analysis   b. Erzincan quake   c. Sinus wave with   d. Sinus wave with
                       record (a_{max}=0.35g)   constant amplitude   increasing-diminishing
                                              (a_{max}=0.36g)   amplitude (a_{max}=0.36g)

Figure 7.31 Yield zone formations around a deep circular opening under different waveforms (partly from Genis and Gercek, 2003).

observational values for the maximum horizontal component of ground accelerations taken at AQK strong motion station in Figure 7.1. The ratio of vertical seismic coefficient to horizontal seismic coefficient was taken as 0.65 in view of measurement results. The roof thickness was assumed to be 300 cm while the surcharge load height varied between 0 cm to 300 cm. The unit weight and tensile strength of roof material were assumed to be $20 \, kN/m^3$ and 560 kPa, respectively. The site observations indicated that a 300 cm deep trench (this value is estimated) was excavated on one side of the sinkhole sites. Therefore, the trench excavations for sewage pipes drastically reduced the effective roof thickness. As expected, the increase of surcharge load drastically reduces the seismic resistance of roof of karstic caves.

7.8   NUMERICAL METHODS

A series of parametric numerical analyses on the shape of underground openings under different high in-situ stress regime and direction and amplitude of earthquake induced acceleration waves was carried out. The details of these numerical analyses can be found in publications by Genis (2002) and Genis and Gercek (2003). Figure 7.31

compares yield zone formations around circular and horse shoe-like tunnels subjected to in-situ hydrostatic stress condition ($P_o = 20$ MPa) under static and dynamic conditions. The reason for such an approach is to eliminate the effects of the in situ stress field on the geometry of failure zone. In order to compare the yield zones determined in either static or dynamic analyses, a simple quantitative measure, i.e. "index of yield zone" or shortly IYZ, was used (Gercek and Genis 1999). IYZ, simply, is the ratio of total area of yielded elements to the cross-sectional area of the opening. In the analyses, the rock mass behaviour is assumed to be elastic brittle-plastic. The properties of the elastic zone and yielded zone were assumed to obey the Hoek-Brown failure criterion with $\gamma = 25$ kN/m^3, $E_m = 27.4$ GPa, $\nu = 0.25$, $\sigma_{cm} = 14.1$ MPa and m $= 3.43$ for elastic zone, and $\sigma_{cr} = 2.84$ MPa and $m_r = 2.23$ for yielded zone. Three different acceleration records were used in these particular analyses. If the maximum amplitudes and dominant frequency characteristics of the earthquake records are almost the same, the yield zones formed under dynamic conditions are same. They are almost circular for the circular tunnels while it is almost elliptical for horse shoe-like tunnels although the acceleration record was uni-directionally applied.

Genis and Aydan (2007) carried out a series of numerical studies for the static and dynamic stability assessments of a large underground opening for a hydroelectric powerhouse. The cavern is in granite under high initial stress condition and approximately 550 m below the ground surface. The area experienced in 1891 the largest inland earthquake in Japan. In the numerical analyses, the amplitude, frequency content and propagation direction of waves were varied (Figure 7.32). The numerical analyses indicated that the yield zone formation is frequency and amplitude dependent. Furthermore, the direction of wave propagation also has a large influence on the yield zone formation around the cavern. When maximum ground acceleration exceeds 0.6–0.7 g, it results in the increase of plastic zones around the opening. Thus, there will be no additional yield zone around the cavern if the maximum ground acceleration is less than these threshold values.

The elasto-plastic dynamic response of an underground shelter in Bukittinggi (Indonesia), which was excavated in a very soft rock in 1942 and experienced the M6.8 Singkarak earthquake in 2007, was chosen as an actual example and its dynamic response and stability were analyzed (Aydan and Genis, 2008b). There was no acceleration record in the vicinity of the underground shelter. However, the inferred maximum ground acceleration was about 0.3 g as shown by Aydan (2007). Therefore sinusoidal waves with frequencies of 0.3, 1 and 3 Hz and amplitude of 0.3 g were applied horizontally to the base of the model. Figure 7.33 shows a three-dimensional perspective view of the underground shelter, acceleration responses at selected positions where some seismic damage was actually observed and displacement and yield zone distributions after the ground shaking disappeared. Although the analyses are limited to simple waveforms, the results can explain possible causes of the damage and its variation in the underground shelter.

Aydan and Genis (2014) have performed three dimensional numerical studies on the response and stability assessment of abandoned mines above a new underground excavation. Figure 7.34 shows a 3D view of the layers, abandoned mine and circular tunnel. Table 7.4 gives the material properties used in numerical analyses. Under the static condition, no yielding occurred. The response of abandoned mine was investigated using the ground motion due to the anticipated M 9 class Nankai, Tonankai

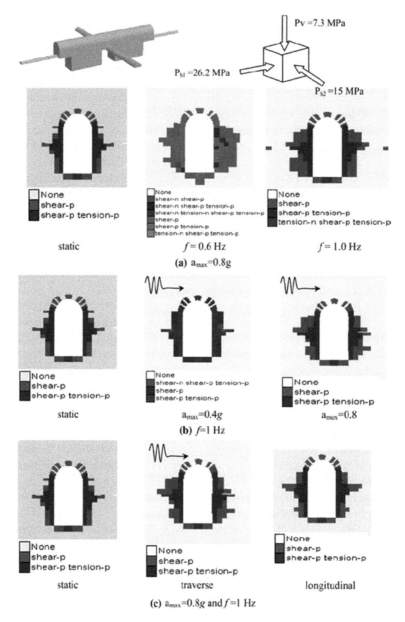

Figure 7.32 Yield zone formation of underground power house for different cases of input ground motions (Aydan *et al.*, 2010a).

and Tokai earthquake at the site under consideration as shown in Figure 7.35 and it is based on the method proposed by Sugito *et al.* (2000). Figure 7.36 shows the yield zone formation in the computational model. As noted from the figure, some yielding occurred at pillars at the deeper section of the model. The results indicated that under dynamic condition, further yielding may occur in the abandoned mines.

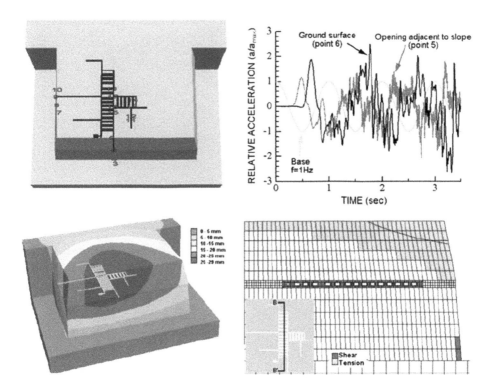

Figure 7.33 Displacement response and yield zone around Bukittinggi underground shelter (arranged from Genis and Aydan, 2008).

Table 7.4 Material properties used numerical analyses.

Layer	γ (kN/m³)	c (kPa)	ϕ (°)	σ_t (kPa)	E (MPa)	ν
Soil	19	50	38	10	270	0.35
Mst-Sst-1	19	700	25	500	750	0.3
Lignite	14	656	45	500	400	0.3
Mst-Sst-2	19	1000	45	700	1073	0.3
Chert	19	3000	45	2000	3647	0.3

Genis and Aydan (2008) carried out a series of elasto-plastic dynamic response analyses of abandoned lignite mine at Mitake town in Japan. Figure 7.37 shows a three-dimensional perspective view of the mine and in-situ stress state inferred from the fault striations using Aydan's method. The material properties used in the analyses are given in Table 7.5. The input ground motion is assumed to be sinusoidal with a chosen period. The responses for actual ground motions induced by nearby earthquakes can be found elsewhere (Aydan and Genis, 2008). Figure 7.38 shows the three-dimensional geological structure on the adopted numerical mesh. For these particular analyses,

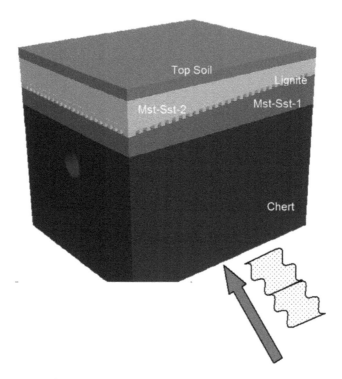

Figure 7.34 Layers of the model and direction of dynamic loading.

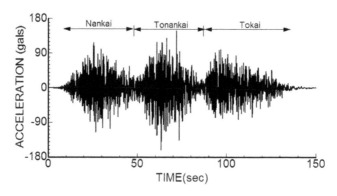

Figure 7.35 Estimated acceleration record due to anticipated Nankai-Tonankai-Tokai earthquake at the site.

the FLAC3D developed by Itasca (2005) is used. The method is based on the finite difference technique utilizing silent boundary conditions and Rayleigh type damping.

The maximum amplitude of the acceleration was selected as 20 and 80 gals with a frequency of 2 Hz. The acceleration is applied for a period of 1.7 s. Figure 7.38 shows the yield zone development for assumed ground motions. While the yielding occurred

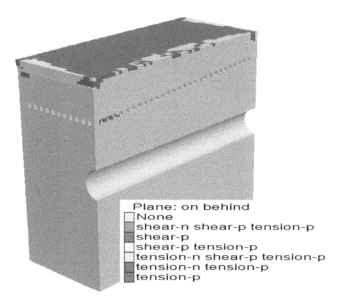

Plane: on behind
None
shear-n shear-p tension-p
shear-p
shear-p tension-p
tension-n shear-p tension-p
tension-n tension-p
tension-p

Figure 7.36 Computed yield zone in the computational model at time step of 120 s of the record shown in Figure 7.33.

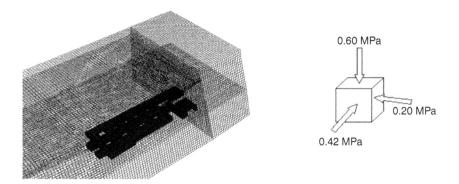

0.60 MPa

0.20 MPa

0.42 MPa

Figure 7.37 Finite difference mesh and assumed in situ stress state.

Table 7.5 Rock material properties used in numerical analyses.

Unit	Uniaxial compressive strength (MPa)	Friction angle (°)	Elastic modulus (MPa)	Unit weight (kN/m³)	Tensile strength (MPa)	Cohesion c_m (MPa)	Shear modulus (MPa)
Lignite	4	53	557	20	0.4	0.669	228.8
Sandstone	2.3	53	257	20	0.06	0.385	102.8
Mudstone	0.96	29	76	20	0.05	0.283	30.4

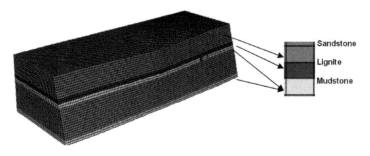

Figure 7.38 A three dimensional view of numerical model of the mine and geologic formations.

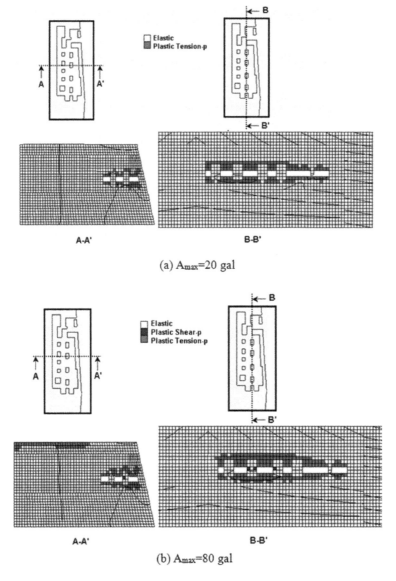

(a) A_{max}=20 gal

(b) A_{max}=80 gal

Figure 7.38 Yield zones around openings and pillars.

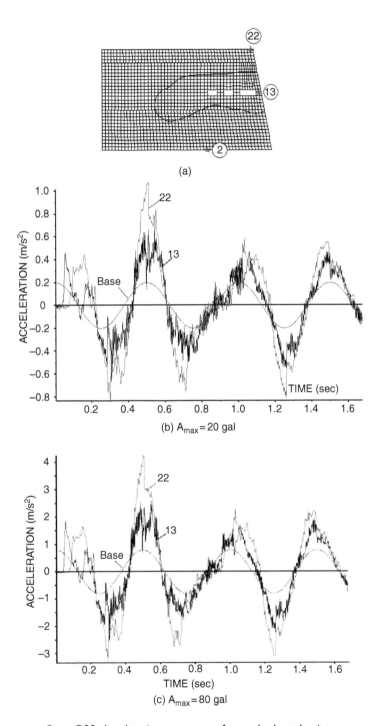

Figure 7.39 Acceleration responses of several selected points.

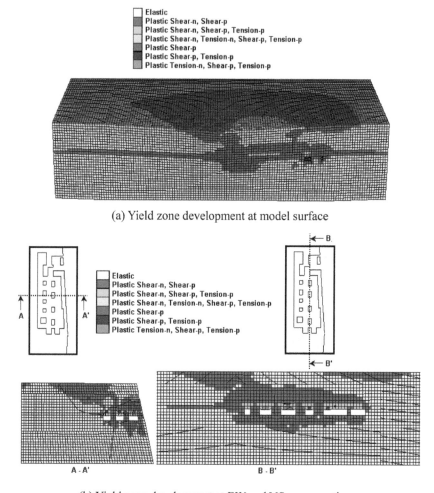

(a) Yield zone development at model surface

(b) Yield zone development at EW and NS cross-sections.

Figure 7.40 Yield zone around mine for ground with reduced strength properties.

due to tensile stresses for the ground motion with an amplitude of 20 gals, the yielding took place due to tensile and shearing for the ground motion with an amplitude of 80 gals. The shear yielding occurred in pillars as seen in Figure 7.38(b).

Figure 7.39 shows the acceleration responses at some selected points. As noted from Figure 7.39, the maximum ground acceleration takes place at the crest of the slope. The ground acceleration at the pillar adjacent to the slope is 3 times higher than that at the base. This tendency is quite similar for ground motions with amplitudes of 20 and 80 gals.

A final computation was carried out the case, in which the properties of rock mass reduced to 1/8 of those of the intact rock while the ground acceleration amplitude was assumed to be 80 gals. This situation corresponds to a state of degradation of rock mass properties with time. Figure 3.39 shows the yield zone development at the model surface and at EW and NS sections. This situation probably corresponds to a state of the total collapse of the abandoned mine. As seen in Figure 3.40, the yielding at ground surface extends to a large area and the yielding in EW (A-A′) cross section resembles a slope failure containing the abandoned mine.

Chapter 8

Dynamic responses and stability of rock slopes

Large inland earthquakes such as 1999 Chi-chi (Taiwan), 2004 Chuetsu and 2008 Iwate-Miyagi (Japan), 2005 Kashmir (Pakistan) and 2008 Wenchuan (Tibet-China) earthquakes caused many large scale rock slope failures in recent years (Aydan, 2006; Aydan and Hamada, 2006; Aydan et al., 2000, 2009a,b, 2012a) (Table 8.1). The slope failures induced tremendous damage to infrastructures as well as to residential areas, and they involved not only cut slopes but also natural rock slopes. Compared to the scale of soil slope failures, the scale and the impact of rock slope failures are very large and the form of failure differs depending upon the geological structures of rock mass of slopes (Figure 8.1) (Aydan, 1989; Aydan et al., 1989, 1992a, 2011). Furthermore, the failure of the rock slope failures may involve both active and passive modes. However, the passive modes are generally observed when the ground shaking is quite large.

8.1 MODEL TESTS

Dynamic tests on model rock slopes are quite rare. Such experiments were first done by Shimizu et al. (1986, 1988) in relation to the investigation of the stability of slopes next to nuclear power plant sites. A shaking table available at the earthquake engineering laboratory of Nagoya University was used during experiments (Figure 8.2). The models of rock mass are hard non-breakable block (wooden blocks, aluminium blocks) representing hard jointed rock mass. The shear resistance of joints is frictional. Figure 8.3 shows examples of slope failures consisting of non-breakable blocks.

A series of shaking table tests was carried out at Tokai University to investigate the initiation of wedge failure and sliding responses of rock wedges under dynamic loads in addition to loading resulting from water and gravitational forces (Kumsar et al., 2000; Aydan and Kumsar, 2010). Dynamic testing of the models was performed using a one-dimensional shaking table, which moves along horizontal plane (Figures 8.4 and 8.5). The waveforms of the shaking table are sinusoidal; saw tooth, rectangular, trapezoidal and triangle. The shaking table has a square shape with 1000 mm long sides. It has a frequency interval between 1 Hz and 50 Hz, a maximum stroke of 100 mm, a maximum acceleration of 6 m/s^2 with a maximum load of 980.7 N. Model frame was fixed on the shaking table to receive the same shaking as that of the shaking table during the dynamic test. The accelerations acting on the shaking table and the models were recorded during the experiment, and saved on a data file as digital data. The displacement responses were recorded using laser displacement transducers.

Table 8.1 A list of major earthquakes, which caused slope failures and rock falls.

Earthquake	Year	Mw	Number of damage and comments
Zenkoji	1847	7.4	A huge slope failure occurred at Mt. Iwakura and a 70 m high landslide dam was created.
Nobibeya	1891	8.0	Many rock slope failures in Neo Valley.
Khorog (Tajikistan)	1911	7.4	567 m high landslide dam created at Sarez. Wedge-like sliding due to conjugate faults.
Diexi	1933	7.4	A huge landslide dam created.
Hebgen Lake	1959	7.4	A huge landslide occurred in Madison River Canyon.
Varto	1968	6.8	Huge landslide at Tepeköy, rockfall at Tapu village.
Izu	1978		Rockfalls along east Izu Roadway. Slope failure at Inatori-Tomoro.
Manjil	1990	7.6	Many rock slope failure and rockfalls. More than 200 mass movements and 51 of them of large scale.
Kobe	1995	6.9	Many small scale slope failure. However, the slope failure at Nikawa.
Dinar	1995	6,2	Wedge failures.
Düzce (Turkey)	1999	7.2	Many slope failures along the fault zone. Slope failure at Bakacak obstructed the highway between Ankara and Istanbul.
Chi-Chi	1999	7.6	Many large scale slope failures and rock falls. Chiufengershan, tsaoling are of great scale and caused landslide dams.
Denali	2002	7.9	Huge rock slope failures occurred in 300 km long and 30 km wide zone along the Denali fault. McGinnes, Black Rapids, West Fork and Susitna Glacier slope failures are of great scale.
Bingöl	2003	6.8	Many slope failures and rock falls at Çiçekdere, Çobantaşı, Göltepe, Soğukçeşme, Yazgülü, Oğuldere, Yolçatı, Kurtuluş.
Off-Sumatra (Aceh)	2004	9.1	Many slope failures and rock falls from Melaboh on the west coast of Sumatra to Andaman. Some of them induced by tsunami waves
Off-Sumatra (Nias)	2005	8.6	Many slope failures and rock falls along entire perimeter of Nias island.
Kashmir	2005	7.5	Many rock slope failures occurred along the fault zone. The slope failure at Hattian created a 180 m high landslide dam. Landslide at Muzaffarabad also created a landslide dam and breached very quickly.
South Sumatra (Bengkulu)	2007	8.2	Many rock slope failures occurred from Bengkulu to Padang along the west coast of Sumatra Island.
Iwate-Miyagi	2008	6.9	Many rock slope failures occurred on the hangingwall side of the earthquake fault. The Aratozawa dam landslide is of great scale.

(Continued)

Table 8.1 Continued.

Earthquake	Year	Mw	Number of damage and comments
Wenchuan	2008	7.5	Many rock slope failures occurred at Yingxiu, Beichuan (Wangjia, Tangjiashan), Donghekou on the hangingwall block of the fault zone. Tangjiashan landslide created 150 m high landslide dam.
L'Aquila	2009	6.4	Slope failures and rock falls at Viapoi, Castel Nouvo.
Suruga Bay	2009	6.9	Slope failures and rock falls at Miho, Kunozan, Abe River Valley, Makinohara, Umeshima, Banjou Waterfall, Matsuzaki.
Padang-Pariaman	2009	7.9	Many slope failures and rock falls at Padang Alai, Tandikat, Padang-Bungus, Padang-Bukittingi Highway.
Haiti	2010	7.0	Many slope failures and rockfalls along the fault zone.
Chirstchurch	2011	6.5	Many rock slope failures and rock falls. Castle Rock, Sumner. Cliff along Goodley Road, Scarborough. Rock fall at Rakaia passed through a house.
Tohoku	2011	9.0	Many slope failures and rock falls at Iwaki, Kesennuma, Rikuzentakada, Shirakawa, Mito, Hitachi.
Erciş	2011	7.2	Many slope failures and rock falls at Gedikbulak, Dağönü, Tirleşin, Yeşilsu, Mollakasım.
Nagano-Hokubu	2014	6.3	Many slope failures in the vicinity of Hakuba, Tachinoma, Kodani, Senkoku.
Gorkha-Nepal	2015	8.1	Gorkha district, Langtan, Melamchi, Sunkoshi, Thomkhola, Kali-Gandaki, Chautara, Nyalam-Tibet.
Kumamoto	2016	7.0	Many rock slope failures and rock falls in the close vicinity of the earthquake fault, which cut through the outer rim of Mt Aso Caldera.

Six special moulds were prepared to cast model wedges (Kumsar *et al.*, 2000). For each wedge configuration, three wedge blocks were prepared. Each base block had dimensions of $140 \times 100 \times 260$ mm. Base and wedge models were made of mortar and their geomechanical parameters were similar to those of rocks. The wedge angles and the initial intersection angles of six wedge blocks numbered as TB1, TB2, TB3, TB4, TB5, TB6 are listed in Table 8.2. Each wedge base block was fixed on the shaking table to receive the same shaking as the shaking table during the dynamic test. The accelerations acting on the shaking table, at the base and wedge blocks, were recorded during the experiment. A laser displacement transducer was used to record the movement of the wedge block during the experiments.

Figure 8.6 shows typical acceleration-time and displacement-time responses for each wedge block configuration. As is noted from the responses shown in Figure 8.6, the acceleration responses of the wedge block indicate some high frequency waveforms on the overall trend of the acceleration imposed by the shaking table. When this type of waveform appears, the permanent displacement of the wedge block with respect to the base block takes place. Depending upon the amplitude of the acceleration waves as well as its direction, the motion of the block may cease. In other words, a step-like behaviour occurs.

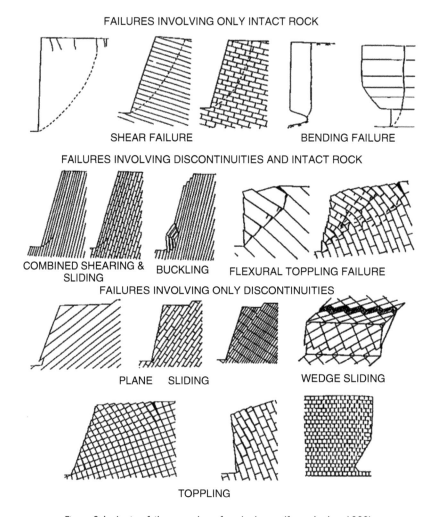

Figure 8.1 Active failure modes of rock slopes (from Aydan, 1989).

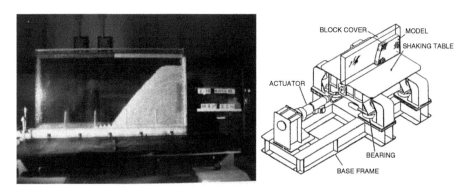

Figure 8.2 Shaking table tests at Nagoya University.

IP-0, 180° **IP-15°** **IP-75°**

CCP-90° **IP-130°** **IP-150°**

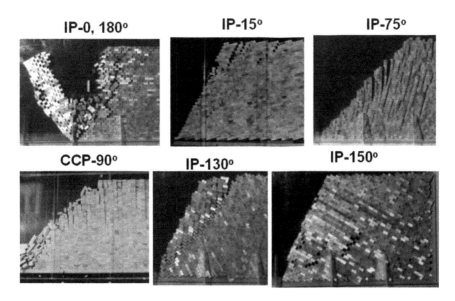

Figure 8.3 Failure modes of rock slope models with non-breakable material.

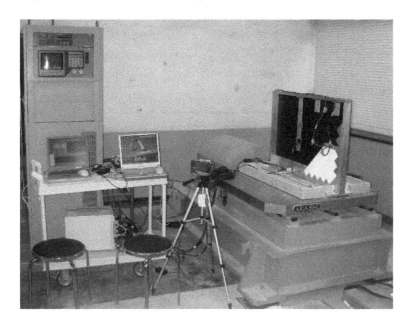

Figure 8.4 A rock slope model on the shaking table.

The motion of the block starts when the amplitude of the input wave acts in the direction of the downhillside and exceeds the frictional resistance of the wedge block. When the direction of the input acceleration is reversed, the motion of the block terminates after a certain amount of relative sliding. As a result, the overall

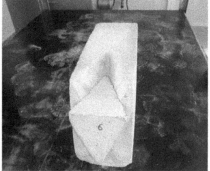

Figure 8.5 Views of model experiments on wedge failure using shaking tables.

Table 8.2 Geometric parameters of the model wedges (after Kumsar et al., 2000).

Wedge no		Intersection angle – i_a (°)	Half wedge angle ω_a (°)
TB1	Swedge120*	29	65
TB2	Swedge100*	29	51
TB3	Swedge90*	31	45
TB4	Swedge71*	27	36
TB5	Swedge60*	30	30
TB6	Swedge45*	30	23

*stands for the wedge block number for each mould (i.e. 1, 2, 3)

displacement response has a step-like shape. Another important observation is that the frictional resistance between the wedge block and base-block limits the inertial forces acting on the wedge block and the base block even though the base-block may have undergone higher inertial forces. The sudden jumps in the acceleration response of the wedge block as seen in Figure 8.6 are due to the collision of the wedge block with the barrier. The initiation of the sliding of the wedge blocks was almost the same as those measured in the first series of the experiments.

Aydan and Amini (2009) carried out a series of experiments on model rock slopes subjected to flexural toppling mode. Model experiments were carried out using breakable blocks. Breakable blocks were made of $BaSO_4$, ZnO and Vaselin oil, which is commonly used in base friction experiments (Aydan and Kawamoto, 1992; Egger, 1983). Properties of block and layers are described in detail by Aydan and Amini (2009). First a series of experiments were carried out on a single column with different height and inclination. Then a series of experiments were carried out for the inclination of layers as 45, 60, 90, 120 and 135 degrees. Before forcing the models to failure in each test, vibration responses of some observation points in the slope were measured with the purpose of investigating the natural frequency of slopes and amplification. Also, deflection of the slope surface and crack initiation were monitored by laser and

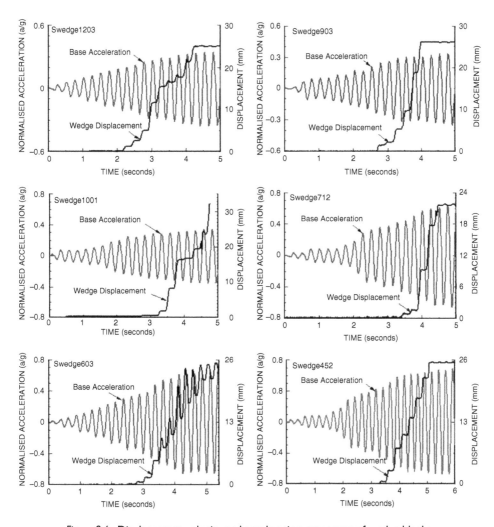

Figure 8.6 Displacement, velocity and acceleration responses of wedge blocks.

acoustic emission sensors. Figure 8.8(a) shows the amplification of motions during a frequency sweeping test with a frequency range of 3–40 Hz for a layer inclination of 60 degrees. It is noted that the acceleration at the crest is amplified by about 5 times the base acceleration. Figure 8.8(b) shows the acceleration, displacement and acoustic emission response for a model slope with a layer inclination of 120 degrees. When rock slopes are subjected to shaking, passive flexural toppling failure mode occurs in addition to active flexural toppling mode (Aydan *et al.*, 2009a, b; Aydan and Amini, 2009). It should be noted that the responses of the model slopes shown in Figure 8.8 are associated with those of slopes subjected to the active and passive flexural toppling modes. Figure 8.9 shows the failure modes of the model slopes with layer inclination of 60, 90 and 120 degrees.

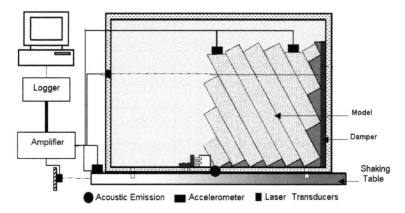

Figure 8.7 An illustration of model experiments and data acquisition system.

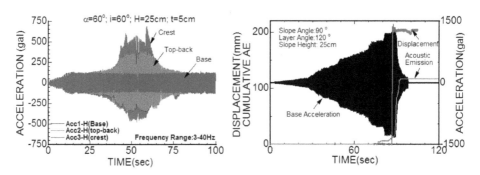

Figure 8.8 Responses of model slopes measured during a sweeping test and failure test.

Figure 8.9 Active and passive failure modes of rock slope models.

Figure 8.10   Failure modes of rock slope models with breakable material.

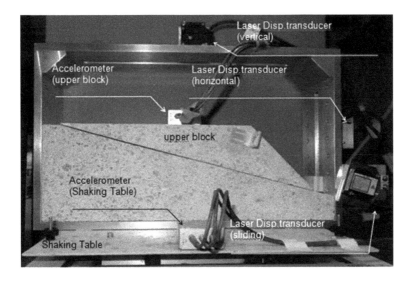

Figure 8.11   View of model rock slope and observation sensor (shaking table test).

These experiments initiated by Aydan and Amini (2009) continued and some experiments were repeated to investigate the effects of cross-joints on response and failure modes (Ohta, 2011; Aydan *et al.*, 2011). Figure 8.10 shows some views of model experiments after failure was observed in these experiments.

Adachi *et al.* (2016) carried out some shaking experiments on the planar sliding of rock slopes as shown in Figure 8.11. The assumed rock slope with a height of 50 m

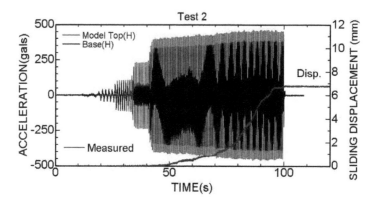

Figure 8.12 Acceleration and displacement responses of the upper-block.

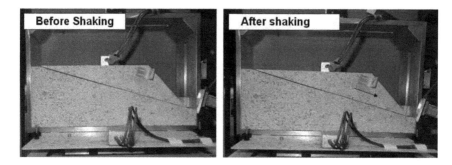

Figure 8.13 Views of a rock slope before and after a shaking table test.

and having a potential plane at an angle of 15 degrees is scaled down to a model with a scale of 1/500. The model material is sandy Ryukyu limestone and locally is known as Awa-ishi. The input-wave, the acceleration of upper block acceleration and relative displacement of the potentially unstable blocks were measured using accelerometers and laser displacement transducers as shown in Figure 8.11. The rock block of the experiment model is made of Ryukyu-limestone. In hard rock, the influence of rock deformation is small and the effect of the slip surface becomes dominant. The static and dynamic friction angle of sliding surface was obtained as 39 degrees and 35 degrees from tilting experiments, respectively. Figure 8.12 shows the acceleration and displacement response of upper-block. Figure 8.13 shows the view of shaking table tests on rock slope. As the natural frequency of the model is very large, the model exhibit a rigid body motion until the block starts to slip. The upper block begins to slip at about 44 seconds and acceleration is about 400 Gal.

Ohta (2011) carried out a new experimental program on the effect of faulting on the stability and failure modes of rock slopes. A series of experiments were carried out on rock slope models with breakable material under a thrust faulting action with an inclination of 45° (Figure 8.14). When layers dip towards valleyside, the ground surface is tilted and the slope surface become particularly steeper. As for layers

Figure 8.14 The effect of thrust faulting on the model rock slopes (from Aydan *et al.*, 2011).

dipping into mountain side, the slope may become unstable and flexural or columnar toppling failure occurs. Although the experiments are still insufficient to draw conclusions yet, they do show that discontinuity orientation has great effects on the overall stability of slopes in relation to faulting mode. These experiments clearly showed that the forced displacement field induced by faulting has an additional destructive effect besides ground shaking on the stability of slopes.

There are overhanging slopes along sea shores or mountainous areas. The overhanging slopes in the vicinity of cliffs next to sea or rivers result from the toe erosion. In mountainous regions, the toes of slopes are excavated and they are often called semi underground openings or half tunnels. Figure 8.15 shows examples of experiments on model half-tunnels in continuous and layered rock mass. The failure occurred due to bending.

8.2 OBSERVATIONS AND CASE HISTORIES

The 1999 Chi-chi, 2004 Chuetsu, 2005 Kashmir, 2008 Wenchuan and 2008 Iwate-Miyagi earthquakes caused many rock slopes and rockfalls (Figures 8.16 and 8.17). These slope failures and rockfalls, in turn, resulted in the destruction of railways, roadways, housing and vehicles. Major slope failures caused by some selected earthquakes are described in this section (Aydan, 2012).

Rockfalls are generally observed in the areas where steep rock slopes and cliffs outcrop. Some rockfalls may induce heavy damage to structures. Nevertheless, rockfalls are results of the falls after the initiation of failure of individual rock blocks in the modes of sliding, toppling or combined sliding and toppling (Aydan *et al.*, 1989). The fall of rock blocks due to sliding mode occurs when the relative sliding exceeds their half width. As for toppling failure, the shaking should be large and long enough to cause the rotation of individual blocks. The height of blocks should be generally larger than their base width.

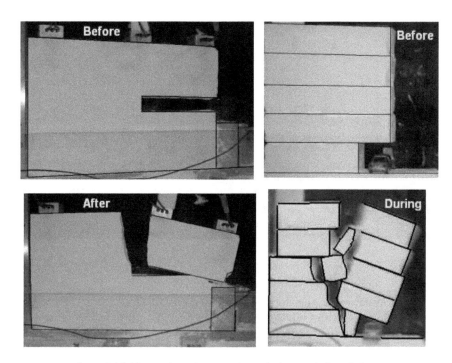

Figure 8.15 Views of overhanging slopes before and after shaking.

Figure 8.16 Views of some slope failures caused by different earthquakes.

(a) Wenchuan earthquake   (b) L'Aquila earthquake   (c) Sinkarak earthquake

Figure 8.17 Rockfalls and their effects.

Figure 8.18 Views of wedge failures in 1995 Dinar earthquake.

8.2.1 1995 Dinar earthquake wedge failures

Dinar earthquake with a magnitude of 6.0 occurred on October 1, 1995 (Aydan and Kumsar, 1997). The earthquake was due to normal faulting and the maximum ground acceleration was more than 300 gals. Many rock slope failures were observed along the surface trace of the earthquake fault. At several locations on the eastern slope of the graben adjacent to the fault scarps, where the rock mass shows up, there were some rock slope failures (Figure 8.18). Rock mass is karstic conglomerate and bedding planes dip towards the south with an inclination of 20°–25°. Most rock slope failures were of small scale and associated with existing joints in rock mass. Slope failures were observed on the eastern side of the graben next to Dinar-Çivril fault. The slope failures shown in Figure 8.18 were wedge failures. Tilting tests indicated that the friction angle of joint surfaces was about 40°.

8.2.2 1999 Chi-Chi earthquake

This earthquake occurred near Chi-chi in the center of Taiwan, about 160 kilometers SSW of the capital city of Taipei, at 01:47 am local time. It was a shallow thrust type

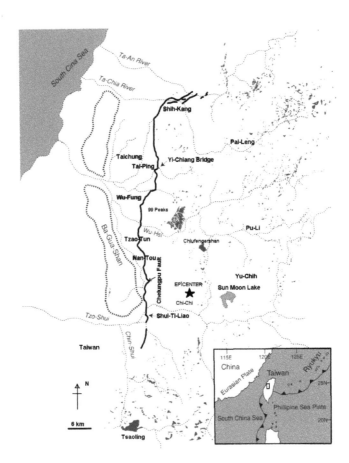

Figure 8.19 Location of slope failures and Chelungpu fault (modified from Kamai *et al.*, 2000).

earthquake, caused by the collision between the Philippine Sea and Eurasian plates. The ground motions were high particularly on the hanging wall side of the earthquake fault and the faulting also induced permanent ground deformations. Numerous rock slope failures were induced by the Chi-Chi earthquake of September 21, 1999 (Figure 8.19) (Kamai *et al.*, 2000). Slope failures mostly occurred on the hanging wall block of the Chelungpu earthquake fault. Among these slope failures, Tsaoling (named also Chaoling) and Chiufengershan (named also Mt. Jio-Fun-Ell-Shan or Nankang) Slope failures are of great interest in view of their geometric dimensions (Figure 8.20).

a) Tsaoling slope failure

The rock mass at this location consists of intercalated sandstone and shale. The overall inclination of the bedding plane was about 14°. A 50–300 m thick, 1800 m wide and 2000 m long mass of sandstone and shale was involved in sliding. The slide surface was somewhat stepped. The estimated volume of the sliding body is about 120 million cubic meters. It is also reported that the earthquakes in 1862 and 1941 (Ishihara, 1985)

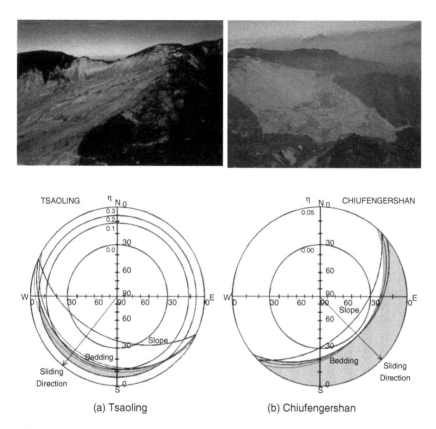

(a) Tsaoling   (b) Chiufengershan

Figure 8.20 Views of large slope failures caused by Chi-chi earthquake and their stereo-net projections (pictures by JGS 2000).

caused partial slope failures at this location. Slides in 1942 and 1979 were associated with heavy rainfalls. Figure 8.20(a) shows the stereo projections of principal geological features with a friction angle of 300. The kinematics analysis implies that the slope should be stable under static conditions. However, a simple analysis with the use of seismic coefficient method would reduce the effective frictional resistance as illustrated in the same figure. In view of the maximum accelerations measured nearby these sites, it is quite plausible that the earthquake induced slope failures along the bedding planes. The sliding masses should have moved almost as monolithic bodies until they struck the valley bottom in view of both theoretical considerations and observations on model rock slopes (Aydan *et al.*, 1989). Reports of non-damaged or slightly damaged structures on the sliding bodies are in agreement with the previous observations of the author. The distance of the slope failure area was 40 km to the earthquake epicenter.

b) Chiufengershan slope failure

The rock mass at this location also consists of intercalated sandstone and shale. The overall inclination of the bedding plane is about 20–27°. A 30–50 m thick, 1000 m wide and 1000 m long mass of sandstone and shale was involved in sliding. The sliding surface was steep as compared with that of Tsaoling slope failure (Figure 8.20(b)). The estimated volume of the sliding body is about 90 million cubic meters and there is no report about any previous sliding at this site. The distance of the slope failure area was about 9 km to the earthquake epicenter. Figure 8.20(b) shows the stereo projections of principal geological features with a friction angle of 300. The kinematics analysis implies that the slope should be stable under a dry-static condition when the cohesion is assumed to be nil. However, if the seismic coefficient method is used, the slope must fail at a horizontal seismic coefficient $\eta = \tan(\phi - \alpha)$ for non-cohesive dry condition, where ϕ and α are friction angle and the inclination of sliding plane, respectively.

8.2.3   2004 Chuetsu earthquake

The Niigata-ken Chuetsu earthquake occurred on October 23, 2004 at 17:56 on JST and it had a magnitude (Mj) of 6.8 on the magnitude scale of Japan Meteorological Agency. The earthquake caused the loss of more than 35 lives and injured more than 2500 people. The Kanetsu Expressway and Hokuriku Shinkansen Line, Joetsu railway line was heavily damaged and the Shinkansen train traveling at a speed of 200 km/h was derailed for the first time in its operation history.

One of the most striking characteristics of this earthquake is extensive slope failures (Figure 8.21). The slope failures caused extensive damage to roadways and expressways as well as destroying homes. The most extensive slope failures were observed in Yamakoshi. The area mainly consists of Neogenic mudstone. This mudstone near ground surface is highly weathered and it had become clayey soil. The slope failure in this area is associated with bedding planes which dip NE with an inclination ranging between 10–25 degrees. Before the earthquake, Typhoon No. 23 passed over the earthquake epicentral area which resulted in very heavy rainfall. Therefore, rock mass and soil layers are expected to be fully saturated at the time of earthquake. The most spectacular slope failures were at Takezawa, Terano and Shiraiwa. The slope failures were categorized as curved deep-seated slope failures, shallow seated slope failures, planar sliding and rock-falls.

(a) Shiraiwa   (b) Terano   (c) Takezawa

Figure 8.21 Views of some slope failures caused by the 2004 Chuetsu earthquake.

8.2.4   2005 Kashmir earthquake

A large devastating earthquake occurred near Muzaffarabad on Oct. 8, 2005. The hypocenter depth of the earthquake was estimated to be about 10 km and it had a magnitude of 7.6. The earthquake killed more than 82000 people in Kashmir, Pakistan (Aydan and Hamada, 2006; Aydan *et al.*, 2009). The earthquake resulted from the subduction of Indian plate beneath Eurasian plate, and the faulting mechanism solutions indicated that the earthquake was due to thrust faulting. The surface expression of the causative fault follows the valley between Bagh to Balakot through Muzaffarabad. The city of Muzaffarabad and Balakot town were the nearest settlements to the epicenter and the fault, and they were the most heavily damaged. Valleys were filled with weakly cemented conglomeratic deposits. Fast-flowing rivers cut through these deposits, resulting in very steep slopes. Many cities, towns and villages were located on this type of terraces. The earthquake caused extensive damage to housing and structures founded on these steep soil slopes. Furthermore, extensive rock slope failures occurred along Neelum and Jhelum Valleys, which obstructed both river flow and roadways.

There were numerous slope failures particularly on the hanging-wall side of the earthquake fault as compared with those on the footwall side of the earthquake fault. Furthermore, the areal extension of the slope failures is much larger on the hanging-wall side than that on the footwall side. They were generally associated with the whitish 100 m thick dolomite layer, which is a highly deformed and fractured rock unit (Figure 8.22a). The wedge-like sliding failure at Hattian was quite large in scale (Figure 8.22b). This slope failure caused a 150 m high slope failure dam with a base length of 1200 m. This slope failure dam created two lakes. The sliding area was about 1.5 km long and 1.0 km wide. Rock mass consisted of shale and sandstone and it constituted a syncline. The estimated wedge angle was about 100° and it was asymmetric. The friction angle of shale from tilting test was more than 35° with an average of 40°. Figure 8.23(a) shows a kinematic analysis of the Hattian slope through the projection of structural planes on equal angle stereo-net together with friction cones. The limiting equilibrium analysis indicated that the safety factor of the slope would be 1.55 under dry static conditions (Aydan and Kumsar, 2010b). However,

(a)   Muzaffarabad   (b) Hattian

Figure 8.22 Views of major slope failures caused by the 2005 Kashmir earthquake.

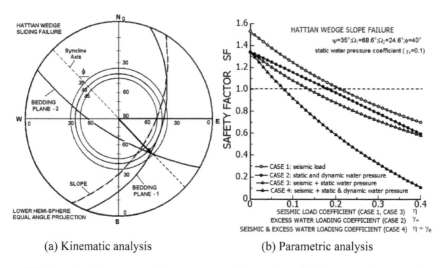

(a) Kinematic analysis   (b) Parametric analysis

Figure 8.23 Results of kinematic and parametric stability analyses for Hattian wedge sliding failure.

the mountain wedge becomes unstable when acceleration is equivalent to the horizontal seismic coefficient of 0.3 and the safety factor becomes 0.9 under such a condition. Some parametric studies for the same wedge sliding at Hattian were carried out and the results are shown in Figure 8.23(b).

8.2.5 2008 Wenchuan earthquake

The 2008 Great Wenchuan (Sichuan) Earthquake with a moment magnitude of 7.9 occurred in Wenchuan County of Sichuan Province of China and caused extensive damage to buildings and infrastructures and it caused the failure of natural slopes as well as cut-slopes. More than 85000 people lost their lives besides the heavy structural damage. The earthquake occurred at the well-known and well-studied Longmen Shan Fault Zone by thrust faulting with dextral component. Preliminary analyses indicated that the rupture process activated a 300 km long fault section. The strong motion records are not available to the scientific community yet. However, the preliminary estimations indicate that the ground motions are high in the epicentral area. The maximum ground acceleration and velocity are expected to exceed 1 g and 100 kine (Aydan et al., 2009b).

One of the most distinct characteristics of 2008 Wenchuan earthquake is the widespread slope failures all over the epicentral area (Figure 8.20). Slope failures, which are fundamentally similar to those observed during the 2005 Kashmir earthquake, may be classified into three categories as soil slope failures, surficial slides of weathered rock slopes and rock slope failures. Rock slope failures are further subdivided into curved or combined sliding and shearing failure, planar sliding, wedge sliding failure and flexural or block toppling. In addition to the active forms of failures, some passive forms of sliding and toppling failures were observed. To initiate passive modes of sliding and toppling failures, very high ground accelerations are necessary. Furthermore, the relative displacement of the block on the sliding plane at the toe of the slope should be greater than its half-length for the collapse of the slope in passive sliding mode. When such relative movements take place, the initial sliding movement changes its character to the rotational mode as seen in Figure 8.3. As a result, the unstable part of the slope rolls down the slope. As for the passive toppling mode, the blocks in the upper part of the slope rotate and roll down. So many rock falls observed during this earthquake were related to these two phenomena.

The satellite images and in-situ investigations indicated that there were very large scale slope failures along the earthquake fault zone. These large scale events took place in Xuankou (southwest end of the fault), Beichuan (central part, three large scale slope failures including the Tangjiashan Slope failure) and Donhekou village (northeast end of the fault). The characteristics of these large scale slope failures are briefly explained in this subsection.

(a) Xuankou slope failure

The Xuankou slope failure (Figure 8.24) involved mainly limestone, which dips toward the valley side with an inclination of 20–25°. Furthermore, there is a fault dipping parallel to the failure surface within the rock mass. The angles of the lower and upper parts of the failed slope are 60° and 30°, respectively. In addition to this new slope failure, one can easily notice the existence of a paleo-slope failure in the foreground. The failure plane of this slope failure is bi-planar and it involved the fault and bedding plane. The fault plane probably assisted the separation of the failed body from the rest of the mountain. The in-situ tilting tests on the limestone joint surfaces yielded the friction angle as 38–42°. If the resistance is purely frictional the expected lateral horizontal seismic coefficient should be greater than 0.36 g. Since this earthquake heavily damaged Yingxiu town, it is more likely that such high ground acceleration did act on this slope.

(b) Beichuan slope failures

There are several large slope failures in Beichuan county and its close vicinity including the Tangjashan Slope failure (Figure 8.24). In association with the motion of the earthquake fault, NW or SE facing slopes failed during this earthquake. There were two large scale slope failures in Qushan town (Beichuan county), which destroyed numerous buildings and facilities. The NW facing slope failure (Jingjiashan) involved mainly limestone while the SE facing slope failure (Wangjiaya) involved sedimentary rocks and quaternary deposits rock unit (Figure 8.24). Limestone layers dip towards the valley side with an inclination of about 30°. Furthermore, there are several faults dipping parallel to the failure surface within the rock mass. The angles of the lower and upper parts of the failed slope are 60° and 30–35°, respectively. The existence of several faults dipping parallel to the slope with an inclination of about 60–65° creates a stepped failure surface. If the resistance is assumed to be purely frictional the expected lateral horizontal seismic coefficient should be greater than 0.18 g.

The SE facing slope (Wangjiaya slope failure, Figure 8.24) may involve a slippage along the steeply dipping bedding plane and shearing through the layered rock mass or sliding along a secondary joint set. In other words, it may be classified as a combined sliding and shearing sliding (Aydan et al., 1992b). The angles of the lower and upper parts of the failed slope are 40–45° and 30–35°, respectively. The layers dip at an

Figure 8.24 Views of some of large slope failures caused by the 2008 Wenchuan earthquake.

angle of 40° towards the valley and shearing plane is inclined at an angle of 20°. Since material properties are unknown for this site as yet, no estimation on possible ground acceleration has been done. Nevertheless, it is more likely that such a high ground acceleration did act on this slope since Qushan town was heavily damaged by this earthquake.

Tangjiashan Slope failure (Figure 8.24), which obstructed Jian River, is located about 2 km NW of Qushan town. The slope failure faces NW. The rock units involved in this slope failure are limestone (top part), and intercalated sandstone and shale (lower part). Rock layers dip towards the valley at an inclination of 40–50°. Although the upper part of the mountain has an slope angle of 40–45°, the lower part is more steeply inclined up to 70° due to the toe erosion by Jian River. The earthquake induced almost a planar sliding of 600 m wide and 450 m long body of rock mass. This slope failure blocked Jian River for a length of 800 m and created a slope failure dam with a maximum height of 124 m.

(c) Donghekou slope failure

This slope failure (Figure 8.24) occurred at Donghekou village of Hongguang Township, Qingchuan County and it is located at the NE tip of the earthquake fault. The upper part of the mountain was covered by limestone while the lower part consists of phyllite, shale and sandstone. This phyllite formation is considerably thick and it is inclined at an angle of 30–45° to NE and it is slightly discordant with upper limestone formation. A paleo slope failure exists to the east of the new slope failure source area. The in-situ tilting tests on the phyllite schistosity (bedding) plane surfaces yielded the

(a) Aratozawa dam   (b) Isawa

Figure 8.25 Views of some slope failures caused by 2008 Iwate-Miyagi intraplate earthquake.

friction angle as 22–24°, which is considerably small. In other words, slopes facing NE cannot be stable even under static conditions if the resistance is only frictional. The initial movement of the unstable part of the mountain should have been in NE direction and then its direction of motion rotated towards SW. Further investigations and laboratory tests on samples from this site should provide essential parameters for investigating the conditions for the failure of the slope at this site.

8.2.6 2008 Iwate-Miyagi intraplate earthquake

2008 Iwate-Miyagi intraplate earthquake occurred on June 14, 2008 at 8:43 AM and had a moment magnitude (M_w) of 6.9 (M_{JMA} is 7.2). While the damage to residential buildings was minimal, it caused extensive slope failures on the hanging wall side of the earthquake fault (Fig. 8.25). The earthquake occurred in a region of upper-plate contraction zone within the complicated tectonics of the Ou Backbone Range, known to have hosted several large earthquakes in historic times. The earthquake was caused by a thrust fault, which was not designated as an active fault in the active fault map of Japan. The observed maximum horizontal accelerations exceeded 1 g at several sites and 0.5 g at about twenty sites. The unusual maximum ground acceleration was measured at IWTH25 station of KIK-NET (2008). However, the non-symmetric nature of the acceleration record at ground surface compared to that in the base rock imply some peculiar amplification of ground motions.

A very large slope failure occurred on the northern side of the reservoir of the Aratozawa dam (Figure 8.25(a)). The total displacement was more than 300 m. The rock mass consists of intercalated loosely cemented sandstone and mudstone (marl-like) and/or siltstone. The inclination of layers was about 10° to eastward. Furthermore, fault traces were observed on the approach road at the right abutment of the dam with about 40 cm upward offset. The high ground acceleration, which was about 1 g at the base of the dam, and high water content of loosely cemented sandstone were probably the major players in the slope failure on a gently bedding surface. The failed rock mass body in Aratozawa Dam slope failure did not slide into the reservoir so that the incident similar to that of the Vaiont dam was luckily not repeated.

Figure 8.26 Wedge failures induced by the 2007 Çameli earthquake.

(a) Castle Rock   (b) Sumner   (c) Raika

Figure 8.27 Views of slope failures caused by 2010 and 2011 New Zealand earthquakes.

8.2.7 Taşcılar wedge failure due to the 2007 Çameli earthquake

An intraplate earthquake struck Çameli town and its vicinity in the southern part of Denizli Province on October 29, 2007 (Kumsar and Aydan, 2014). Planar and wedge sliding failures took place in rock slopes consisting of marl units. The expected ground motions to induce wedge failures are generally higher than those for planar sliding (Kumsar *et al.*, 2000) due to the wedging action. The wedge angle seen in Figure 8.26 is about 80°–90° and its intersection angle is inclined at angle of about 40°. As reported by Kumsar and Aydan (2014), the strong motion records taken at Çameli are sufficient to induce rock wedge sliding failure.

8.2.8 2010 and 2011 New Zealand earthquakes

In addition to extensive ground liquefaction, failures of retaining walls and numerous slopes occurred in both earthquakes at hilly sites in the vicinity of Lyttleton and the volcanic rocks of the Port Hills (Aydan *et al.*, 2012). In the 2010 earthquake, a limited number of slope failures, such as a slope failure at the Rakaia Gorge, which blocked State Highway 77 and small rock falls at Castle Rock (Figure 8.27a), Lyttelton and Summit Road were observed. While in the 2011 Christchurch earthquake, major rock

Figure 8.28 Views of slope failures and rockfalls induced by the Padang-Pariaman earthquake.

falls, wedge and toppling failures occurred, particularly in Castle Rock, Shag Rock, Redcliffs, Sumner, Taylor Mistakes, Scarborough Hills, Godley Head, Dyers Pass, Ferry Mead, McCormack and Rapaki-Governor's Bay, which resulted in a number of fatalities and damaged structures (Figure 8.27b–c). This is probably due to the very large vertical ground accelerations produced by the 2011 earthquake as well as topographic amplification effects that increased the ground motions. In addition, the source of some rock falls may be rock blocks loosened in the 2010 event. The rocks forming the slopes that failed are volcanic rocks mainly comprised of jointed basaltic and trachytic lavas, agglomerates and tuffs (Aydan *et al.*, 2012).

8.2.9   2009 Padang-Pariaman earthquake

Extensive slope failures were observed in Padang Alai and Tandikat (Figure 8.28). The area is covered with very loosely compacted and layered volcanic ash. The grain size of layers varies from the clay size to gravel size. Large fragments are made of pumice and their unit weight is very light. The sandy and gravel sized layers are porous and rich in water content. The site investigation indicated that layers have inclination more than 10 degrees towards to slope side. The failure of slopes took place along the interface of clayey and sandy layers due to intense ground shaking, which is expected to be more than 0.3 g and the failed body becomes fluidized during the motion. More than 200 people were killed in a slope failure at Tandikat. Furthermore, the hard rocks are andesite or basalt and they are jointed due to cooling joints. There were numerous rockfalls in mountainous regions consisting of hard rocks. The spectacular rock falls

Figure 8.29 Views of slope failures (from Aydan *et al.*, 2017).

and slope failures particularly southern inner rim of Maninjau Crater (Figure 8.28). The debris created some landslide dams and the breach of the landslide dam caused huge debris flow 15 hours after the earthquake and causing some structural damage (Figure 8.28).

There was extensive damage to roadways due to slope failures in the mountainous areas such as Padang Alai and Tandikat. This damage was of great scale when roadways run along the thin ridge of the mountains. Roadways were damaged by rock falls. These failures were observed along Padang-Bukittinggi Highway and Padang-Bungus. The fallen rock blocks were more than 5 m in diameter (Figure 8.28).

8.2.10   2016 Kumamoto earthquakes

Many rock slope failures and rock falls occurred in the close vicinity of the earthquake fault, which cut through the outer rim of Mt Aso Caldera. The slope failures are associated with jointing and they may be broadly classified as toppling, planar sliding, wedge sliding and combined toppling & sliding as seen in Figures 8.29 and 8.30.

The protection measures such as rockbolts, frames and wire-meshes with fences were not effective, generally. Although rockbolts and rock anchors are generally effective against toppling type slope failures, their length was not sufficiently long. The shotcreting is not an effective support measure except preventing the water entry into the slopes. Many rock slopes, which were shotcreted, failed easily. Rockfalls were wide spread and wire-meshes with fences used as a counter measure were not effective in many locations. In many cases, rock blocks flew over the meshes and/or bent steel fences.

Figure 8.30 Views of some rockfalls (from Aydan *et al.*, 2017).

8.3 EFFECTS OF TSUNAMIS ON ROCK SLOPES

The effects of tsunamis on slopes along seashores are almost unknown and it did not receive any attention in the field of rock mechanics and rock engineering. Aydan (2012) was first to point out the effects of the tsunamis on rock slopes. In this section a brief description on the effects of tsunamis of the 2004 Aceh (Off-Sumatra) earthquake and the 2011 Great East Japan earthquake on rock slopes along seashores explained.

8.3.1 M9.3 2004 Aceh (Off-Sumatra) earthquake

The seashore between Banda Aceh and Lamno is particularly mountainous and slopes are steeply inclined. Furthermore, the rock layers are folded and inclined towards the sea. The failure of rock slopes was mainly due to sliding on bedding planes of hard sedimentary rocks (i.e. limestone, sandstone). The slopes facing the causative fault plane failed mainly (Fig. 8.31). There were also some toppling failures observed near Lhonga.

8.3.2 M9.0 2011 Great east Japan earthquake

The seashore between Ishinomaki (Miyagi Prefecture) and Noda (Iwate Prefecture) is particularly mountainous and slopes are steeply inclined. Furthermore, the rock layers are folded and inclined towards the sea. The failure of rock slopes was mainly due to sliding on bedding planes of hard sedimentary or metamorphic rocks and the slopes facing the causative fault plane (Figures 8.32–8.34). Tsunami waves first apply shock

Figure 8.31   Failures of rock slopes induced by the 2004 Aceh earthquake.

Figure 8.32   Effects of the tsunami on rock slopes with horizontal layering.

Figure 8.33 Views of the effects of the tsunami on rock slopes with inclined layering.

Figure 8.34 Views of the effects of the tsunami on jointed rock slope failures.

Table 8.3 Parameters of Eq. (8.1) for disrupted and coherent Slope failures.

Condition	A	B	C
Disrupted	0.3	0.8	25
Coherent	0.2	0.8	30

type forces with tremendous amplitude on rock slopes along shores and climb up. Then they pull back and apply drag forces on already disturbed rock mass by shaking due to the preceding earthquake and shock waves by the tsunami, which results in the failures of rock slopes in different modes depending upon its structural geologic features. Particularly, the effective stress is drastically reduced due to the rapid drawdown of the seawater and insufficient drainage of saturated rock mass by the seawater. The situation is very close to the effective stress variations when the water level of reservoirs is rapidly lowered (Aydan *et al.*, 2012). The author also observed this situation in his experiments on rock blocks subjected to planar failure subjected to rapid rising and subsequent lowering of water in reservoirs.

8.4 EMPIRICAL APPROACHES FOR DYNAMIC SLOPE STABILITY ASSESSMENT

Keefer (1984) studied slope failures induced by earthquakes in USA and other countries, and he proposed some empirical bounds for slope failures, which are classified as disrupted or coherent. The empirical bounds of Keefer is not specifically given as formula. Aydan (2007) compiled many slope failures caused by recent worldwide earthquakes according to Keefer's classifications. Modifying the previous equation of Aydan (2007), the following empirical equation was proposed by Aydan *et al.* (2012) for the maximum hypocentral distance of disrupted and coherent slope failures as a function of earthquake magnitude and fault orientation:

$$R_o = A^*(3 + 0.5 \sin\theta - 1.5 \sin^2\theta) * e^{B \cdot M_w} - C \tag{8.1}$$

Constants A, B and C of Eq. (8.1) for disrupted and coherent landslides are given in Table 8.3. Since ground accelerations differ according to the location with respect to fault geometry, the empirical bounds proposed herein can also provide some bases for the scattering range of observations. Figure 8.35 compares case history data with the bounds given by Eq. (8.1) and Keefer's bound.

8.5 LIMITING EQUILIBRIUM APPROACHES

The assessment of the stability of natural rock slopes against earthquakes is very important and one of our urgent issues is how to address it and how to devise the methods and technology for mitigation. It should be also noted that the failure forms induced

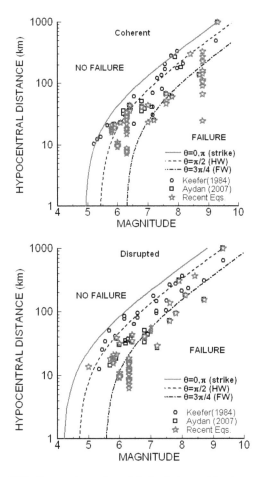

Figure 8.35 Comparison of empirical relations with observations.

by earthquakes might involve passive modes in addition to failures modes classified by Aydan *et al.* (1989).

Aydan and his research group (i.e. Aydan, 1989, 2006a; Aydan and Kawamoto, 1992; Aydan *et al.*, 1989, 1991, 1992; Aydan and Kumsar, 2010; Aydan and Ulusay, 2002; Kumsar *et al.*, 2000) proposed an integrated system based on several limiting equilibrium methods with the consideration of possible failure modes to determine the limiting stable slope angle under the given seismic, geometrical and physical conditions. The fundamental concept of limiting equilibrium methods for the considered failure modes is illustrated in Figure 8.36. The details of these limiting equilibrium methods can be found in the respective articles. Figure 8.37 shows a plot of the slope angle of various rock slopes versus the inclination of the thoroughgoing discontinuity set whose strike is parallel or nearly parallel to the axis of the slope. Stable slopes are denoted by S while failed slopes by F. The plotted data include the data on presently stable natural rock slopes and rock slopes failed due to earthquakes. Most of the data are compiled by

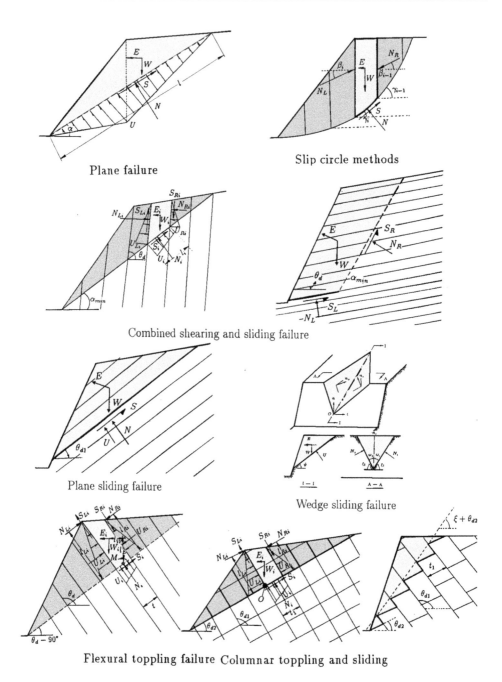

Plane failure

Slip circle methods

Combined shearing and sliding failure

Plane sliding failure

Wedge sliding failure

Flexural toppling failure Columnar toppling and sliding

Figure 8.36 Fundamental concept of limiting equilibrium methods for each failure mode of rock slopes.

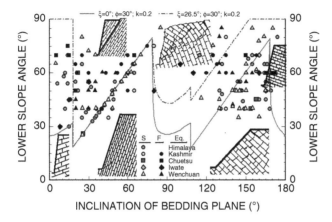

Figure 8.37 The relation between slope angle and bedding plane angle for stable (S) and failed (F) case histories in recent earthquakes.

Aydan *et al.* (2011). In the plots, the stability charts of a slope with a ratio of t/H:1/75 for cross continuous and intermittent patterns ($\xi = 26.5°$) for $\eta = 0.0$ are also included to have a qualitative insight rather than a quantitative comparison (Aydan *et al.*, 1989). The chosen value of t/H is arbitrary and may not correspond to the ratios of slopes plotted in the figure. It is also interesting to note there are almost no failed slopes when the slope angle is less than 25–30° and most of the failed slopes have a slope angle greater than 25–30°. This is in accordance with the conclusion of Keefer (1984). Nevertheless, it is also noted that there are a great number of stable slopes having slope angle greater than 25–30°. This implies that the angle and the height of slopes cannot be only parameters determining the overall stability of natural rock slopes. Therefore, the orientations of discontinuity sets, their geometrical orientations with respect to slope geometry and their mechanical properties and loading conditions must also play a great role in determining the stable angles of natural rock slopes. The results shown in Figure 8.37 may serve as guidelines for a quick assessment of the stability of natural rock slopes and how to select the slope-cutting angle in actual restoration of the failed slopes.

8.6 NUMERICAL METHODS

As explained in Chapter 2, there are various numerical methods to analyse the response of rock slopes during dynamic loading as well as upon failure. In this section, some applications of these methods are described. It should be noted that the equivalent rock mass approach may not be appropriate for rock slopes as they have numerous discontinuities. Therefore, the emphasis is given to discontinuum approaches herein.

8.6.1 Discrete Element Method (DEM)

Discrete Element Approach (DEM) or Universal Discrete Element Code (UDEC) is probably the first numerical technique to be applied to rock slope engineering

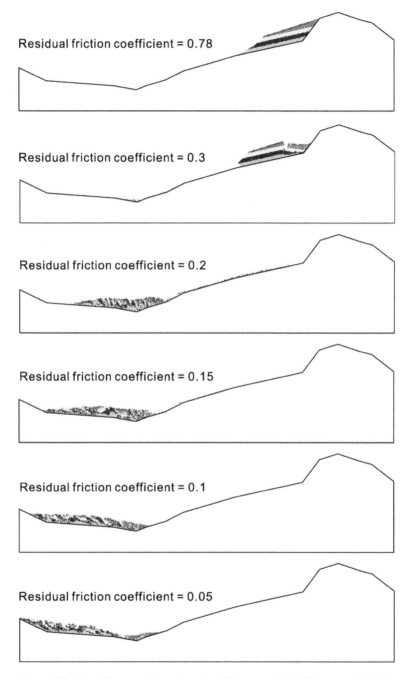

Figure 8.38 Post failure configurations for different residual friction angle (160s).

problems. The method itself is based on the hyperbolic type formulation and it is quite easily adopted for evaluating the dynamic response and stability of slopes. Now the UDEC has some options to consider the effect of seismic shaking. The most difficult aspect with the UDEC is how to determine the geometry of the failing mode and appropriate parameters for failing body. Many numerical experiments have to be carried out to simulate the post-failure geometry of the failed slope. The chosen parameters may be quite different from those determined from laboratory or in-situ experiments. There are several applications of DEM (UDEC) to simulate the post-failure motions of major rock slope failures observed during earthquakes. The computations carried out by Tang *et al.* (2009) for the Tsaoling landslide are chosen for an application example of DEM (UDEC) to earthquake induced slope failures. Figure 8.38 shows several computational results for the post-failure configuration of failed body for different residual angle values at the computational time 160s.

Tang *et al.* (2009) decided that the residual friction angle coefficient of 0.15 yields the post-failure configuration, which is close to the observed configuration. However, it should be noted that this simulation assumed the initial rupture configuration rather than its simulation.

8.6.2   Displacement Discontinuity Analyses (DDA)

Displacement Discontinuity Analyses (DDA) has been updated to simulate the post failure motion of rock slopes. Wu *et al.* (2008) utilized DDA to simulate the post-failure motion and configuration of Tsaoling landslide. Figure 8.39 shows the results for scenarios 1 and 3. When results by DDA and DEM are compared with each other, the results are somewhat different from each other. Furthermore, the properties used do not correspond to those determined from experiments.

8.6.3   Discrete Finite Element Method (DFEM)

DFEM are applied to various rock stability problems (Aydan, *et al.*, 1996a,b; Mamaghani *et al.*, 1994; Mamaghani *et al.*, 1999; Tokashiki *et al.*, 1997). DFEM coded in three different versions, namely, elliptical (pseudo-dynamic), parabolic and hyperbolic. The current codes are appropriate for the initiation and propagation of failure in initial stages. If the post-failure motions are to be simulated, the codes have to be updated with appropriate considerations. Here several example of applications are given.

First the dynamic stability of square and rectangular blocks on a plane with an inclination of 30° was analyzed by the DFEM. The rectangular block was assumed to have a height to breadth ratio $h/b = 1/3$. The friction angles for square and rectangular blocks are $\phi = 25°$ and $\phi = 35°$, respectively. Figure 8.40 shows computed configurations of the square block of size 4 m × 4 m and a rectangular block of size 12 m × 4 m. The square block slides on the incline (time step $\Delta t = 0.04$ sec) while the rectangular block topples (time step $\Delta t = 0.01$ sec). These predictions are consistent with the kinematic conditions for the stability of a single block in the previous example as well as with the experimental results reported by Aydan *et al.* (1989). It should, however, be noted that the discretization of the domain, mechanical properties of blocks and contacts, and time steps may cause superficial oscillations and numerical instability. It is also worth

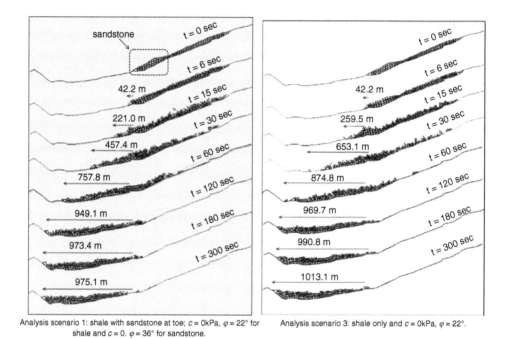

Analysis scenario 1: shale with sandstone at toe; $c = 0\text{kPa}$, $\varphi = 22°$ for shale and $c = 0$. $\varphi = 36°$ for sandstone.

Analysis scenario 3: shale only and $c = 0\text{kPa}$, $\varphi = 22°$.

Figure 8.39 Post failure configurations for scenarios 1 and 3 by DDA.

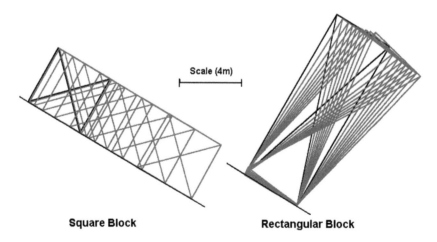

Figure 8.40 Dynamic stability of a block on an incline.

noting that any hyperbolic type equation system requires a certain kind of damping (viscosity) to attain stationary solution, which requires information on the viscous characteristics of rocks and discontinuities. Since time-dependent characteristics of discontinuities and intact rocks are less studied and experimental data are still limited, the inertia term is neglected in the computations reported hereafter using the DFEM.

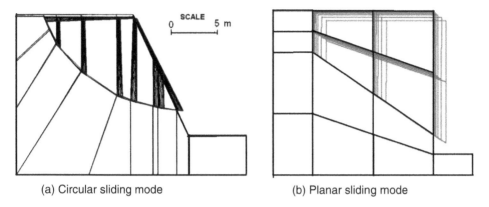

(a) Circular sliding mode   (b) Planar sliding mode

Figure 8.41 Numerical simulation of various slope failure modes.

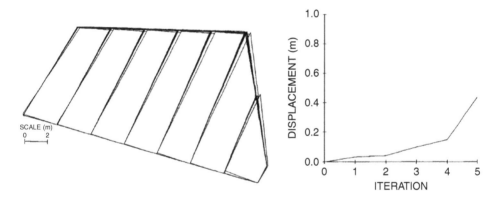

Figure 8.42 Simulation of toppling failure by DFEM.

The next two examples are concerned with circular sliding and planar sliding of rock slopes. The material properties and boundary conditions were given in detail elsewhere (Aydan *et al.*, 1997). Figure 8.41 shows the configurations of the slope for each failure mode at each computation step. The displacement of the failure body for a circular sliding tends to become asymptotic to a certain value following the rapid motion as the inclination of the sliding surface decreases after each computation step. As for planar sliding failure, a separation at the vertical discontinuity occurs and sliding along the inclined discontinuity takes place. The displacement of a sliding block increases as the computation step number increases as the failure plane remains the same.

Aydan *et al.* (1996a,b) applied DFEM to simulate rock slopes subjected to toppling failure and buckling failure. Figure 8.42 and 8.43 show the computational results for these two failure modes of slope failures. Buckling failure simulation is actually a simulation of buckling failure of the floor of the Elbistan open-pit mine in Turkey.

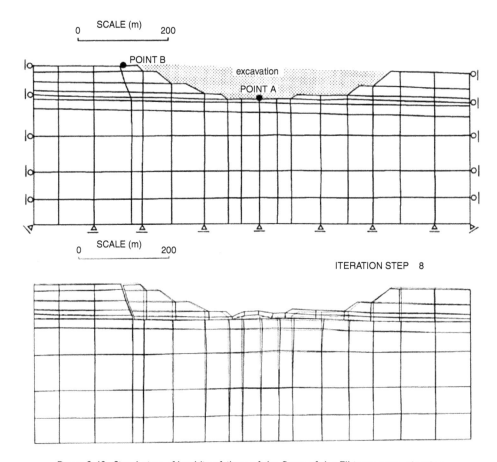

Figure 8.43 Simulation of buckling failure of the floor of the Elbistan open-pit mine.

8.7 ESTIMATIONS OF POST FAILURE MOTIONS OF SLOPES

The estimation of travel distance of natural slopes upon failure and their effect on engineering structures as well as on the natural environment is also of great importance (Aydan, 2006a; Aydan and Ulusay, 2002; Aydan and Kumsar, 2010; Aydan *et al.*, 2006b). The travel distance may be of tremendous scale and it may cause severe damage to settlements and structures. Although this issue is well known and some simple methods are available (i.e. Aydan *et al.*, 1992; Tokashiki and Aydan, 2010), the present numerical methods are still insufficient to model post-failure motions. Back analyses of post-failure travel distances of selected slope failures using the methods proposed Aydan (Aydan and Ulusay, 2002; Aydan *et al.*, 1992b) are presented in this section. Figure 8.44 illustrates the fundamental concept of the method.

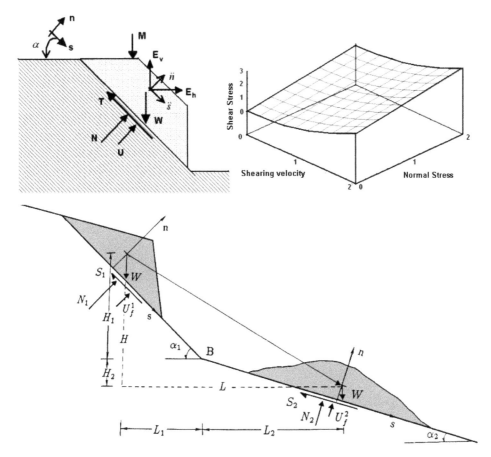

Figure 8.44 An illustration of the fundamental concept of the method for post-failure motions of rock slopes (modified from Aydan and Ulusay (2002) and Aydan et al. (1992b)).

(a) Chiufengershan slope failure

The accelerations records at nearby station denoted TCU089 (CWB, 1999) were used in computations. The computed response of displacement and velocity of the mass center and its path are shown in Figure 8.45. The material and geometrical properties used in computation are also given in Figure 8.45. As noted from the figure, the path of the mass center during motion is well estimated by the mathematical model for chosen material properties.

(b) Shiraiwa slope failures

Computations carried out for Shiraiwa slope failure, which is also called Myoken, Uragara or Yokowatashi, are only reported herein. Rock mass is andesitic tuff with a layered structure. The slope failure took place in the form of plane sliding and hit the railway lines. The acceleration records at Ojiya, which is about 4 km south of the site,

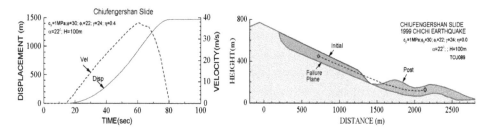

Figure 8.45 Comparisons of computed results for Chuifengershan landslide.

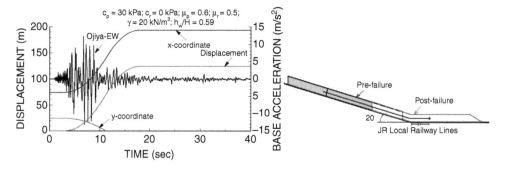

Figure 8.46 Comparisons of computed results for Shiraiwa slope failure.

were used in computations. Figure 8.46 shows the absolute displacement, horizontal and vertical position of the slope failure mass center in time space together with EW component of Ojiya strong motion record released by K-NET (2004) and shear strength properties of the failure surface. The comparison of the computed position of the mass center with that obtained from the geometry of the slope failure body is almost the same.

(c) Hattian slope failure

The post failure motion of the Hattian slope failure is analyzed by using the Abbotabad acceleration record by multiplying a factor of 1.8 so that the seismic coefficient is 0.3. The computational results are shown in Figure 8.47. Since the profile of the failure plane was assumed to be linear, the velocity of the failed body is linear following the initial stage. If the topographical data of the slope failure area becomes accessible, it will be desirable to carry out further analyses for the failed body and potentially unstable zones within the slope failure area.

(d) Aratozawa dam slope failure

The maximum ground accelerations at Aratozawa dam were 865, 810 and 1024 gals for EW, NS and UD components, respectively (Aydan, 2015). The slope failure body was 1350 m long and 800 m wide. Although the thickness ranges from place to place, the highest slope failure scarp was about 100 m. There are loosely cemented sandstone

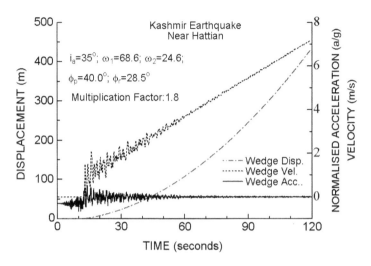

Figure 8.47 Comparisons of computed results for Hattian slope failure.

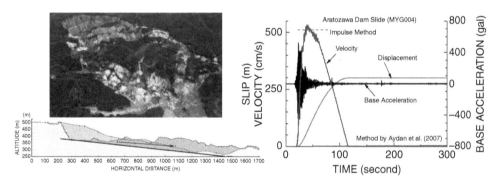

Figure 8.48 Comparisons of computed results for Aratozawa dam slope failure.

layers of volcanic origin, sandwiched by relatively impermeable marl-like mudstone. These sandstone layers played a major role in the slope failure. The author carried out some laboratory tests on samples collected from the site. The computational results are based on the laboratory results and topographical maps before and after the slope failures. The computational results are shown in Figure 8.48. The failed rock mass body in Aratozawa Dam landslide shown in Figure 8.48(e) by the 2008 Iwate-Miyagi Earthquake luckily did not move into the reservoir so that an incident similar to that of the Vaiont dam did not occur.

(e) Wedge sliding

Aydan and Kumsar (2010) proposed a dynamic limiting equilibrium method to estimate both the initiation of wedge failure and their responses in time-domain under dynamic loading. Theoretical estimations are compared with the measured sliding

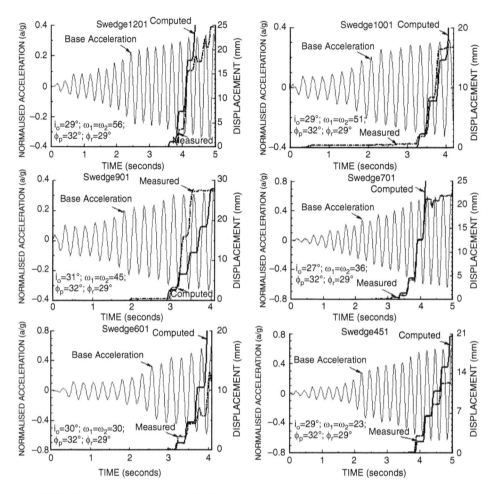

Figure 8.49 Comparison of computed displacement response with measured displacement responses model wedges.

motion of the wedge block throughout the experiments. A series of computational analyses were carried out. Figure 8.49 compares computational results for all wedge blocks with measured sliding responses. In computations, the peak friction angle of discontinuity planes reported in the previous publication (Kumsar *et al.*, 2000) was reduced by 3° in order to take into account the slight damage to surfaces due to multiple utilization of wedge blocks. As noted from the measured and computed displacement responses, the results are remarkably similar to each other. Although some slight differences exist, these may be associated with the variation of the non-linear surface friction between the base-block and wedge block and the negligence of viscous effects in computations.

Chapter 9

Dynamic responses and stability of historical structures and monuments

Historical structures are mainly masonry structures, which are composed of blocks made of natural stones, bricks or both, and they are assembled in different patterns with or without mortar. Furthermore, masonry houses presently constitute more than 60 percent of the residential buildings all over the world and they have a very long building history. In spite of widespread utilization of masonry structures in building history all over the world, there are a few studies on their seismic response and stability.

The author has been involved with the model studies on the seismic response and stability of masonry structures such as arches, castle walls, dams, retaining walls, bridges, pyramids, towers and buildings. Furthermore, he had a chance to be a part of reconnaissance groups for earthquakes in Turkey, Japan, Indonesia, Italy, Taiwan, China, India, Kashmir and was able to investigate the response and stability of various masonry and historical structures.

The author and his group was involved with the restoration of some masonry structures in the Okinawa Island of Japan. These structures were the famous Shuri Castle, Nakagusuku Castle, Gushikawajo Castle, an arch bridge in Iedonchi royal garden and Yodore royal mausoleum of the Ryukyu Imperial period. In addition the author and his group was involved with the static and dynamic stability assessment of some natural rock structures such as Perry Banner Rock, Wakariji Rock and Himeyuri mausoleum.

In this chapter, the observations of damage to actual masonry and historical structures, the shaking table experiments, available limiting equilibrium and numerical methods for estimating their responses are presented.

9.1 OBSERVATIONS

Observations of damage to masonry and historical structures by earthquakes are described in this section. It should be also noted that damage to historical structures reflects the past seismicity of regions.

9.1.1 Examples from Turkey

There is much evidence of strong earthquakes, which affected many antique settlements in Western Turkey such as Smyrna (old İzmir presently known as Bayraklı district), Agora, Efes, Milet, Magnesia, Aphrodisias, Hierapolis, Laodekia, Collosea,

Figure 9.1 Views of some damaged antique settlements in western Aegean region.

Metropolis, Tralles, Nysa, Tripolis, Sardis, Aizonoi, Sagalosos, Aspendos, Pergamon, Troy, Knidos etc. in the region in the past. Figure 9.1 shows views of some of these antique settlements. For example the antique city called Efes (Ephesus) was destroyed by earthquakes in the past and the city was abandoned following an earthquake in the 10th century. The recent archaeological excavations clearly showed the traces of such events of past earthquakes in some antique settlements (i.e. Altunel 1998). The author recently found a relative slip of about 450 cm in Sardis remains. If this relative slip is associated with the AD 17 earthquake (Ergin *et al.*, 1967; Soysal *et al.*, 1981), its estimated moment magnitude and rupture length would be about 7.6 and 104 km from the empirical relation of Aydan (1997, 2007, 2012) between the earthquake magnitude and maximum relative slip with the consideration of faulting sense.

(a) 1998 Adana-Ceyhan earthquake

An earthquake of magnitude 6.3 on the Richter Scale occurred at 16:56 (13:56 GMT) local time on June 27, 1998 in Adana province in Southern Turkey, which was subjected to big earthquakes in the past ($M_s = 6.1$ and $M_s = 5.2$ at Ceyhan-Misis region on March 20, 1945 and October 22, 1952, respectively). This earthquake is officially called *Adana-Ceyhan Earthquake*. Turkish Earthquake Research Department (ERD)

Figure 9.2 Damage to Lokman Hekim bridge over Ceyhan river.

estimated that the epicenter of the main shock was about 38 kilometers away from the centre of Ceyhan town. The peak ground accelerations were 0.28 g in E-W, 0.22 g in N-S, and 0.08 g in vertical directions (ERD). The most severe damage was observed at a stone masonry bridge, called Lokman Hekim Bridge (Figure 9.2(a)) and built in AD 6 century. This is a 10-arched masonry bridge and it is 132.7 m long and 6.50 m wide. The stone blocks are sandstone. There is a permanent relative displacement of

Figure 9.3 A view of damaged Mihrimah Sultan Mosque.

the bridge between its abutments. As a result, fracturing and separation of blocks were observed (Figures 9.2(b)).

(b) 1999 Kocaeli earthquake

The minarets of Beşiktaş Mosque and Kabataş Mosque indicated some relative movements and sliding among limestone blocks. The damage was mainly observed at the location where their rigidity is discontinuous. The most severe damage was observed at Mihrimah Sultan Mosque. This mosque has one hundred and sixty-one windows, built by Sinan for Mihrimah Sultan who was the daughter of Süleyman the Magnificent. The damage was concentrated at the NE corner of the structure. The separation and relative sliding in this corner resulted in the fall of limestone blocks from the outer arch of the main dome. Nevertheless, the main reason is associated with the settlement of ground at that corner before the earthquake resulted in the loosening of the walls at those specific locations. The ground shaking caused further loosening of the walls, which subsequently caused the falling of block walls from the dome (Figure 9.3).

The Istanbul city walls, once an impenetrable fortification, stretch seven kilometers from the Sea of Marmara to the Golden Horn. Restored recently, and many times

Figure 9.4 Partial collapse of İstanbul City Wall (Surlar).

previously, these walls date from the fifth century and the reign of Emperor Theodosius II. UNESCO has declared the city walls, and the area which they enclose, to be world cultural heritage. The walls were heavily damaged in 1766 and 1894 earthquakes and were repaired. During this earthquake, partial collapses of the walls were observed at several locations (Figure 9.4).

East of Izmit, Sakarya is the provincial capital of Adapazari, an important agricultural and industrial region. The Beşköprü Bridge was built by the Byzantine emperor Justinian in 553, and stretches for 429 meters across the Çarksuyu over the old course of Sakarya river. Eight arches connect the two shores (Figure 9.5). Some damage to this arch bridge was observed.

(c) Buldan earthquake

An earthquake with a magnitude of 5.6 occurred in Buldan area of Denizli city on 26 July, 2003. The earthquake caused some damage to city walls of the antique city of Tripolis built in 160 BC and to catacombs near Yenice town (Aydan and Kumsar, 2005). Some blocks of travertines had fallen down or were displace as seen in

Figure 9.5 A view of Beşköprü Bridge over Çarksuyu.

Figure 9.6 Views of fallen or displaced blocks of travertine at Tripolis Antique City.

Figure 9.6. Some part of slopes of marl-like soft rock saw a number of catacombs collapse as seen in Figure 9.7. A open-crack was developed as a result of ground shaking, which resulted in the failure of catacombs.

(d) Erciş earthquake

Two devastating earthquakes with moment magnitudes of 7.2 and 5.6 occurred on October 23, 2011 (Erciş earthquake) and November 9, 2011 (Edremit earthquake), respectively, in the Van Province of the eastern Turkey. The first earthquake caused

Figure 9.7 Collapsed catacombs near Yenice Town (from Aydan and Kumsar, 2005).

(a) before   (b) after

Figure 9.8 Heavily damaged Kara Yusuf Pasha Kümbet (Aydan *et al.*, 2012).

some damage to historical structures from Karakoyunlu Türkmen State, which governed Azerbaijan and Iraq during a period between 1375 and 1468. Erciş served as the capital of Karakoyunlu Türkmen State and there are many historical remains from that period in the Erciş and Van and Bitlis provinces. There are a number of different mausoleums (Kümbet in Turkish) built for the khans (king) and/or their aristocratic families. They are dodecahedral or cylindrical built on a cubic base. Two kümbets were damaged by this earthquake. Particularly the Kara Yusuf Pasha Kümbet (also locally known Zortul), which was built near Çatakdibi village of Erciş during the period of Cihan Şah, suffered very heavy damage by the earthquake as seen in Figure 9.8.

(a) before   (b) after

Figure 9.9 Damaged Kadem Paşa Hatun Kümbet (from Aydan *et al.*, 2012).

Figure 9.10 Stone monuments (toppled) Van Museum (from Aydan *et al.*, 2012).

The second kümbet damaged by the earthquake is Kadem Paşa Hatun. This küm-bet is about 1–2 km to the east of Erciş, at the junction of the Erciş–Van and Patnos–Van roads. It was also built in 1458 for the wife (hatun) of Kadem Paşa during the period of Cihan Khan (Shah). The damage was due to relative movement among blocks, which caused the opening of vertical joints between stone blocks of the structure as seen in Figure 9.9.

It was also reported that some stone monuments with cuneiform (originally devel-oped by Sumerians, who are of Central Asian origin) inscriptions of Urartu (probably related to Ural-Altaic origin) period in the Van museum toppled as seen in Figure 9.10.

Figure 9.11 Toppled wall of Urartu castle in Çelebibağ village.

(a) before   (a) after

Figure 9.12 Views of Kayaçelebi mosque before and after the earthquake (from Aydan et al., 2012).

The museum building also suffered some damage, some clay tablets were broken and some collections belonging to Urartu period were also damaged.

A stonewall of castle remains from the Urartu period collapsed in Çelebibağ village. This wall collapsed towards the south as seen in Figure 9.11.

There are also some historical mosques and other historical structures to the south of the castle. There are three historical mosques, namely, Kayaçelebi, Hüsrevpaşa and Süleyman Han mosques. Hüsrevpasa mosque was built in 1567 by the famous Turkish architect Mimar Sinan. The Kayaçelebi mosque was built in 1660 and lost its minaret during the October 23, 2011 earthquake (Figure 9.12). The November 9,

Figure 9.13 Partial collapse of a masonry bridge at Kono (from Murai and Matsuda, 1975).

2011 earthquake caused more damage to the main compounds and minarets and the inner plastering of walls fell down.

9.1.2 Examples from Japan

It is also known that earthquakes caused tremendous damage to historical structures in Japan. The damage to some historical structures in the last 100 years is detailed briefly in this sub-section.

(a) 1975 Central Oita Prefecture earthquake

On April 21, 1975 an earthquake with a magnitude of 6.4 occurred in Central Oita Prefecture. This earthquake caused damage to some stone masonry structures such as bridges and retaining walls in addition to damage to other structures (Murai and Matsuda, 1975).

(b) 1995 Kobe earthquake

The Great Hanshin earthquake or Kobe earthquake, occurred on Tuesday, January 17, 1995, at 05:46 JST (January 16 at 20:46 UTC) in the southern part of Hyōgo Prefecture, Japan. It measured 6.8 on the moment magnitude scale and Mj7.3 (adjusted from 7.2) on JMA magnitude scale. The total earthquake rupture time was 20 seconds and the hypocenter was 16 km deep. Besides tremendous damage and heavy casualties many masonry type structures were affected by this earthquake. Figure 9.14 shows the dislocation of granitic blocks of a temple in Kobe City.

Figure 9.14 Dislocated granitic blocks of a temple in Kobe City.

Figure 9.15 The collapse of some parts of Imabari Castle due to the 2001 Geiyo earthquake.

(c) 2001 Geiyo earthquake

The 2001 Geiyo (or Akinada) earthquake occurred with a moment magnitude of 6.7 on March 24 at 15:27 local time near Hiroshima, Japan. Some part of Imabari Castle walls collapsed as seen in Figure 9.15.

(d) 2007 Noto earthquake

The Noto Peninsula (Noto-Hanto) earthquake occurred at 9:42 JST on March 25, 2007 and it had a magnitude (Mj) of 6.9 on the magnitude scale of the Japan

Figure 9.16 Toppled fort-gate of Soji temple.

Meteorological Agency. Damage to temples was generally observed at the fort-gate (torii) made of granitic columns and beams. One spectacular example was observed at Soji Temple next to the NTT building. The granitic columns were uprooted from their base and toppled as seen in Figure 9.16. The main compounds were generally undamaged due to superior earthquake-resistant-ability.

(e) 2007 Kameyama earthquake

The Kameyama earthquake occurred at 12:19 JST on April 15, 2007 and it had a magnitude (Mj) of 5.3 on the magnitude scale of the Japan Meteorological Agency. The earthquake injured 12 people and caused some structural damage (Aydan *et al.*, 2007). The strong ground motions were quite high in the epicentral area with high frequency components although its moment magnitude was only 5.0.

The strong motion networks of the Japan Meteorological Agency and K-Net and Kik-Net of NIED recorded very high ground accelerations in the epicentral area although the moment magnitude (Mw) of this earthquake was only 5.0. The maximum ground acceleration was observed at Geino strong ground motion station of KIK-NET, which was 3 km away from the epicenter, and it was about 850 gal at ground surface while the maximum acceleration was 338 gal at base (rock), implying an amplification factor of 2.3–2.5. The maximum ground acceleration in the Kameyama station was 771 gal. The highest acceleration was NS component with an amplitude of 716 gals. In spite of the small magnitude of this earthquake, the maximum ground acceleration and velocity exceed those estimated from empirical equations.

Kameyama castle was built in 1590 on a 10 m high hill. The inclination of south, east and west walls is about 50° while the inclination of the north wall is much steeper, about 70–75°. The earthquake caused the collapse of the NW corner of the 5 m high northern wall, where a masonry stone stair is located. The collapsed section was 2 m wide and this section was actually damaged by a typhoon in 1972 and it was repaired

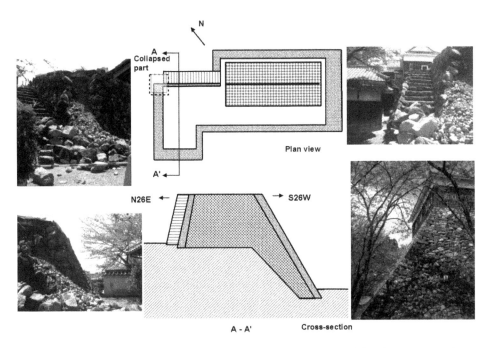

Plan view

Cross-section

A - A'

Figure 9.17 Plan and cross sectional view of Kameyama Castle (not to scale) and some views of intact and collapsed parts of the castle walls (modified from Aydan *et al.*, 2007).

(Aydan *et al.*, 2007). The stones used at the NW corner were typically 30 cm wide, 30 cm high and 55 cm long and the rock itself is either andesite or diorite. Figure 9.17 shows plan and cross-section illustrations of the castle together with some views of the intact and collapsed sections of the castle.

(f) 2009 Suruga Bay earthquake

The Suruga Bay earthquake occurred at 05:07, 11 August 2009 with a magnitude of 6.5 (Mj) (Mw 6.2) in the area of the anticipated Tokai earthquake according to the Japan Meteorological Agency (JMA). However, the hypocenter of the earthquake is inferred to be in the subducting Philippine Sea Plate beneath Euro-Asia plate. The earthquake occurred on an unknown fault and the focal mechanism implied that it was due to thrust faulting, whose strike was perpendicular to that of the anticipated Tokai earthquake. The ground motions were very high all around the Bay. The maximum ground acceleration was greater than 0.7 g at Omaezaki strong motion station of the Japan Meteorological Agency (JMA). Ground amplifications were also very high, particularly in Shizuoka.

Damage to temples and shrines generally occurs at gates and buildings. The gates of temples are either made of granitic stone or wood. Furthermore, damage to monuments and stone lanterns occurs along the alleys leading to the main compound. The damage to such structures occurred in temples and shrines situated over soft ground or hilly ground with ridges (Figure 9.18). The damage to main buildings and auxiliary

Yaizu   Sagara

Figure 9.18 Damage to temples and shrines.

structures was mainly due to the displacement of roof tiles and/or the relative sliding of the wooden columns over the foundation stones. In some temples and shrines, uneven settlement of ground due to liquefaction caused some distressing in the main buildings.

The most publicized retaining wall failure was observed at the remains of Sunpu Castle (Figure 9.19). The retaining walls of about 8–9 m high failed at three locations in N-S direction, indicating the directivity effect of the strong ground motions. Two failures occurred at the south and north side of the outer moat (soto-bori) and one failure occurred in the south side of the inner moat (uchi-bori). The size of corner stones is generally large (longest side length is about 200 cm) and placed in an intermittent pattern with an average inclination of 70°. The sizes of blocks at other parts are very variable. Nevertheless the longest side of the stones cut into pyramid shape is generally more than 80 cm and it is placed into the inside of the wall (Figure 9.20). Old stones are generally made of andesite and porphyrite while the newly restored parts consist of basalt blocks. It is well known that the retaining walls suffered heavy damage in the past. It is reported that all retaining walls collapsed during the 1854 Ansei earthquake. Some parts of the restored retaining wall failed again in the 1935 Shizuoka earthquake. Although the configuration of the retaining walls of Japanese castles is earthquake-resistant, the ground motions are generally expected to be large enough to make these retaining walls fail.

SUNPU CASTLE

Figure 9.19 An aerial view of Sunpu castle and views of failed retaining walls.

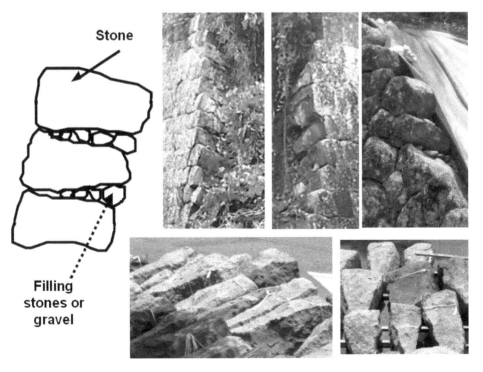

Figure 9.20 Masonry technique for building retaining and castle walls of Japan and views of stones used in Sunpu castle retaining walls.

(g) 2010 Okinawa earthquake

An earthquake with a magnitude of 6.9 on Japan Meteorological Agency Scale occurred near Okinawa Island of Japan. The focal depth of the quake, which occurred at 5:31 a.m. local time (2031 GMT Friday), was about 40 km under the sea 107 km east off Naha, capital of Okinawa (Aydan and Tokashiki, 2011). The damage reported after the quake were ruptured water pipes, fallen or cracked water tanks, fall of roof tiles and the collapse of some parts of walls of Katsuren Castle. In addition some slope failures also occurred.

The collapse of some of castle walls at Katsuren Castle, which is designated as a world heritage site, occurred. The castle is located over a hill in Uruma City and the nearest strong motion station of the K-NET strong motion network is Gushikawa. The NW corner of the castle wall with a height of 4m collapsed and there were numerous dislocations blocks and rotation of blocks in the castle as seen in Figure 9.21. The typical size of the blocks ranges between 50 to 60 cm as seen in Figure 9.22. As the castle is situated on the top of the hill, it is likely that ground motions were amplified.

(h) 2016 Kumamoto earthquakes

Kumamoto prefecture has numerous cultural masonry structures. The Kumamoto Castle is one of prominent structure and suffered heavy damage at various locations (Figure 9.23). The damage to castle retaining walls was quite heavy. The Kumamoto Castle also suffered heavy damage in the 1899 earthquake. The damage was quite heavy where the re-construction was implemented. It seems that the workmanship in re-constructing the damaged parts in 1889 earthquake was not appropriate. When one closely inspects the failure of the castle retaining walls, it can be found that rounded cobbles were used as backfill material. The low friction angle of the backfill material may be one of the prime causes of the failure besides strong ground shaking.

There are also famous masonry arch bridges in the earthquake affected area. One of the Futamata bridges failed due to out-of-plane loading while the other one suffered almost no damage (Figure 9.23). The side-wall of the bridge facing the epicenter failed during the earthquake. Nevertheless, the arch structure was stable despite some small relative displacements of the blocks occurred.

The temples in Japan have gates called TORII in Japanese. These gates are generally built using granite. At several locations where the ground shaking was heavy, the gates were either toppled or partially fallen down as seen in Figure 9.23.

9.1.3 Examples from Italy

The 2009 L'Aquila earthquake with moment magnitude of 6.3 occurred in the Abruzzi region of Central Italy. The earthquake caused heavy damage to cultural and historical buildings of larger scale such as domes, towers and facades. The failure mechanism of these buildings is fundamentally quite similar to that of ordinary masonry buildings. The non-existence of tie beams make these buildings quite vulnerable to heavy damage to collapse. The dome of Santa Maria del Suffragio collapsed although the use of wooden tie beams were used. As the wooden tie beams were not continuous over the perimetry of the dome, its effect was disastrous as seen in Figure 9.24(a).

Figure 9.21   Collapse of the castle wall and dislocation or rotation of blocks and walls.

The masonry cylindrical tower at Stefano di Sessanio, which is about 20 km from the epicenter, was completely collapsed (Figure 9.25). It was important to notice that another tower with almost same height located at the foot of the same hill survived the earthquake. This may be related to the difference in the amplification characteristics

Figure 9.22 Views of block layout and typical blocks at the collapsed section.

Figure 9.23 Views of damage to masonry structures (from Aydan et al., 2017).

of the top and foot of the hill as well as the shape and shaking characteristics of the towers.

A 2 m long 2.5 m high stone masonry arch bridge collapsed and was repaired by filling crashed lime stone into the collapsed section as seen in Figure 9.26. The southwest side of abutments probably moved towards west and this movement resulted in the loss of arching effect of this masonry arch bridge and collapsed during the earthquake. Furthermore, the infill cover of this bridge was thin. This may be also an additional cause for the collapse of this small arch bridge.

(a) Santa Maria del Suffragio   (b) Paganica Church

Figure 9.24 Failure of domes and facades of some large scale historical buildings.

(a) before   (b) after

Figure 9.25 Views of the tower of Stefano di Sessanio before and after the earthquake.

9.1.4   Examples from Egypt

Egypt has a historical record of earthquake activity extending over the past 4,800 years. It was found that, for Egypt, a total of 58 earthquakes are reported felt, with intensities of V–IX, during the period 2200 BC–1900 AD. Some of these earthquakes are reported with poor information regarding the epicentral area; some have locations outside the Egyptian border. Altogether, 22 of the earthquakes have reliable information concerning the location. 11 of these earthquakes caused destruction (Table 9.1) (Figure 9.27).

Abu Simbel Temples are rock temples in Nubia Region of Egypt on the west bank of Lake Nasser and it is about 230 km from Aswan City. The temples were originally were in the mountainside during the reign of Pharaoh Ramesses II in the 13th century

Figure 9.26 Collapsed arch bridge and damaged road surfacing.

Table 9.1 Earthquakes damaged historical structures in Egypt.

Year	Io	Mb	Remarks
365		8.3	Aegean Subduction Zone, Tsunami, Cleopatra's palace disappeared into sea, liquefaction, 5000 people killed in Alexandria
1326			Alexandria, Pharos light house ruined, Kayit Bey (Qaitbay) built his castle over this structure
1356			Cairo, casing stones of pyramids fallen off
1992	VIII	5.4	Cairo (Dashour Earthquake) 70 people killed, damage at burial room and passages to Saqqara step pyramid. Rock block falls from Kufhu pyramid, collapse of minarets in Cairo

BC, as a monument for himself and his queen Nefertari. Four colossal 20 meter statues of the pharaoh with the double Atef crown of Upper and Lower Egypt decorate the facade of the temple, which are 38 meters wide and 33 meters high. The colossal statues decorate the facade of the temple with a height of 33 m and width of 38 m was sculptured directly from the rock in which the temple was located. All statues represent Ramesses II. The statue to the left of the entrance was damaged by an earthquake, leaving only the lower part of the statue still intact (Figure 9.27). The head and torso can still be seen at the statue's feet. Although it is difficult to find any official record for the earthquake to topple this colossal, it is supposed to be caused by an earthquake after its construction (1279–1213 BC). Ramesses II was said to be still alive at that time and he was never told about it. The temple complex was relocated due to the possibility of inundated by the reservoir lake of Aswan High dam in its entirety in 1968, on an artificial hill made from a domed structure, high above the reservoir.

There are a number of pyramids built during the era of Pharaohs. The stepped pyramid built by King Zoser at Sakkara is probably the first pyramid. It was during the fourth dynasty that the royal pyramids were evolved and the largest pyramids were built at Giza. Khufuo built the Kufhu great pyramid. The second and third pyramids are called Khafra and Menkaura. Many other pyramids followed in other locations, chiefly Abusir and Saqqara for the Kings of the fifth and sixth dynasties. The pyramids are mainly made of limestone blocks put together in a three-dimensional intermitted pattern so that mechanical interlocking was achieved. This feature of the pyramids

Abu Simbel   Giza-Pyramids

Alexandria   Luxor-Karnak   Giza - Menkaure Pyramid

Figure 9.27 Traces of earthquakes in the well-known monuments of Egypt.

made them quite resistant against earthquakes for several millenniums. The pyramids were covered with limestone or Aswan granite case-stones. The case stones of the pyramids at Giza were fallen off due to the earthquake in 1356 AD. Therefore, some case stones at the great pyramid are still visible while the fallen case stones of the Menkaura Pyramid made of granite of Aswan can be seen at the foot of the pyramid (Figure 9.27). The 1992 Dashfur (Cairo) earthquake damaged the burial room and passages to Saqqara step pyramid. It was reported some rock block falls occurred from Kufhu pyramid.

9.1.5   Examples from India

Kutch (Kaatch) region in Gujarat State of India was severely shaken by a powerful earthquake at 8:46 am on 26 January 2001 of the India Standard Time, which has been the most damaging earthquake in the last five decades in India. The M7.9 quake caused a large loss of life and property. Several cities and large towns such as Bhuj, Anjaar, Vondh and Bhachau sustained widespread destruction and casualties. There are many monuments, historical structures such as castles and places and mosques and temples. Most of these structures built using mainly sandstone and they are generally of dry-masonry type. The author visited most of these structures during the reconnaissance of the damage induced by the 2001 Kutch earthquake. Especially tower like-masonry structures such as minarets and clock and watch towers located at far-distant locations such as Ahmedabad were highly affected by the earthquake (Figure 9.28).

Ghandidham   Ahmedabad   Wankaner   Bhuj

Morbi   Morbi   Bhuj

Figure 9.28 Damaged masonry type historical structures by the 2001 Kutch earthquake.

9.1.6 Examples from Kashmir

On October 8, 2005 at 8:50 (3:50 UTC), a large devastating earthquake occurred in Kashmir region of Pakistan. The depth of the earthquake was estimated to be about 10 km and it had the magnitude of 7.6. The maximum ground acceleration for Balakot was inferred to be greater than 0.9 g from overturned vehicles in the direction parallel to the axis of the valley. This probably represents the largest ground acceleration in the epicentral area. Besides widespread slope failures all over the epicentral area, the Kashmir earthquake of October 8, 2005 caused extensive damage to housing and structures founded on sloping soil deposits. One of the most important historical masonry structures is Dome Fort. This fort suffered extensive damage in the form of collapse of retaining walls, castle walls and arches as seen in Figure 9.29. The walls were constructed using rounded large stones brought probably by Jhelum and Neelum Rivers from adjacent steep mountains.

9.1.7 Examples from Portugal

The 1755 Lisbon earthquake, also known as the Great Lisbon Earthquake, occurred in the Kingdom of Portugalon Saturday, 1 November 1755 at around 9:50. The moment magnitude of this earthquake is estimated to be around 8.5–8.8 and the resulting tsunami arrived at Lisbon approximately 40 minutes later inducing heavy casualties

Figure 9.29 Damaged masonry structures of the Dome Fort by the 2005 Kashmir earthquake.

Figure 9.30 Effects of the 1755 Lisbon earthquake at Carmo Convent

in addition to those from ground shaking. Heavy ground shaking caused the collapse of the walls of Covilha Castle, Churches and major towers in the City. Traces of the earthquake can be still observed at Carmo Convent on a hill in the Lisbon City as seen in Figure 9.30.

Figure 9.31 Damage to Washington Monument caused by the August 23, 2011 earthquake.

9.1.8 Examples from USA

The most historical monument for American having European origin is the Washington Monument. This monument was constructed at two stages, 1848–1854 stage and 1876–1884 stage in the shape of Egyptian Obelisk. It is about 166 m high and the ratio of height to width is 10. The monument was built using marble blocks with mortar on blue gneiss foundation. On August 23, 2011 an earthquake with a magnitude of 5.8 occurred. Although the earthquake was about 200 km away from the monument was heavily shaken and caused substantial damage to the monument (Figure 9.31). When the author visited the monument in May 2012, it was closed to the visitors and repair works was not commenced. The earthquake caused the opening of joints and spalling and relative sliding and cracking of stones. In addition to the heavy damage to the Washington Monument, many historical masonry structures suffered extensive damage in Washington City as well as in other sites near the epicentral area. The damage at National Cathedral in Washington DC was also heavy.

The 1500 km long San Andreas fault runs along the west coast of USA and cause many earthquakes from time to time. One of the largest events occurred in 1906 near San Francisco and caused heavy damage to masonry structures. Similar situations were also observed in other major cities along the west coast.

9.2 MODEL EXPERIMENTS ON MASONRY STRUCTURES

A series of model experiments on the seismic response and stability of masonry structures such as arches, castle walls, dams, retaining walls, bridges, pyramids, towers

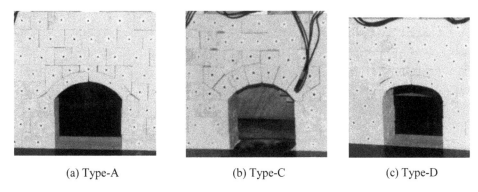

| (a) Type-A | (b) Type-C | (c) Type-D |

Figure 9.32 Model arch types.

and buildings were performed using a shaking table (see Aydan *et al.*, 2003, 2005) for details). In these model studies, masonry structures were constructed using Ryukyu limestone blocks as building materials.

9.2.1 Experiments on arches

Five arch configurations denoted as Type-A, B, C, D and Type-E, four (Type-A, B, D, E) of which are commonly used in Shuri Castle in Okinawa Island, Japan were tested (Figure 9.32). The remaining arch form (Type-C) is quite common almost all over the world. The arches of Shuri Castle generally consist of two monolithic blocks in the form of a semi-circle or an ovaloidic shape while the Type-C arch consists of several blocks and has a semi-circular shape. As the shaking table was uniaxial, the effect of the direction of input acceleration wave was investigated by changing the longitudinal axis of the arches.

Figure 9.33 shows the failure state of Type-A arch for shaking directions of 0°, 45° and 90°. The experimental results indicated that the common form of failure for all arch types for a shaking direction of 0° is sliding at abutments and inward rotational fall of arch blocks subsequently. As for 90° shaking, the arch failed in the form of toppling. The failure for 45° shaking was a combination of sliding and toppling. The experiments clearly indicated that the amplitude of acceleration waves to cause failure was the lowest for 90° shaking while it was the maximum for 0° shaking. Figure 9.34 shows measured acceleration and displacement responses of Arch Type-C for 0° direction.

9.2.2 Experiments on pyramids

Both 2D and 3D models of pyramids consisting of limestone blocks were tested. In all experiments, the governing mode of failure was due to relative sliding among the layers of blocks as seen in Figure 9.35. If the motion of the blocky layer is not obstructed by the roughness of the block interfaces, the pyramids keep their original configuration during motion. However, if the inter-block sliding is obstructed by some asperities, then

(a) Shaking direction of 0°

(b) Shaking direction of 45°   (c) Shaking direction of 90

Figure 9.33 Failure modes of arch type-A.

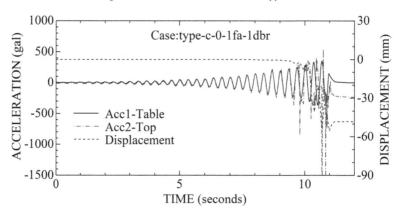

Figure 9.34 Measured acceleration and displacement responses for arch type-C for shaking direction of 0°.

block separation starts to take place gradually as shown in Figure 9.36. The failure of pyramids was purely governed by the frictional properties of block interfaces.

9.2.3   Experiments on castle walls

The walls of many castles and historical cities all over the world are built either vertically or inclined. Furthermore, the outer shells of the walls are built with neatly placed blocks of rocks while the core generally consists of rubble material. A series of

Figure 9.35 Failure modes of 2D and 3D pyramids for horizontal shaking.

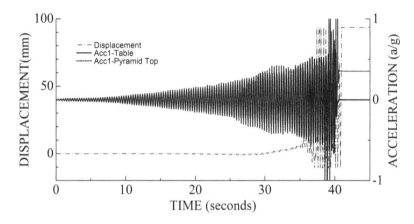

Figure 9.36 Measured acceleration and displacement responses of a 3D pyramid built with Ryukyu limestone blocks.

experiments were carried out to see the effect of the inclination of outer shell with the consideration of the basal inclination of the foundation blocks. The inclinations of the castle wall were 73°, 84° and 90° with 0° and 7° basal inclinations of the foundation block. Figure 9.37 shows the failure of castle walls with an inclination of 84° for 0° and 7° basal inclinations of the foundation block. Figure 9.38 shows the acceleration records for an experiment on the castle walls with inclination of 84°. The experiments showed that the decrease in wall inclination results in higher resistance against shaking. Furthermore, walls with a 7° basal inclination have a higher resistance against shaking as compared with walls with a 0° basal inclination. The fundamental mode of failure is toppling. However, some inter-block sliding may be caused and the gap may be filled by backfill material, which may result in more unstable wall configuration after each wave passed through.

9.2.4   Experiments on retaining walls

In model tests, the aim was to investigate the height/width ratio, the inclination of retaining walls and back-filling material. In tests, the ratio of width to height was varied between 0.25 and 0.625. The wall inclinations were 73°, 84° and 90° with 0° and 7° basal inclinations. The back filling material was sand N7 and sandy gravel.

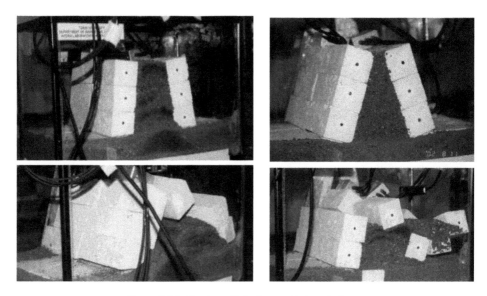

Figure 9.37 Castle walls before and after shaking.

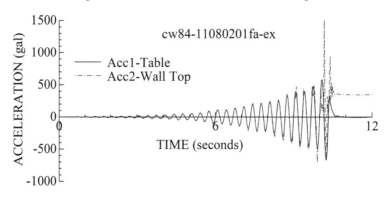

Figure 9.38 Acceleration records for an experiment on the castle walls with inclination of 84°.

Figure 9.39 shows some views of the model experiments on retaining walls for two types of back-filling materials. Figure 9.40 shows measured accelerations records. The fundamental model of failure was toppling (rotational failure) for walls with greater height/width ratios. After each shaking cycle, the interface gap was filled with a wedge of back-filling material, which resulted in unstable wall configuration. If the walls are inclined and/or have a 7° basal inclination, they can resist high acceleration amplitudes. This may also explain why such walls are more stable as compared with walls with 90° inclination during earthquakes.

9.2.5   Experiments on houses

The experiments were carried out on single story houses having a heavy roof. In experiments, plastic blocks were used. The model houses were built in such a way that two

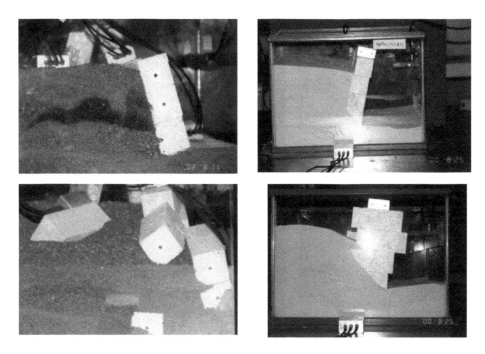

Figure 9.39 Failure modes of retaining walls.

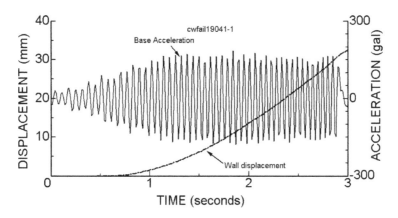

Figure 9.40 Measured acceleration and displacement responses for a model wall numbered cwfail19041-1.

sidewalls would be parallel to ground shaking while the other two walls would be subjected to out-of plane loading. Figure 9.41 shows some views of the experiments. Figure 9.42 shows measured accelerations records. The experiments clearly showed that the walls, which are subjected to out-of plane loading, tend to collapse first while the sidewalls parallel to shaking tend to fail by inter-block sliding. The ground shaking

Figure 9.41 Some views of experiments on model houses.

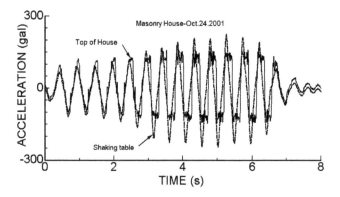

Figure 9.42 Measured acceleration responses for a model house.

to cause total collapse of the side-walls parallel to the direction of shaking should be such that the accumulated relative displacement of the inter-block sliding should exceed the half length of the block in the respective direction. The experiments also showed that the corners of the buildings are quite prone to fail first due to the concentration of two failure modes at such particular locations.

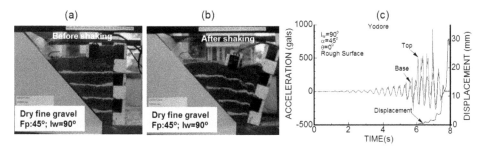

Figure 9.43 A view of the physical model before and after the shaking.

9.2.6 Some special structures

(a) Yodore Royal Mausoleum

Yodore Royal Mausoleum is located in Urasoe City. It was recently restored as part of a restoration project of masonry historical structures in Ryukyu Islands. A very high retaining wall of about 12 m was constructed as part of the overall project. The slope behind the retaining wall was cut at an angle of 45°. The wall was stable under gravitational load. However, it failed during a heavy season following backfilling. The authors investigated the causes of this failure through some physical model tests (Figure 9.43). A physical model of retaining wall with a back filling was constructed. The model was stable under dry condition. When the model was fully saturated and the retaining wall had no drainage holes, it was found that the retaining wall would fail. Next, the stability of the retaining wall was investigated under dynamic conditions. Figure 9.43 shows an example of a physical model before and after shaking together with displacement response of the wall. As noted from the figure, the retaining wall starts to exhibit non-linear behaviour when the acceleration level is about 100 gals. After each cycle of shaking, the displacement of the wall increases. This deformation mode involves both relative slip and rotation of the wall with respect to its base.

(b) Gushikawa Castle

Two karstic caves exist beneath Gushikawa Castle remains in Itoman-shi in Okinawa Island of Japan. These caves are large (more than 20 m wide) and they may result in their collapse by the continuing erosion process (Geniş *et al.*, 2009). As it is difficult to model an actual structure in a reduced scale of the geometry, stress conditions and constitutive parameters of materials, the model experiments presented in this sub-section are intended to illustrate what one should expect under natural conditions and to understand the underlying mechanism of the response and stability of natural underground openings subjected to earthquakes. A two-dimensional model with a scale of 1/100 of the actual configuration of Cave A and its surrounding was prepared (Figure 9.44a). The model rock mass is 400 mm long, 250 mm high and 50 mm thick. The model material was a mixture of $BaSO_4$, ZnO and Vaseline oil, which were used in base-friction model tests.

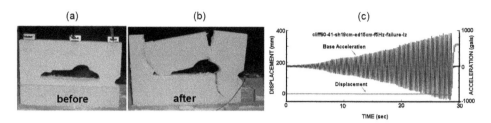

Figure 9.44 Views of Cave A before and after shaking and the measured response during shaking.

The cavity was first excavated and its deformation to the excavation procedure was monitored. The opening was stable under static conditions. Following the frequency response experiments under a base acceleration of 100 gals, the model was subjected to horizontal shaking along the longitudinal axis of the cave A until the failure occurred. The frequency of the applied base acceleration was 5 Hz and its amplitude was about 600 gals when the model with the cave failed. Figure 9.44a,b shows a shaking model experiment in relation to the effects of karstic caves beneath Gushikawa castle remains. Figure 9.44c shows the measured displacement in relation to the applied base acceleration. As noted from the figure, the elastic displacement is very small to be discernible and the displacement becomes very large during the failure stage suddenly.

The cavity was filled with a very weak material (sponge), similar to that in actual conditions, to see the effect of the filling of the cavity. As in the previous case, the same procedure was followed and the model was subjected up to 1 g, which was the maximum allowable capacity of the shaking table. Although the filling material was very weak, the opening remained stable compared to that in the unfilled state.

(c) Arch bridge of Iedonchi imperial garden

Iedonchi imperial garden is very close to Shuri Castle. There are many stone structures carved into different artistic configurations in this garden. It has been recently found that that one of the arch bridges is dilapidating. Tokashiki *et al.* (2014) have been investigating the stability of the arch bridge by model experiments using the base-friction apparatus and shaking table. The preliminary results indicated that the bridge must be stable under gravitational loading condition. Figure 9.45 shows views of the model arch bridge before and after shaking and the measured responses for the experiment numbered CASE4D shown in Figure 9.45 also. The vibration, which may result from nearby constructions, bombing during the Second World War and/or earthquakes, causes the loss of the arching effect and may result in the collapse of the bridge. However, it was experimentally found that the increase of contact strength of block interfaces through grouting dramatically increases the resistance of the arch bridge against horizontal shaking (Tokashiki *et al.*, 2014).

(d) Perry banner rock

A natural rock block exists in Nakagusuku Village. There is a drawing depicting the banner on the rock block at the time the famous Black Ship visited Japan in 1853.

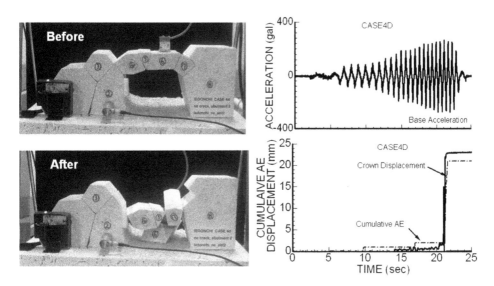

Figure 9.45 Views of the arch bridge model experiment CASE4D before and after shaking and its response.

The vicinity of the rock block shown in Figure 9.46(a) has been recently designated as a touristic spot and the authorities are concerned with its static and dynamic safety. The authors have been asked to investigate the stability of this rock block. There is a stepped discontinuity surface on which the block sits. Tokashiki *et al.* (2014) developed a physical model of this block with a similar density and studied its stability using a shaking table (Figure 9.46). Depending upon the direction of the shaking the block starts to exhibit non-linear behaviour at about 100 gals for the direction of 90 degrees while it is about 230 gals for the direction of 0 degree. The block becomes unstable when the base acceleration exceeds 350 gals. The authors also studied how to increase the dynamic resistance of the block against large earthquakes using an agent to bond the block to its base. The experimental results clearly indicated that the increase of the resistance was possible and the overall seismic resistance of the block increases.

9.3   LIMIT EQUILIBRIUM APPROACHES

Aydan and Tokashiki have been investigating the stability of masonry structures under both static and dynamic conditions in Ryukyu Islands using experimental, analytical and numerical techniques (Aydan *et al.*, 2001, 2002; Tokashiki *et al.*, 2006, 2014). Several mechanical models developed during their collaborative study and they are validated through shaking table experiments. In this section, some of these methods are briefly presented.

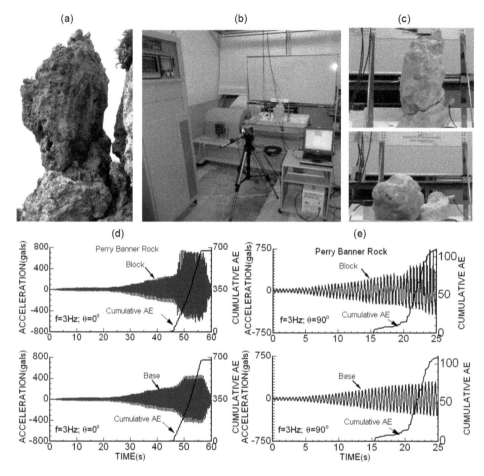

Figure 9.46 Views of the Perry Banner Rock, its physical model tests using a shaking table and measured acceleration and AE count responses.

9.3.1 Retaining walls

There are four different situations for the failure of retaining walls, namely, 1) Stable; 2) Sliding; 3) Toppling; 4) Toppling and sliding. If the seismic coefficient method (pseudo-dynamic) is employed the initiation of sliding failure can be obtained in the following forms (Figure 9.47):

a)   Transition from stable to sliding mode

$$\eta_s = \frac{a}{g} = \frac{(\sin\theta + \cos\theta\tan\phi_{ws}) - K\dfrac{\gamma_s}{\gamma_w}\dfrac{h}{t}\dfrac{\cos^2\theta}{2}(\cos\theta - \sin\theta\tan\phi_{ws})}{(\cos\theta - \sin\theta\tan\phi_{ws}) + K\dfrac{\gamma_s}{\gamma_w}\dfrac{h}{t}\dfrac{\cos^2\theta}{2}(\cos\theta - \sin\theta\tan\phi_{ws})} \tag{9.1}$$

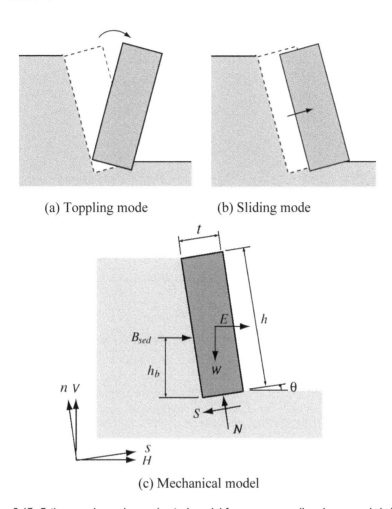

(a) Toppling mode   (b) Sliding mode

(c) Mechanical model

Figure 9.47 Failure modes and a mechanical model for masonry wall under ground shaking.

b)   Transition from stable to toppling mode

$$\eta_t = \frac{a}{g} = \frac{(h\sin\theta + t\cos\theta) - K\dfrac{\gamma_s}{\gamma_w}\dfrac{h}{t}\cos^2\theta\left(\dfrac{h}{3}\cos\theta - t\sin\theta\right)}{(h\cos\theta - t\sin\theta) - K\dfrac{\gamma_s}{\gamma_w}\dfrac{h}{t}\cos^2\theta\left(\dfrac{h}{3}\cos\theta - t\sin\theta\right)}$$ (9.2)

where a: maximum horizontal acceleration; g: gravitational acceleration; θ: wall inclination and base inclination; ϕ_{ws}: friction angle between wall and backfill soil; γ_s: unit weight of backfill soil; γ_w: unit weight of wall; h: length (height) of wall; t: width of wall; K: lateral force coefficient resulting from backfill

c)   Transition from stable mode to combined sliding and toppling failure mode requires that both equations (9.1) and (9.2) must be simultaneously satisfied.

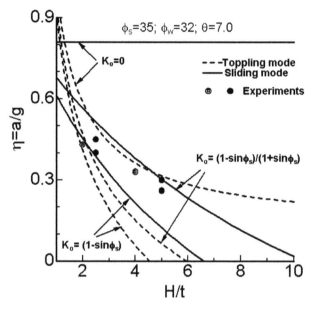

Figure 9.48 Comparison of computed stability chart with experimental results.

Figure 9.48 compares the computed results with measured results for the dynamic stability of the retaining wall for sliding and toppling modes using the dynamic limit equilibrium method (D-LEM) (Tokashiki *et al.*, 2007).

It is also possible to model the rigid body translation and rotation of walls with the consideration of inertia term. In this type of formulation the earthquake force is assumed to be proportional to the mass of retaining masonry wall. The differential equations can be obtained for sliding mode and toppling mode as (Tokashiki, 2011)

Sliding Mode

$$\frac{d^2s}{dt^2} = g[A(t)(\cos\theta - \sin\theta\tan\phi_{ws}) - (\sin\theta + \cos\theta\tan\phi_{ws})] \qquad (9.3)$$

Toppling Mode

$$\frac{d^2\alpha}{dt^2} = \frac{3g}{R}\left[B(t)\left(\cos\theta\frac{b}{3t} - \sin\theta\right) + \frac{a(t)}{g}\left(\cos\theta\frac{b}{t} - \sin\theta\right) - \left(\sin\theta\frac{b}{t} + \cos\theta\right)\right] \qquad (9.4)$$

where $B(t) = \frac{\gamma_s}{\gamma_w}\frac{b}{t}\frac{\cos^2\theta}{2}K_o\left(1 + \frac{a(t)}{g}\right); A(t) = \left(B(t) + \frac{a(t)}{g}\right); R = \sqrt{\left(\frac{b}{t}\right)^2 + 1}; K_o = \frac{1-\sin\phi_s}{1+\sin\phi_s}$
or $K_o = 1 - \sin\phi_s$ (Jacky load coefficient).

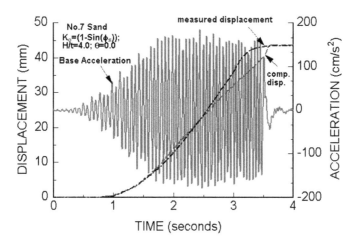

Figure 9.49 Comparison of measured responses with computed responses.

One can integrate the above equation in time domain and compute the response of retaining walls. This approach is called dynamic-limiting equilibrium method (D-LEM).

Aydan *et al.* (2002, 2003) and Tokashiki *et al.* (2006, 2007, 2008) have carried out numerous experiments on retaining walls with different configurations and backfilling materials. Figure 9.49 compares the measured response of retaining walls with those obtained from D-LEM for retaining walls.

As explained in section 9.1.2, the collapsed north wall of Kameya castle is about 5 m high with an inclination of about 70° and block sizes range between 50–60 cm. The block size of older parts of the castle walls is more than 100 cm and their inclination is about 50°. Furthermore, the other walls of the castle are more than 10 m high. The friction angles of backfill soil, friction between the wall and backfill soil are not measured. Nevertheless, they would be greater than 30° in view of past experiences. In the computations for dynamic response and stability of the wall, the NS component of the acceleration records taken at Kameyama strong motion station of K-NET was used and the responses of the castle wall during shaking for sliding and toppling failure modes were computed (Aydan *et al.*, 2007). The computed displacement responses shown in Figure 9.50 correspond to those of the wall mass center. The sliding mode indicates that the wall would be displaced about 160 cm while the toppling mode implies rotation of about 10° (45/250). These results imply that the earthquake shaking was sufficient to induce both sliding and toppling modes of failure. Nevertheless, the effect of sliding mode is more dominant. Since the displacement exceeds the wall width, it may be inferred that the failure of the castle wall was a natural consequence of earthquake shaking.

As also explained in Section 9.1.2, the 2010 Okinawa earthquake caused the collapse of some parts of castle walls at Katsuren Castle, which is designated a world heritage site (Figures 9.21 and 9.22). The castle is located over a hill in Uruma City and the nearest strong motion station of the K-NET strong motion network is Gushikawa.

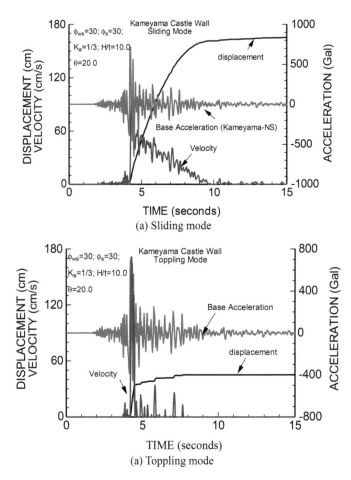

Figure 9.50 Computed displacement and velocity response of mass center for sliding and toppling modes of failure.

The NW corner of the castle wall with a height of 4 m collapsed and there were numerous dislocations and rotation of blocks in the castle as seen in Figure 9.22. A series of analyses were carried out using the dynamic limit equilibrium method (D-LEM) and the acceleration records at Chinen and Gushikawa strong motion stations of the K-NET strong motion network. The typical size of the blocks ranges between 50 to 60 cm as seen in Figure 9.22. The analyses were carried out to back-analyze the collapse of the wall using the strong motion records taken at Gushikawa and Chinen. The wall is stable against toppling mode for strong motions at Gushikawa and Chinen. If the records taken at Gushikawa are used, the relative sliding cannot be greater than 10 cm, which implies that the wall should be stable although some slip might take place. However, if the records taken at Chinen are used the relative sliding can be greater than 60 cm for $\theta = 5°$, which exceeds the half size of the block and this implies that the wall should collapse (Figure 9.51). The bulging of the wall and inclination of the foundation rock strongly supports that this condition would

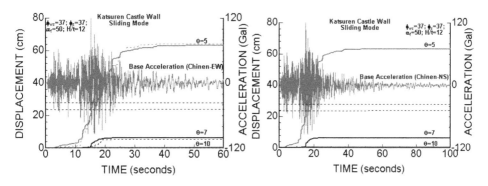

Figure 9.51   Sliding responses of the castle wall for Chinen record.

FAILURE MODE OF A WALL

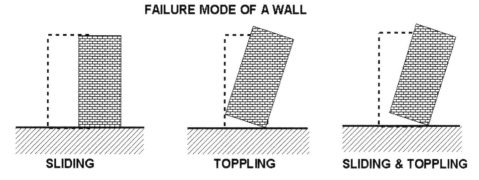

SLIDING   TOPPLING   SLIDING & TOPPLING

Figure 9.52   Failure modes of a wall.

be prevailing at the location of the collapse. As the castle is situated on top of the hill, it is likely that ground motions might have been amplified also.

9.3.2   Castle walls

When walls are vertical with little bonding agent, the seismic resistance of the masonry walls against toppling will mostly depend upon the wall geometry while the seismic resistance against shear will be frictional. The conditions for different modes of failure shown in Figure 9.52 can be derived for a horizontal seismic coefficient as follows (Aydan *et al.*, 1989, Aydan, 2001):

Toppling condition

$$\frac{a}{g} > \frac{t}{h} \tag{9.5}$$

Sliding Condition

$$\frac{a}{g} > \tan \phi \tag{9.6}$$

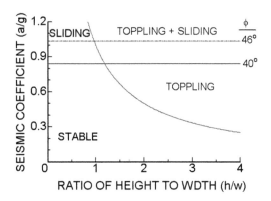

Figure 9.53 Pseudo-dynamic stability condition for a wall.

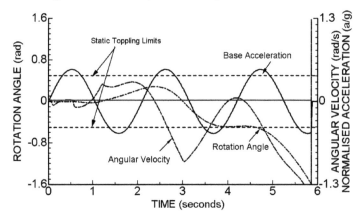

Figure 9.54 Dynamic response of toppling block under sinusoidal base shaking.

Toppling & Sliding Condition

$$\frac{a}{g} > \frac{t}{h} \quad \text{and} \quad \frac{a}{g} > \tan\phi \tag{9.7}$$

It should be noted that the transition between toppling bound to toppling-sliding bound may be slightly different if the dynamic equilibrium equations are taken into account (Aydan *et al.*, 1989). Nevertheless, the conditions from Eq. (9.5) to (9.7) are sufficient enough to evaluate the seismic resistance of the walls.

Figure 9.53 shows the relation between wall height to width ratio and lateral seismic coefficient. As noted from Figure 9.53, the walls are quite vulnerable to toppling rather than sliding in view of the conventional ratio of the wall height to wall width. Eqs. (9.3) and (9.4) can be simplified to obtain the dynamic response of walls by omitting the components of forces resulting from back-filling. Figure 9.54 shows a dynamic solution of Eq. (9.4) without any back-filling pressure for a wall with a height to width ratio of 2 under a sinusoidal base shaking. Provided that the friction coefficient is greater than 0.5, the wall becomes unstable during the second cycle.

9.3.3 Arch bridge (Iedonchi)

The physical situation of the arch bridge of Iedonchi can be modeled as shown in Figure 9.55. For the assumed conditions, the force equilibrium for horizontal direction can be written as

$$\sum H_A = H_A - H_B - k_H W_a = 0 \tag{9.8}$$

$$H_A = H_B + k_H W_a \tag{9.9}$$

As for vertical direction, one can derive the following relation

$$\sum V = V_A + V_B - W_a = 0$$

$$\Rightarrow V_A + V_B = W_a \ \Rightarrow \ V_A = W_a - V_B = W_a \left(\frac{1}{2} + k_H \frac{h}{L} \right) \tag{9.10}$$

Furthermore, the vertical reaction force V_B at point B can be obtained from moment equilibrium as

$$\sum M_{(A)} = W_a \cdot \frac{L}{2} - V_B \cdot L - k_H W_a \cdot h = 0 \ \Rightarrow \ V_B = W_a \left(\frac{1}{2} - k_H \frac{h}{L} \right) \tag{9.11}$$

where k_H: horizontal seismic coefficient, W_a: weight of arch, W_{ab}: weight of abutment block, L: arch span, h: arch height.
The horizontal reaction force H_A at point A can be obtained as

$$\sum H = H_A - H_C - k_H \frac{W_a}{2} = 0 \ \Rightarrow \ H_A = H_C + k_H \frac{W_a}{2} \tag{9.12}$$

From the following moment equilibrium

$$\sum M_{(A)} = \frac{W_a}{2} \cdot \frac{L}{4} - H_c \cdot h - k_H \frac{W_a}{2} \cdot \frac{h}{2} = 0 \tag{9.13}$$

Horizontal reaction forces H_A and H_C are obtained as

$$H_C = \frac{W_a}{8} \frac{L}{h} - k_H \frac{W_a}{4} \ \Rightarrow \ H_A = \frac{W_a}{8} \cdot \frac{L}{h} - k_H \tag{9.14}$$

Trust in the arch can be obtained from the horizontal force equilibrium as

$$\sum H = T - H_A - k_H W_{ab} = 0 \ \Rightarrow \ T = H_A + k_H W_{ab} \tag{9.15}$$

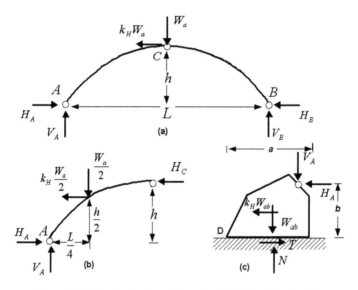

Figure 9.55 Mechanical modeling of arch bridge of ledonchi.

Thus the shear force acting at the base of abutment takes the following form

$$T = \frac{W_a}{8} \frac{L}{h} + k_H \frac{W_a}{4} + k_H W_{ab} \qquad (9.16)$$

From the force equilibrium for vertical direction, one can write the following

$$\sum V = N - V_A - W_{ab} = 0 \implies N = V_A + W_{ab} \qquad (9.17)$$

The vertical reaction is obtained as

$$N = \frac{W_a}{2}\left(1 + k_H \frac{2h}{L}\right) + W_{ab} \qquad (9.18)$$

If the shear resistance is purely friction the stability condition can be written as

$$\frac{T}{N} \le \tan\phi \qquad (9.19)$$

From above relations, one finally gets the following condition for the stability

$$\frac{W_{ab}}{W_a} \ge \frac{\dfrac{L}{8h} + \dfrac{k_H}{4} - \left(\dfrac{1}{2} + k_H \dfrac{h}{L}\right)\tan\phi}{\tan\phi - k_H} \qquad (9.20)$$

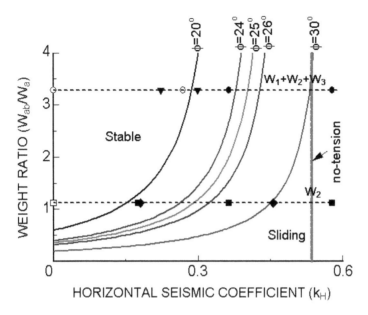

Figure 9.56 Comparison of stability conditions with experimental results.

Furthermore, the stability condition against overturning of the abutment block at point A is obtained through the moment equilibrium given by

$$\sum M_{(D)} = -H_A \cdot b + V_A \cdot a + W_{ab}\frac{a}{2} - k_H W_{ab}b \cdot \frac{b}{2} \geq 0 \tag{9.21}$$

as

$$\frac{W_{ab}}{W_a} \geq \frac{\left(\frac{1}{8}\frac{L}{b} + \frac{k_H}{4}\right) - \left(\frac{1}{2} + k_H\frac{b}{L}\right)\frac{a}{b}}{\frac{a}{2b}(1 - k_H)} \tag{9.22}$$

In addition the horizontal reaction at point B, can be obtained from the following force equilibrium as

$$H_B = \frac{W_a}{4}\left(\frac{L}{2b} - 3k_H\right) \tag{9.23}$$

For $H_B = 0$, the seismic coefficient for the arch to collapse is obtained as

$$k_H = \frac{L}{6b} \tag{9.24}$$

Figure 9.56 compares the computational results with measured limiting accelerations to induce failure.

9.4 NUMERICAL METHODS

In literature, the applications of discrete type numerical methods such as DEM and DDA to masonry and historical structures made of stones are almost none and most discrete-type numerical modeling involves the use of DFEM. Therefore, several applications of the DFEM to fully dynamic and pseudo-dynamic analyses of masonry arch structures are described in this section (Aydan, 1998; Aydan *et al.*, 1996, 2001, 2003, 2011; Tokashiki *et al.*, 1997, 2001).

9.4.1 Fully dynamic analyses

The acceleration waves shown in Figure 9.57 and material properties for blocks and interfaces given in Table 9.2 and 9.3 are used in the analyses presented in this section.

(a) Masonry tower or wall (out-of plane)

Figure 9.58 shows initial and deformed configurations of a dry masonry tower or wall (out-of-plane) at the time steps of 22 and 50 (4.4 and 10 seconds) and responses of a nodal point at the top of the structure with time. As seen from the figure, there is a relative sliding at the base of the wall and separation and rotation of blocks within the wall occur. Furthermore, the wall does not return its original position at the end of shaking.

(b) Wall (in-plane)

In the analysis, the foundation of the structures was subjected to the acceleration waveform 1 horizontally in plane shown in Figure 9.57. Figure 9.59 shows the initial and deformed configurations at time steps 22 and 50. The displacement response of the top of a masonry wall is also shown in the same figure. As noted from the figure, there is a relative sliding at the base of the wall and separation and rotation of blocks occur within the wall. Particularly the block separation and movement is quite amplified at the top of the wall. Furthermore, the wall does not return to its original position at the end of shaking and some permanent relative displacements occur between the wall base and its foundation.

Table 9.2 Material properties of blocks.

Unit weight (kN/m^3)	λ (MPa)	μ (MPa)	λ^* (MPa · s)	μ^* (MPa · s)
25	30	30	30	30

Table 9.3 Material properties of contacts.

λ (MPa)	μ (MPa)	λ^* (MPa · s)	μ^* (MPa · s)	h (mm)	c (kPa)	σ_t (kPa)	tan ϕ
5	2.5	5	2.5	5	0	0	0.7

(c) Arch

Figure 9.60 shows the initial and deformed configurations of a masonry arch at the time steps of 23 (4.6 seconds) and 50 (10 seconds) subjected to the Acc. No. 1 and Acc. No. 2, respectively. For plotting the deformed configurations, in this example,

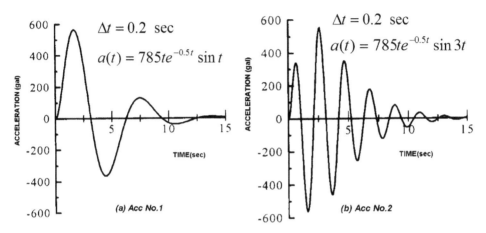

Figure 9.57 Imposed horizontal acceleration waves on foundations (from Aydan, 1998 and Mamaghani et al., 1999).

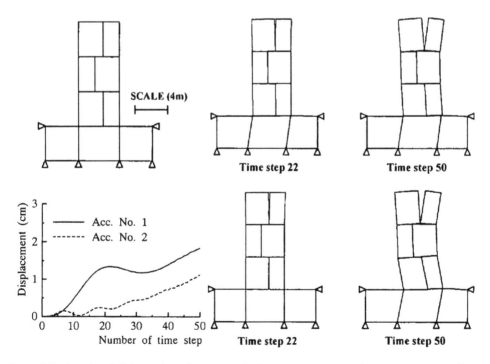

Figure 9.58 Initial and deformed configurations displacement response of a masonry tower (from Aydan, 1998 and Mamaghani et al., 1999).

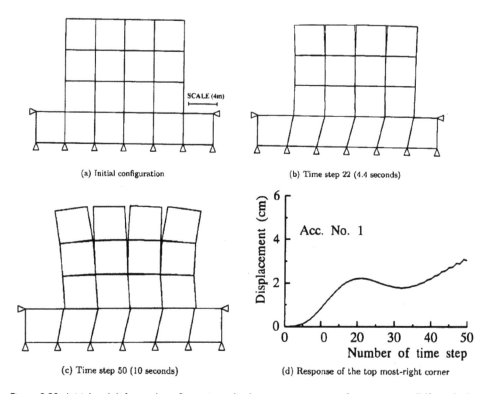

(a) Initial configuration

(b) Time step 22 (4.4 seconds)

SCALE (4m)

(c) Time step 50 (10 seconds)

(d) Response of the top most-right corner

Acc. No. 1

Figure 9.59 Initial and deformed configurations displacement response of a masonry wall (from Aydan, 1998 and Mamaghani et al., 1999).

the displacement in the deformed configurations is amplified 50 times to make more visible the mode of failure (deformed configuration) from the initial configuration. Figure 9.60b shows that the arch has slid at the base at the time step 23 under Acc. No. 1 and the crown blocks of the arch start to fall apart while the side columns are still stable. Figure 9.60d shows that, under Acc. No. 1 at the time step 50, the arching action disappears and the crown blocks fall apart. The columns slide relative to the base, and they tend to topple in two opposite directions. The blocks tend to separate within the side columns.

Figure 9.60d shows that, under Acc. No. 2 at time step 23, there is no slide at the base of the arch while the crown blocks are separated and tend to fall apart. At time step 23, the side columns of the arch exhibit relatively stable behaviour under Acc. No. 2 as compared with Acc. No. 1. However, under Acc. No. 2 at time step 50 (10 seconds), the side columns of the arch slide at the base and the arching action disappears while the blocks start to fall apart. As expected, the toppling (failure) modes of the side columns of the arch differ, depending on the nature of the imposed form of acceleration waves.

Figure 9.60c shows the displacement responses with time of a nodal point at the top most-right corner of the arch corresponding to Acc. No. 1 and Acc. No. 2. The results

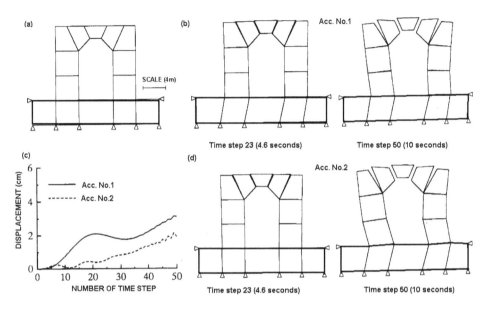

Figure 9.60 Initial and deformed configurations and displacement response of the top of a masonry arch (from Aydan, 1998 and Mamaghani et al., 1999).

in Figure 9.60c indicate that, as expected, the displacement of the side column of the arch with time is much more severe under Acc. No. 1 as compared with Acc. No. 2, especially in the early stage of loading. Figures 9.60b and 9.60d show that, under both of the imposed acceleration waves, the reaction of the toppled columns forces the crown block to move upward. This is because of the geometrically symmetric configuration of the structure and outward inclination of the crown block contact interfaces at the center of symmetry (Figure 9.60a). As can be realized by examining the displacement response curves in Figure 9.60b, the real value of the upward displacement is very small as compared to the dimension of the crown block. It should be noticed that in Figures 9.60b and 8d, the displacement is amplified using the illustration scale factor (50 times the actual scale) to make the failure mode of the whole structure more visible.

9.4.2   Pseudo-dynamic analyses

As the mechanical properties for dynamic analyses are not well-known, the inertia term in the DFEM formulation is ignored and pseudo-dynamic type computations are implemented in this sub-section. Material properties of blocks and contacts are given in Table 9.4.

(a) Arch structure

A masonry arch structure is considered and the physical and mechanical properties given in Table 9.4 are used. The friction coefficient among blocks is 0.7. When the dead weight of the blocks is imposed, the arch remains stable. Analyzing the arch as a

Table 9.4 Material properties of blocks and contacts.

Unit weight (kN/m³)	E (MPa)	ν	E_n (MPa)	G_s (MPa)	h (mm)	c (kPa)	σ_t (kPa)	$\tan\phi$
25	50	0.2	5	0.5	5	0	0	0.7

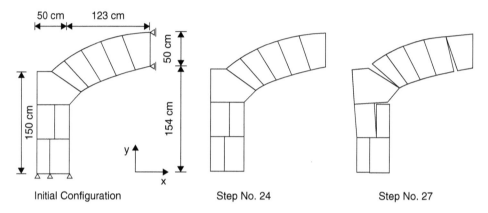

Figure 9.61   Failure mechanism of a masonry arch (from Aydan, 1998).

three-hinged simple structure can validate this result. It is still stable when the distributed traction less than 150 kgf/m² is imposed on the arch. However, if the traction reaches that level, then the arch starts to be unstable. Figure 9.61 shows the configurations of the arch at different iteration steps. The failure mode of the arch is quite similar to what one would observe in actual tests.

(b) Pyramid

In the next example, a pyramid structure was analyzed. The physical and mechanical properties of blocks and contacts were the same as those given in Table 1. This structure is stable under its own weight and for small values of horizontal load as it can be predicted from a simple limiting equilibrium analysis. However, when the horizontal loads are increased, the pyramid becomes unstable for horizontal forces $F_1 = 500$ kgf and $F_2 = 13000$ kgf, and the failure mechanism is shown in Figure 9.62. As seen from the figure, blocks slide and some of the contacts are separated with the increasing number of iterations.

(c) Arch gate of Shuri Castle

A series of analyses using the discrete finite element method (DFEM) were performed on arch gates of Shuri Castle (Tokashiki et al., 2002; Aydan et al., 2005). The details of the analyses can be found in the respective article. Figure 9.63 shows a computational example on the arch gate under a horizontal seismic load and gravity. As noted from Figure 9.63, the failure occurs due to sliding of the sidewall and subsequent rotation and fall of the arch blocks.

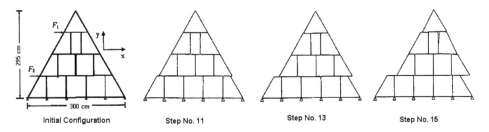

Figure 9.62   Failure mechanism of a masonry pyramid (from Aydan, 1998).

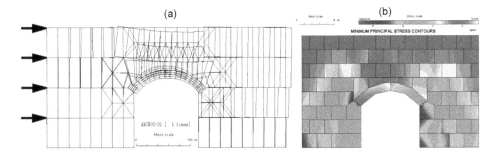

Figure 9.63   Numerical DFEM analyses for arches and computational results (arranged from Tokashiki et al., 2002 and Aydan et al., 2005).

(d) Retaining wall

Tokashiki *et al.* (2001) analyzed the bulging phenomenon of retaining wall using the DFEM. The details of the analyses can be found in the respective article. The stability analysis of this masonry wall was first carried out under gravitational loading. Figure 9.64 shows how the retaining wall is modeled in the numerical analyses. The computed configuration is barrel-like and it is quite close to the actual observations. A pseudo-dynamic simulation of the same retaining-wall was carried out. Figure 9.65(b) shows the deformed configuration of the retaining wall for a horizontal seismic coefficient of 0.2. As expected, the horizontal displacement of the wall is larger than that for gravitational loading, and the effect of wall height on its deformation under gravitational and seismic loading. As expected the deformation of wall increases as a result of seismic loading and some relative displacements among blocks occur.

(e) Himeyuri Mausoleum

There is a huge karstic cave beneath the Himeyuri monument in Okinawa Island. The enlargement of the monument was considered and the authors were consulted if the karstic cave would be stable upon the enlargement. Figure 6.66 shows the beam modelling of overhanging part. Table 9.5 and Figure 6.67 compares the maximum tensile and compressive stresses computed from beam theory and FEM. Despite some slight differences, the results are quite similar. If the dynamic load resulting from the earthquakes is considered using the seismic coefficient method, the stresses become much higher as noted in Table 9.5 (see details in Aydan *et al.*, 2011).

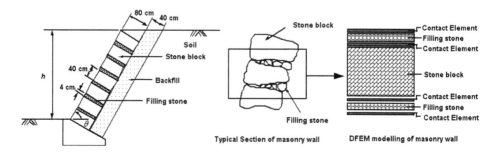

Figure 9.64 A typical masonry retaining wall and its DFEM representation.

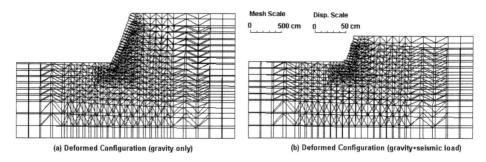

(a) Deformed Configuration (gravity only)   (b) Deformed Configuration (gravity+seismic load)

Figure 9.65 Comparison of deformed configurations for gravity only and gravity and seismic loading.

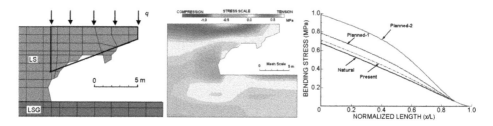

Figure 9.66 Computational model and comparison of stresses from the bending theory and FEM.

(e) Karstic caves beneath Gushikawa castle remains

In addition to 2D and 3D static numerical analysis, a series of dynamic analyses were carried out (Geniş et al., 2009). The Gushikawa castle remains are approximately 120 km away from both Ryukyu Trench and Okinawa Trough. The area experienced a very strong earthquake with a magnitude of 7.2 on February 27, 2010 (Aydan and Tokashiki, 2010). The maximum ground accelerations nearby the castle remains ranged between 70–120 gals (0.07–0.12 g). Therefore, the consideration of dynamic condition is very important for the stability of Karstic caves even after filling the cavities.

Table 9.5 Computed maximum tensile and compressive stresses from bending theory and FEM.

Loading Condition	Analysis Method	Max. Tensile Stress (MPa)	Max. Compressive Stress (MPa)
Natural	FEM	0.557	−1.363
	Theory	0.677	−0.677
Present	FEM	0.631	−1.402
	Theory	0.713	−0.713
Planned-2	FEM	0.770	−1.478
	Theory	0.991	−0.991

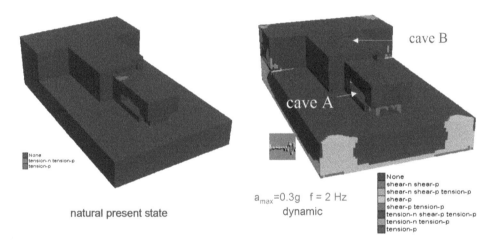

natural present state

$a_{max}=0.3g$ f = 2 Hz
dynamic

Figure 9.67 Plastic zone formation for static and dynamic conditions.

The mesh used in 3D static analyses was used with the introduction of viscous boundary conditions following the static loading. The main purpose was to assess the stability of karstic caves in the natural state and filled state. The input waves were assumed to be horizontal and perpendicular to the longitudinal axis of the numerical mesh. Since there is no strong motion record for this area, sinusoidal acceleration waves with amplitude of 0.3 g and a frequency of 2 Hz to simulate the worst scenario of dynamic loading were used.

Figure 6.67 shows the plastic zone formation in the computational model for the non-filled state of caves under static and dynamic conditions. Unexpectedly, plastic zone development occurred at the boundaries and corners of the 3D model. If attention is given to the vicinity of the karstic caves, it is easily noticed that the plastic zone formation is very extensive for both caves and the geometry of plastic zone implies how the total collapse of the caves would occur under dynamic conditions. The results indicate that Cave-A would fail by toppling towards the west side while Cave-B will collapse as a circular sinkhole (i.e. doline).

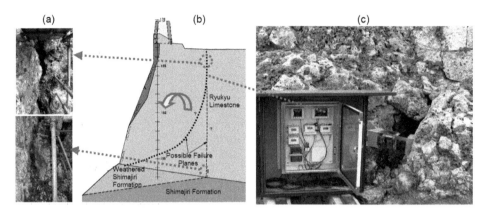

Figure 9.68 Views of monitoring location and instrumentation.

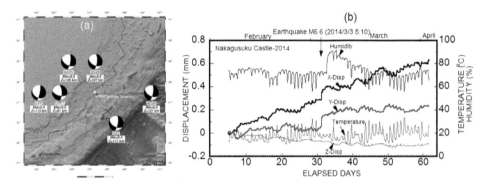

Figure 9.69 Preliminary monitoring results during February and March, 2014.

9.5 MONITORING AT NAKAGUSUKU CASTLE

There is a continuous crack in Ryukyu limestone layer (about 20 m thick) at the south-east side of the Nakagusuku Castle Remains designated a World Heritage Site by UNESCO (Figure 13). The authorities are concerned with the stability of this part of the castle remains particularly during earthquakes as some part of the Katsuren Castle remains collapsed during the M7.2 2010 Okinawa earthquake. The authors established a long-term multi-parameter (displacement, acoustic emission, inclination, tempera-ture, humidity, air pressure and accelerations) monitoring system (Figure 6.68). This system utilizes solar power as the energy source for instruments and it is environmen-tally friendly. An earthquake with a moment magnitude of 6.5 occurred at 5:10 AM on 2014 March 13 (JST) in East China Sea at a depth of 120 km on the western side of Okinawa Island (Figure 6.66a). Another earthquake occurred at 11:27 AM on the same day near Kumejima Island. Although the magnitude of the earthquake was inter-mediate and far from the location, some permanent displacement occurred as seen in Figure 6.69b.

Chapter 10

Dynamics of loading and excavation in rocks

10.1 DYNAMICS OF LOADING

10.1.1 Uniaxial tensile loading experiment

It is very rare to see discussion or experimental results on the load-displacement-time or stress-strain-time responses. The author devised an experimental set-up. The experimental set-up consists of an acrylic bar attached to a strain gauge and fixed to a support above. The diameter and length of the acrylic bar were 8 mm and 200 mm respectively. An object with a given weight was instantaneously applied to the lower end of the bar. The strain response was monitored using WE7000 dynamic data acquisition system with a sampling interval of 10 ms. Figure 10.1 illustrates the experimental set-up. In the first stage, 500 gf was applied and then the load was increased by 1543 gf. Figure 10.2 shows the strain variation with time. As noted from the figure, the strain fluctuates and become asymptotic to the static strain level for the applied stress level. Although the experiment is very simple, it is clearly shown that the loading of samples and structures as well as excavation of rock engineering structures should be treated as a dynamic phenomenon.

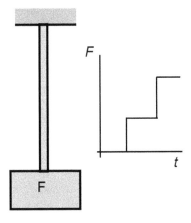

Figure 10.1 Schematic illustration of the experimental set-up.

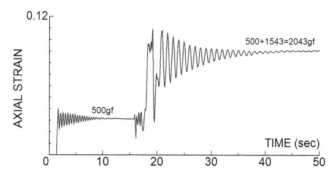

Figure 10.2 Strain response of an acrylic bar subjected to the weight of an object.

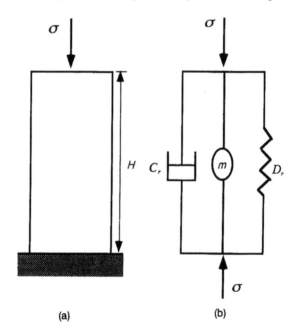

(a)   (b)

Figure 10.3 (a) Uniaxial Compression Test, (b) Its mechanical model.

10.1.2 Uniaxial compression loading

Aydan (1997) proposed a method to model the dynamic response of rock samples during loading. In this subsection, this method and several examples of its applications to some typical situations are presented.

(a) Theoretical formulation

Let us consider a sample under uniaxial loading as shown in Figure 10.3(a). The force equilibrium of such a sample can be written in the following form (Figure 10.3(b)):

$$\sigma = D_r \varepsilon + C_r \dot{\varepsilon} + \rho H \ddot{u} \tag{10.1}$$

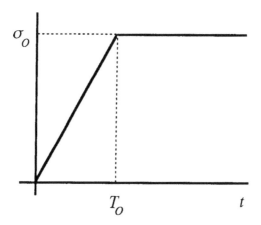

Figure 10.4 Time-history of uniaxial loading.

where D_r, C_r, ρ, H are elastic modulus, viscosity coefficient, density and sample height. If acceleration $\ddot{u}$ is uniform over the sample and its strain ε is defined as

$$\varepsilon = \frac{u}{H} \tag{10.2}$$

Eq. (10.1) becomes

$$\sigma = D_r \varepsilon + D_r \dot{\varepsilon} + \rho H^2 \ddot{\varepsilon} \tag{10.3}$$

Let us assume that stress is applied onto the sample in the following form (Figure 10.4)

for $0 \leq t \leq T_0$

$$\sigma = \frac{\sigma_0}{T_0} t \tag{10.4}$$

for $t \geq T_0$

$$\sigma = \sigma_0 \tag{10.5}$$

The solutions of this ordinary differential equation are:

Case 1: Roots are real

$$\varepsilon = C_1 e^{\lambda_1 t} + C_2 e^{\lambda_2 t} + \varepsilon_p \tag{10.6}$$

where

$$\lambda_1 = \frac{1}{2\rho H^2}\left(-C_r + \sqrt{C_r^2 - 4D_r \rho H^2}\right), \quad \lambda_2 = \frac{1}{2\rho H^2}\left(-C_r - \sqrt{C_r^2 - 4D_r \rho H^2}\right)$$

for $0 \leq t \leq T_0$

$$\varepsilon_p = \frac{\sigma_0}{\rho H^2 T_0} \frac{1}{\lambda_1^2 \lambda_2^2}[\lambda_1 \lambda_2 t + (\lambda_1 + \lambda_2)]$$

for $0 \leq t \leq T_0$

$$\varepsilon_p = \frac{\sigma_0}{\rho H^2} \frac{1}{\lambda_1 \lambda_2}$$

Case 2: Roots are same

$$\varepsilon = [C_1 + C_2 t]e^{\lambda t} + \varepsilon_p \tag{10.7}$$

where

$$\lambda = -\frac{C_r}{2\rho H^2}$$

for $0 \leq t \leq T_0$

$$\varepsilon_p = \frac{1}{\lambda^3} \frac{\sigma_0}{\rho H^2 T_0}$$

for $t \geq T_0$

$$\varepsilon = \frac{1}{\lambda^2} \frac{\sigma_0}{\rho H^2}$$

Case 3: Roots are complex

$$\varepsilon = e^{pt}[A \cos qt + B \sin qt] + \varepsilon_p \tag{10.8}$$

where

$$p = -\frac{C_r}{2\rho H^2}, \quad q = \sqrt{4D_r \rho H^2 - C_r^2}$$

for $0 \leq t \leq T_0$

$$\varepsilon_p = \frac{1}{(p^2 + q^2)^2} \frac{\sigma_0}{\rho H^2 T_0}[(p^2 + q^2)t + 2p]$$

for $t \geq T_0$

$$\varepsilon_p = \frac{1}{p^2 + q^2} \frac{\sigma_0}{\rho H^2}$$

Integration constants C_1 and C_2 can be determined from the following initial conditions:

for $0 \leq t < T_0$

$$\varepsilon = 0 \quad \text{at } t = 0$$
$$\dot{\varepsilon} = 0 \quad \text{at } t = 0 \tag{10.9}$$

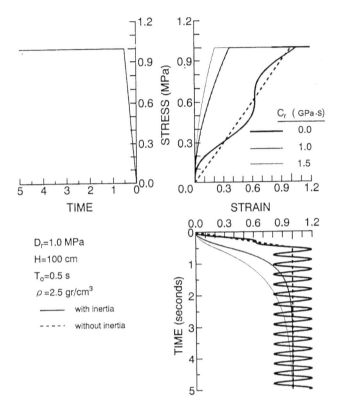

Figure 10.5 The effect of viscosity coefficient on dynamics response of a sample subjected to uniaxial compression.

for $t \geq T_0$

$$\begin{aligned} \varepsilon &= \varepsilon_0 \quad \text{at } t = T_0 \\ \dot{\varepsilon} &= \dot{\varepsilon}_0 \quad \text{at } t = T_0 \end{aligned} \qquad (10.10)$$

Integration constants can be easily obtained for above conditions for each case. However, their specific forms are not presented as they are too lengthy.

(b) Applications

Several applications of the theoretical relations derived in the previous section are given herein to investigate the effects of viscosity coefficient, elasticity coefficient, loading rate, and sample height.

1   *The effect of viscosity coefficient*: Figure 10.5 shows the effect of viscosity coefficient on the deformation responses of a sample. It is of great interest that when rock is elastic, an oscillating behaviour must be observed. Furthermore, stress-strain relation is not linear and it also oscillates as the applied stress is

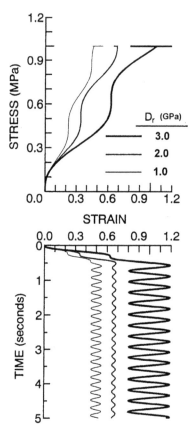

Figure 10.6 The effect of elastic coefficient on dynamics response of a sample subjected to uniaxial compression.

linearly increased. However, this oscillating behaviour is suppressed as the viscosity coefficient increases.

2 *The effect of elasticity coefficient:* Figure 10.6 shows the effect of elasticity coefficient on the deformation responses of a sample with a viscosity coefficient of 0 GPa·s. The amplitudes of oscillating part and stationary part of strain decrease as the value of elasticity coefficient increases. Nevertheless, the oscillating behaviour is apparent for each case.

3 *The effect of loading rate:* Figure 10.7 shows the effect of loading rate on the deformation responses of a sample with a viscosity coefficient of 0 GPa·s. While the amplitude of stationary part of strain remain the same, the amplitude of oscillating part of strain decreases as the loading rate decreases. Although the oscillating behaviour could not be suppressed, the effect of oscillation tends to become smaller.

4 *The effect of sample height:* Figure 10.8 shows the effect of sample height on the deformation responses of a sample with a viscosity coefficient of 0 GPa·s. The amplitudes of oscillating part and stationary part of strain remain the same

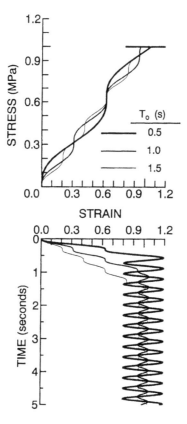

Figure 10.7 The effect of loading rate on dynamics response of a sample subjected to uniaxial compression.

while the period of oscillations becomes larger as the value of sample height increases.

10.2 DYNAMICS OF EXCAVATIONS

In this section, several examples for the dynamic response of excavations are given. Although the assumed excavation geometries are simple, the examples should be sufficient to illustrate the fundamental aspects of dynamics of rock excavations.

10.2.1 Loading of inclined semi-infinite slabs

A semi-infinite slab on an incline is considered and it is assumed to be subjected to instantaneous gravitational loading. The original formulation was developed by Aydan (1994) and it is adopted herein for assessing the dynamic response of a semi-infinite layer on an incline, which is a very close situation to the slope stability assessment of Terzaghi (1960).

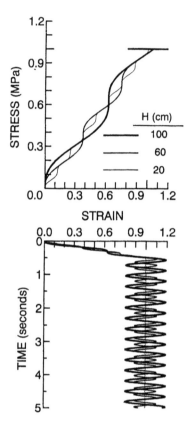

Figure 10.8 The effect of sample height on dynamics response of a sample subjected to uniaxial compression.

Let us assume that the deformation is purely due to shearing under gravitational loading and the slab behaves in a visco-elastic manner of Kelvin-Voigt type given by (Figure 10.10):

$$\tau = G\gamma + \eta\dot{\gamma} \tag{10.11}$$

where G is elastic shear modulus and η is viscous shear modulus. This model is known as the Voigt-Kelvin model (Eringen, 1980). When $G = 0$, then it simply corresponds to a Newtonean fluid. On the other hand, when $\eta = 0$, it corresponds to a Hookean solid.

Let us consider an infinitesimal element within an inclined infinitely-long layer as illustrated in Figure 10.11. The governing equation takes the following form by considering the equilibrium of the element by applying Newton's 2nd Law (Eringen, 1980):

$$\frac{\partial\tau}{\partial y} - \frac{\partial p}{\partial x} + \rho g \sin\alpha = \rho\ddot{u} \tag{10.12}$$

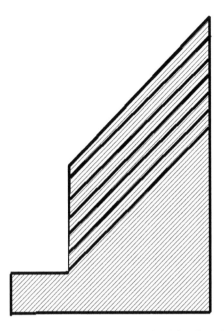

Figure 10.9 An illustration of semi-infinite slope (modified from Terzaghi (1960)).

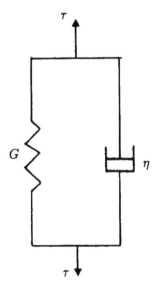

Figure 10.10 Constitutive model.

If the thickness of the layer does not vary with x and the medium consists of a same material, then $\partial p/\partial x = 0$ and the above equation becomes:

$$\frac{\partial \tau}{\partial y} + \rho g \sin \alpha = \rho \ddot{u} \qquad (10.13)$$

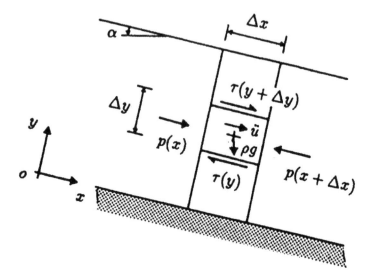

Figure 10.11 Mechanical model for shearing of semi-infinite slope.

(c) Closed form solutions

Assuming that shear strain and shear strain rate can be defined as

$$\gamma = \frac{\partial u}{\partial y}, \quad \dot{\gamma} = \frac{\partial \dot{u}}{\partial y} \tag{10.14}$$

and introducing the constitutive law given by Eq. (10.11) into Eq. (10.13) yields the following partial differential equation

$$\rho \frac{\partial^2 u}{\partial t^2} + \eta \frac{\partial^2}{\partial y^2} \left(\frac{\partial u}{\partial t} \right) + G \frac{\partial^2 u}{\partial y^2} = \rho g \sin \alpha \tag{10.15}$$

Let us assume that the solution of this partial differential equation is given as (i.e. Kreyszig, 1983; Zachmanoglu and Thoe, 1986),

$$u(y, t) = Y(y) \cdot T(t) \tag{10.16}$$

As a particular case, $Y(y)$ is assumed to be of the following form by considering an earlier solution of the equilibrium equation without inertial term for semi-infinite slab with free-surface boundary conditions (Figure 10.12(a)):

$$Y(y) = y \left(H - \frac{y}{2} \right) \tag{10.17}$$

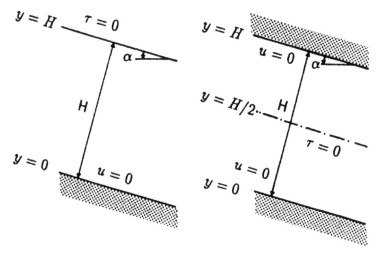

(a) Free surface condition   (b) Constrained surface condition

Figure 10.12 Boundary conditions.

Inserting this relation into Eq. (10.15), we have

$$\rho \frac{\partial^2 T}{\partial t^2} y \left(H - \frac{y}{2} \right) + \eta \frac{\partial T}{\partial t} + GT = \rho g \sin \alpha \tag{10.18}$$

Integrating the above equation with respect to y for bounds $y = 0$ and $y = H$ results in the following second order non-homogeneous ordinary differential equation

$$\frac{\partial^2 T}{\partial t^2} - \frac{3\eta}{\rho H^2} \frac{\partial T}{\partial t} + \frac{3G}{\rho H^2} T = \frac{3g \sin \alpha}{H^2} \tag{10.19}$$

The solutions of this differential equation are:

Case 1: Roots are real

$$T = C_1 e^{-\lambda_1 t} + C_2 e^{-\lambda_2 t} + \frac{1}{\lambda_1 \lambda_2} \frac{3g \sin \alpha}{H^2} \tag{10.20}$$

where

$$\lambda_1 = \frac{1}{2\rho H^2} \left(-3\eta + \sqrt{9\eta^2 - 12G\rho H^2} \right), \quad \lambda_2 = \frac{1}{2\rho H^2} \left(-3\eta - \sqrt{9\eta^2 - 12G\rho H^2} \right)$$

Case 2: Roots are same

$$T = [C_1 + C_2 t] e^{-\lambda t} + \frac{1}{\lambda^2} \frac{3g \sin \alpha}{H^2} \tag{10.21}$$

where

$$\lambda = -\frac{3\eta}{2\rho H^2}$$

Case 3: Roots are complex

$$T = e^{pt}[A \cos qt + B \sin qt] + \frac{1}{p^2 + q^2}\frac{3g \sin \alpha}{H^2} \qquad (10.22)$$

where

$$p = -\frac{3\eta}{2\rho H^2}, \quad q = \sqrt{12G\rho H^2 - 9\eta^2}$$

Integration constants C_1 and C_2 can be determined from the following initial conditions:

$$\begin{aligned} u(y, t) &= 0 \quad \text{at } t = 0 \\ \dot{u}(y, t) &= 0 \quad \text{at } t = 0 \end{aligned} \qquad (10.23)$$

For the above initial conditions, the integration constants for each case are

Case 1: Roots are real

$$C_1 = -\frac{1}{\lambda_1(\lambda_2 - \lambda_1)}\frac{3g \sin \alpha}{H^2}, \quad C_2 = \frac{1}{\lambda_2(\lambda_2 - \lambda_1)}\frac{3g \sin \alpha}{H^2} \qquad (10.24)$$

Case 2: Roots are same

$$C_1 = -\frac{1}{\lambda^2}\frac{3g \sin \alpha}{H^2}, \quad C_2 = \frac{1}{\lambda^2}\frac{3g \sin \alpha}{H^2} \qquad (10.25)$$

Case 3: Roots are complex

$$C_1 = -\frac{1}{p^2 + q^2}\frac{3g \sin \alpha}{H^2}, \quad C_2 = \frac{p}{q} \cdot \frac{1}{p^2 + q^2}\frac{3g \sin \alpha}{H^2} \qquad (10.26)$$

Integration constants for constrained boundary conditions can be obtained in a similar manner. This situation may be quite relevant to the response of soft layers sandwiched between two relatively rigid layers in underground excavations and trap-door experiment used particularly for underground openings in soil (Terzaghi, 1946).

(b) Finite element formulation

Taking a variation on δu, introducing the constitutive relation given by Eq. (10.11) into Eq. (10.13) and employing the usual finite element discretization technique yields the following

$$M\ddot{U} + C\dot{U} + KU = F \qquad (10.27)$$

where

$$\mathbf{M} = \rho \int_V \mathbf{N}^T \mathbf{N} dV, \quad \mathbf{C} = \eta \int_V \mathbf{B}^T \mathbf{B} dV, \quad \mathbf{K} = G \int_V \mathbf{B}^T \mathbf{B} dV, \quad \mathbf{F} = \rho g \sin \alpha \int_V \mathbf{N}^T dV$$

Choosing a linear type shape function for the space y as:

$$\mathbf{N} = [N_i, N_j] \tag{10.28}$$

where

$$N_i = \frac{y - y_i}{L}, \quad N_i = \frac{y_j - y}{L}, \quad L = y_j - y_i$$

and carrying out integrations, the above matrices are specifically obtained as:

$$\mathbf{M} = \frac{\rho A_e L}{6} \begin{bmatrix} 2 & 1 \\ 1 & 2 \end{bmatrix}, \quad \mathbf{C} = \frac{\eta A_e}{L} \begin{bmatrix} 1 & -1 \\ -1 & 1 \end{bmatrix}, \quad \mathbf{K} = \frac{G A_e}{L} \begin{bmatrix} 1 & -1 \\ -1 & 1 \end{bmatrix},$$

$$\mathbf{F} = \frac{\rho g \sin \alpha L A_e}{2} \begin{Bmatrix} 1 \\ 1 \end{Bmatrix} \tag{10.29}$$

where A_e is the area of an element. Using the central difference technique for discretizing time, we have the following:

$$\mathbf{K}^* \mathbf{U}_{n+1} = \mathbf{F}^*_{n+1} \tag{10.30}$$

where

$$\mathbf{F}^*_{n+1} = \mathbf{M}_K \mathbf{U}_n - \mathbf{M}_C \mathbf{U}_{n-1} + \mathbf{F}_n, \quad \mathbf{K}^* = \frac{1}{\Delta t^2} \mathbf{M} + \frac{1}{2\Delta t} \mathbf{C}, \quad \mathbf{M}_K = \frac{2}{\Delta t^2} \mathbf{M} - \mathbf{K},$$

$$\mathbf{M}_C = \frac{1}{\Delta t^2} \mathbf{M} - \frac{1}{2\Delta t} \mathbf{C}$$

Figure 10.13 compares the solutions obtained from closed-form solution and FEM for a 4 m thick semi-infinite slab.

Using the solutions presented in this section, one may compare the expected responses under different circumstances. Such a comparison has been already done by Aydan (1994–1998). The negligence of inertia component in Eq. (10.11) results in a parabolic partial differential equation. If the viscous effect is neglected in the resulting equation it would result in a differential equation of elliptical form. Figure 10.14 compares the responses obtained for three situations of the differential equation. As noted from the figure, all solutions converge to the solution obtained from the elliptical form (static case). The inertia component implies that displacement as well as resulting stresses and strains responses would be greater than those of the elliptical-form.

10.2.2   Excavation of circular underground openings

Excavation of tunnels is done through drilling-blasting or mechanically such as TBM and/or excavators. The most critical situation on stress state is due to the drilling-blasting type excavation since the excavation force is applied almost impulsively.

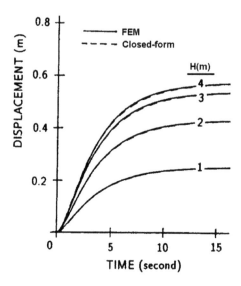

Figure 10.13 Dynamic response of 4 m thick semi-infinite slab under instantaneously applied gravitation load.

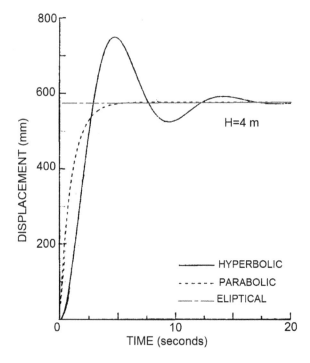

Figure 10.14 Comparison of responses obtained for hyperbolic, parabolic and elliptical forms of the differential equation for a 4 m thick slab under instantaneous gravitational loading.

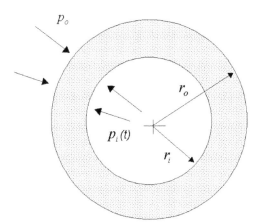

Figure 10.15 Illustration of dynamic excavation of a circular opening under hydrostatic in-situ stress condition.

The dynamic response of circular tunnels during excavations under hydrostatic in-situ stress condition can be given as (Figure 10.15)

$$\frac{\partial \sigma_r}{\partial r} + \frac{\sigma_\theta - \sigma_r}{r} = \rho \frac{\partial^2 u}{\partial t^2} \tag{10.31}$$

where $\sigma_r, \sigma_\theta, u, \rho$ and r are radial, tangential stresses, radial displacement, density and distance from the center of the circular cavity, respectively. Let us assume that the surrounding rock behaves in a visco-elastic manner of Kelvin-Voigt type, which is specifically given as

$$\begin{Bmatrix} \sigma_r \\ \sigma_\theta \end{Bmatrix} = \begin{bmatrix} D_1 & D_2 \\ D_2 & D_1 \end{bmatrix} \begin{Bmatrix} \varepsilon_r \\ \varepsilon_\theta \end{Bmatrix} + \begin{bmatrix} C_1 & C_2 \\ C_2 & C_1 \end{bmatrix} \begin{Bmatrix} \dot{\varepsilon}_r \\ \dot{\varepsilon}_\theta \end{Bmatrix} \tag{10.32}$$

where $\varepsilon_r, \varepsilon_\theta$ and $\dot{\varepsilon}_r, \dot{\varepsilon}_\theta$ are radial and tangential strain and strain rates, respectively. The strain and strain rates are related to the radial displacement in the following form:

$$\varepsilon_r = \frac{\partial u}{\partial r}; \quad \varepsilon_\theta = \frac{u}{r} \quad \text{and} \quad \dot{\varepsilon}_r = \frac{\partial \varepsilon_r}{\partial t}; \quad \dot{\varepsilon}_\theta = \frac{\partial \varepsilon_\theta}{\partial t} \tag{10.33}$$

Eringen (1961) developed closed-form solution for Eq. (10.31) under blasting loads. In order to deal with more complex boundary and initial conditions and material behaviour, Eq. (10.31) is preferred to be solved using a dynamic finite element code. The discretized finite element form of Eq. (10.31) together with the constitutive law given by Eq. (10.32) takes the following form:

$$\mathbf{M\ddot{U}} + \mathbf{C\dot{U}} + \mathbf{KU} = \mathbf{F} \tag{10.34}$$

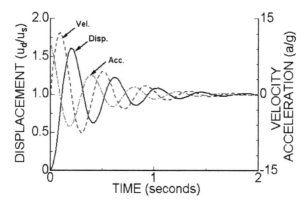

Figure 10.16 Responses of displacement, velocity and acceleration of the tunnel surface.

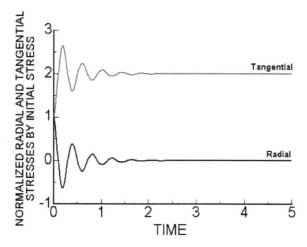

Figure 10.17 Responses of radial and tangential stress components nearby the tunnel surface (25 cm away from the perimeter).

Eq. (10.34) has to be discretized in time-domain and the resulting equation would take the following form

$$K^*U_{n+1} = F^*_{n+1} \qquad (10.35)$$

The specific form of matrices in Eq. (10.35) may change depending upon the method adopted in the discretization procedure in time-domain. For example, if the central difference technique is employed, the final forms would be the same as those given in sub-section 10.3.1. A finite element code has been developed by the author and used in the examples presented in this sub-section.

In this sub-section, the dynamic response of a circular tunnel under the impulsive application of excavation force is presented. The results were initially reported in the publication by Aydan (2011).

Figure 10.16 shows the responses of displacement, velocity and acceleration of the tunnel surface with a radius of 5 m. As noted from the figure, the sudden application of the excavation force, in other words, sudden release of ground pressure results in 1.6 times the static ground displacement at the tunnel perimeter and shaking disappears almost within 2 seconds. As time progress, it becomes asymptotic to the static value and velocity and acceleration disappear.

The resulting tangential and radial stress components nearby the tunnel perimetry (25 cm from the opening surface) are plotted in Figure 10.17 as a function of time. It is of great interest that the tangential stress is greater than that under static condition. Furthermore, very high radial stress of tensile character occurs nearby the tunnel perimeter. This implies that the tunnel may be subjected to transient stress state, which is quite different than that under static conditions. However, if the surrounding rock behaves elasticallly, they will become asymptotic to their static equivalents. In other words, the surrounding rock may become plastic even though the static condition may imply otherwise.

Chapter 11

Blasting

11.1 BACKGROUND

Blasting is the most commonly used excavation technique in mining and civil engineering applications. Blasting induces strong ground motions and fracturing of rock mass in rock excavations.

The excavation of rocks in mining and civil engineering applications by blasting technique is the most common technique since chemical blasting agents developed for centuries (e.g. Hendron, 1977; Hoek and Bray, 1981). However, the development of the modern blasting techniques is after the invention of dynamite. Blasting induces high ground motions and fracturing of rock mass adjacent to blastholes (e.g. Thoenen and Windes, 1942; Attewell et al., 1965; Siskind et al., 1980; Kutter and Fairhurst, 1971). Particularly, high ground motions may also induce some instability problems of rock mass and structures nearby (Northwood et al., 1963; Tripathy and Gupta, 2002). Furthermore, it may cause some environmental problems due to noise as well as vibrations of structures near the populated areas (e.g. Aydan et al., 2002).

Models for attenuation of ground motions induced by blasting are generally based on velocity type attenuation following the initial suggestions pioneered by United States Bureau of Mines (USBM) (Thoenen and Windes, 1942) and many models follow the footsteps of the USBM model (e.g. Attewell et al., 1965; Tripathy and Gupta, 2002 etc.). These models are often used in an empirical manner to assess the environmental effects on structures and human (Northwood et al., 1963; Hendron, 1977; Siskind et al., 1980). Although it is mathematically possible to relate ground motion parameters with each other, it is not always straightforward to do so (Aydan et al., 2002). Depending upon the sampling intervals of ground motion records as well as superficial effects particularly during integration process, the ground motion parameters and records may be different if they are measured by velocity-meters or accelerometers. However, it should be noted that the acceleration records are essential when they are used in stability assessments.

11.2 BLASTING AGENTS

11.2.1 Dynamite

Dynamite is an explosive material of nitroglycerin, using diatomaceous earth, or another absorbent substance such as powdered shells, clay, sawdust, or wood pulp.

Table 11.1 A summary of basic parameters of commonly used explosives.

Explosive	Detonation Velocity (m/s)	Density (g/cm³)	Detonation Pressure (GPa)
Dynamite	4500–6000 (7600)	1.3, 1.593 (1.51)	6–13.6 (22.0)
ANFO	2700–3600	0.882–1.10	0.7–9.0

The Swedish chemist and engineer Alfred Nobel invented dynamite in 1867. Dynamite is usually in the form of cylinders about 200 mm long and about 32 mm in diameter, with a weight of about 186 g. Dynamite is generally used in underground excavations as it produces less harmful gases during explosion.

11.2.2 Ammonium Nitrate/Fuel Oil (ANFO)

ANFO (Ammonium Nitrate/Fuel Oil) is a widely used bulk industrial explosive mixture. It consists of 94% porous prilled ammonium nitrate (NH_4NO_3) (AN) that acts as the oxidizing agent and absorbent for the fuel and 6% fuel oil (FO). ANFO is widely used in open-cast coal mining, quarrying, metal mining, and civil construction as it is low cost and easy to use matter among other conventional industrial explosives. The initiation of blasting is achieved using primer cartridges.

11.2.3 Blasting pressure for rock breakage

The detonation pressure is empirically related to the density (ρ_o) and detonation velocity (D) of explosives and the following formula is generally used for estimating the intrinsic pressure of explosives

$$p = \frac{1}{1+\gamma}\rho_o D^2 \tag{11.1}$$

where γ is the parameter related to the intrinsic properties of explosives. Its value is generally 3 (Persson *et al.*, 1994). Table 11.1 summarizes detonation pressure of several explosives used in rock breakage. However, the actual blasting pressure acting on the wall of holes is much less than the detonation pressure due to the gap between explosives, thickness of the casing, deformability characteristics and fractures in rock mass.

11.3 MEASUREMENT OF BLASTING VIBRATIONS IN OPEN-PIT MINES AND QUARRIES

The author has conducted measurements of blasting-induced motions in open-pit mines and quarries. The results of measurements conducted by the author are briefly explained in this sub-section.

Figure 11.1 A general view of Orhaneli open-pit mine, Gümüşpınar village behind the pit and the major fault plane on the left side.

11.3.1   Orhaneli open-pit lignite mine

Measurements were carried out in Orhaneli open-pit and Gümüşpınar village, separately (Aydan, 2002). Figure 11.1 shows a general view of open-pit mine and nearby Gümüşpınar village. The village is about 1 km from the open-pit mine. In the open-pit mine, a major normal fault and several minor strike-slip faults were observed. The minor strike-slip fault is observed near the ground surface and they are limited to near-surface layers. The rock mass above the lignite seam consists of tuff, sandstone, breccia and marl.

Figure 11.2 shows the layout of the instrumentation employed at the open-pit mine blasting test. The blasting hole is 9.65 m deep with a diameter of 25 cm. It is a single hole with two deck charges. Both upper and lower deck charges consist of 50 kg of ANFO and 1 kg of cap-sensitive emulsion explosive (dynamite). Initiation was done by non-electric shock tube (NONEL) detonator with 25 ms delay. The upper deck was fired first. Figure 11.3 shows a view of the blasting operation experiment. The distance to the blasting hole was 66 m from Acc-3.

Figure 11.3 shows the records of accelerometer denoted as Acc-1 during blasting. As seen from the figure, the magnitude of the vertical component of the acceleration waves is the largest while that of the traverse components is the least among other components. Furthermore, the peak of the vertical component appears a few milliseconds before the others. The reason for differences between vertical and horizontal components may be related to the damage and weakening caused by the previous blasts to the top 1–1.5 m of the bench on which the instruments were located. Therefore the weakened top part of the bench can be regarded as low velocity layer as compared with the rest of the bench below. This low velocity (damaged) layer may cause the attenuation of horizontal components while its effect on the vertical component is less since the

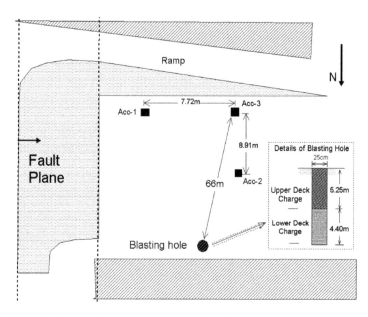

Figure 11.2 The layout of instrumentation employed in the open-pit mine blasting test.

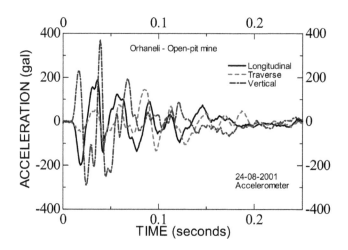

Figure 11.3 Acceleration records from Acc-1 accelerometer.

vertical component wave travels mostly through undisturbed marl beds. Furthermore, the interface between coal and marl, which is just 8–10 m below the ground surface, may act as a good reflecting surface so that the vertical component is enhanced in amplitude.

Figure 11.4 shows the vertical acceleration records of accelerometers denoted as Acc-1, Acc-2 and Acc-3 during blasting. Although the magnitudes of initial peaks follow the order of distance to the blasting hole, the magnitude of the farthest accelerometer denoted as Acc-3 is larger than the others. This may be caused by some

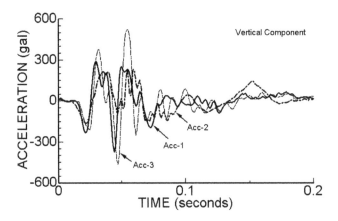

Figure 11.4 The vertical components from accelerometers denoted as Acc-1, Acc-2 & Acc-3.

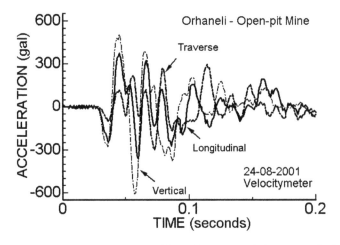

Figure 11.5 Acceleration components obtained from the numerical derivation of velocity records.

slight variation of fixation and ground conditions beneath the accelerometers. The peak value exceeds 500 gal and the waves attenuate as time goes by within 0.2 seconds following the blasting. The ground nearby accelerometers was blasted previously. The accelerometer locations could be disturbed at varying degrees depending upon the distance to the previous blastholes. Therefore, the attenuation of records of accelerometer Acc-2 is greater than the others and long period waves become dominant.

Figure 11.5 shows the acceleration obtained from the numerical derivation of velocity records using sampling interval of the velocity-meter (denoted Vel-3) next to the accelerometer Acc-3 (see Figure 11.2 for location). Although records are very similar to each other, the accelerations computed from velocity-meters are larger in amplitude as compared with those from accelerometers. For example, for the vertical component, the maximum amplitude of an acceleration wave was 525 gal from the accelerometer as compared with 609 gal computed from the velocity record of the velocity-meter. The amplitude obtained from the velocity-meter records is about 1.16 times that of the true acceleration records. The difference may arise from one or all of the following reasons: slight variation of fixation of instruments, the errors inherent in numerical derivation arising from digital wave forms or the frequency dependence of directly measured acceleration value. The validity of the first and second reasons must be checked by further measurements and analysis. If, especially, the second reason holds true, the empirical damage criteria based on the peak particle velocity of the ground, could not be employed straightforwardly in the case of integration of directly measured acceleration values, or vice versa. Hence a different damage criterion should be developed for the case of direct monitoring of acceleration.

The amplitude of the vertical component of acceleration waves caused by blasting is larger than that of other components. The amplitude of the acceleration waves is in the order of vertical, longitudinal (radial), traverse (tangential). However, the response spectra imply that amplifications are in the reverse order. Fourier spectra of longitudinal, traverse and vertical components of acceleration records of accelerometer and velocitymeter indicated that dominant frequencies of the waves observed at 8–10 Hz and 30–40 Hz, into account the fundamental vibration mode. The results indicate that structures having a natural period less than 0.05s could be very much influenced. The effect of blasting should be smaller for structures having natural periods greater than 0.1 s.

11.3.2 Demirbilek open-pit lignite mine

Aydan *et al.* (2013) performed ground motion measurements in the open-pit mine near Demirbilek village during blasting, in which the attenuation of ground motions and the effect of existing fault were investigated, using both velocity-meters and accelerometers simultaneously. YOKOGAWA WE7000 modular high-speed PC-based data-acquisition system was used. It can handle 16 channels simultaneously. The sampling interval was set to 10 ms during measurements. AR-10TF accelerometers with three components were used and this device can measure accelerations up to 10 g. X and Y components of the accelerometer were aligned to radial and tangential directions with respect to the blasting location. Z-direction measured UD component of ground motions. Figure 11.6 shows one of the layouts for ground motion observations and a view of blasting.

A typical blasthole in lignite mines of Turkey is generally 8–10 m deep with a diameter of 25 cm. It consists of 50–75 kg of ANFO and 0.5–1 kg of cap-sensitive emulsion explosive (dynamite). One third of each blast-hole stemmed with soil and initiation was done by non-electric shock tube detonator with a 25–50 ms delay.

The characteristics of ground motions induced by blasting depend upon the amount of explosives, the layout of blastholes and benches, delays as well as geomechanical properties of rock mass. Figure 11.7 shows an example of acceleration responses during the blasting experiment numbered 47 with a 15.5 kgf ANFO explosive. The distance of blasted hole was approximately 40 m away from the monitoring location on the same bench level. Typical rocks observed in the lignite mine are lignite itself, marl, sandstone, mudstone and siltstone.

As mentioned in the introduction, attenuation relations used in the evaluation of the effects of blasting are mostly of velocity type. There are very few attenuation

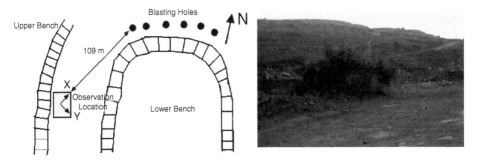

Figure 11.6 A layout of ground motion observations and a view of blasting.

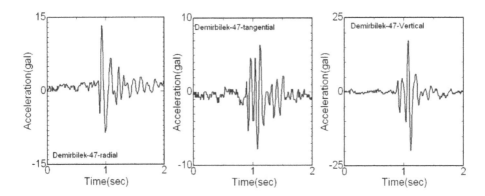

Figure 11.7 Acceleration responses measured during the in-situ blasting experiment numbered 47.

relations for blasting-induced accelerations. Dowding (1985) proposed an empirical attenuation relation for maximum acceleration. Wu *et al.* (2003) also developed an empirical relation using results of small-scale field blast tests involving soil and granite. However, the empirical relations particularly overestimate maximum accelerations within a distance of 100 m to the blast location. Nevertheless, the empirical relation proposed by Dowding performs better than that by Wu *et al.* (2003). Therefore, Aydan *et al.* (2013a) developed their attenuation relations for ground conditions typical in lignite mines of Turkey.

The attenuation of ground acceleration may be given in the following form as a convolution of three functions F, G and H in analogy to the attenuation relation of ground motions induced by earthquakes.

$$a_{max} = F(V_p)G(R_e)H(W) \tag{11.2}$$

where V_p, R_e and W are elastic wave velocity, distance from the explosion location and weight of explosives. The units of V_p, R_e and W are m/s, meter (m) and kilogram force (kgf) while the unit of acceleration is gal. In analogy to the spherical attenuation

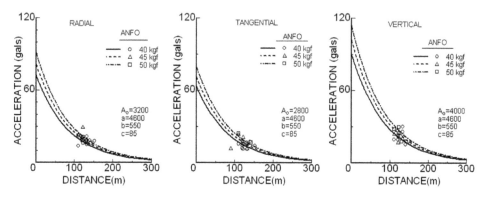

Figure 11.8 Attenuation of maximum ground acceleration with distance for a single hole blasting experiment.

relation proposed by Aydan (1997, 2001, 2007, 2012; Aydan and Ohta, 2011), the functions F, G and H: may be assumed to be one of the following forms:

$$F(V_p) = A_o(e^{V_p/a} - 1)$$ (11.3a)

$$H(W) = (e^{W/b} - 1)$$ (11.3b)

$$G(R_e) = Ae^{-R_e/c}$$ (11.3c)

The coefficients of a, b and c are found to be approximately 4600, 550 and 85, respectively, for the observation data obtained from single-hole blasting experiments at Demirbilek open-pit mine as shown in Figure 11.8 together with values of coefficients of Eq. (11.3) for each component. The value of coefficient A_o is found to be 3200, 2800 and 4000 for radial, tangential and vertical components, respectively. However, the value of coefficient c may be different for each component of ground acceleration in relation to explosive type. The value of coefficient c is applicable to ANFO explosives, which are commonly used in Turkish lignite mines. The empirical relation by Dowding (1985) follows a similar approach in order to take into account the effect of ground conditions in attenuation relations.

11.3.3   ELI Işıkdere open-pit mine

(a) 2010 measurements

2010 measurements were carried out using a single accelerometer of G-MEN type. The blasting holes were 7 m deep and two rows of the holes with a separation distance of 7 m were drilled parallel to the bench face. For each hole, the amount of ANFO was 50–75 kg and stemming was 1/3 of the total depth of the hole. Each sag has 25 kg ANFO. The delay between the front and back row of the holes was 25 ms. Rock mass blasted consists of mainly marl. Figure 11.9 shows views of blasts while Figure 11.10

Figure 11.9 Views of blasting on August 23, 2010 at ELI Işıkdere open pit mine.

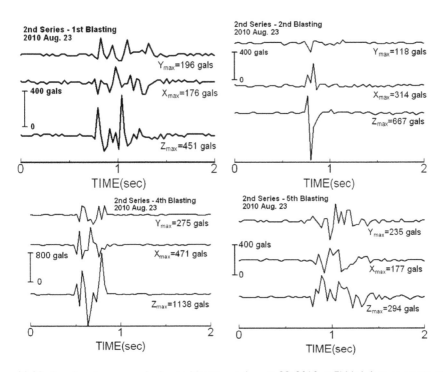

Figure 11.10 Accelerations records due to blasting on August 23, 2010 at ELI Işıkdere open pit mine.

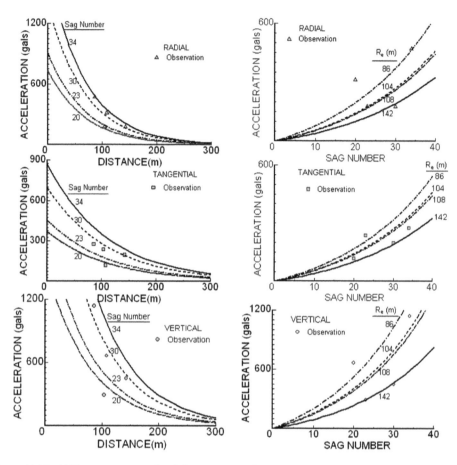

Figure 11.11 (a) The attenuation of radial, tangential and vertical accelerations and (b) the increase of the maximum acceleration with respect to number of sags.

the acceleration records for different blasts. As noted from the figure, the acceleration records are not symmetric with respect to time axis.

Figure 11.11 shows the attenuation of radial, tangential and vertical accelerations and the increase of the maximum acceleration with respect to sag number in accordance with functional forms of Eq. (11.3b,c). As expected, the maximum acceleration decreases as a function of distance while the amplitude of the maximum ground acceleration increases as the amount of ANFO increases.

(b) 2011 measurements

The 2011 measurements were carried out using 7 QV3-OAM stand-alone accelerometers with trigger mode and 2 G-MEN type accelerometers. The first series of investigations were aimed to see the attenuation of accelerometers. The nearest station to the blasting point was 10 m. Figure 11.13 shows some of the acceleration records.

(a) First series　　　　　　　　　　　　(b) Second series

Figure 11.12 Views of blasting experiments before and during blasts.

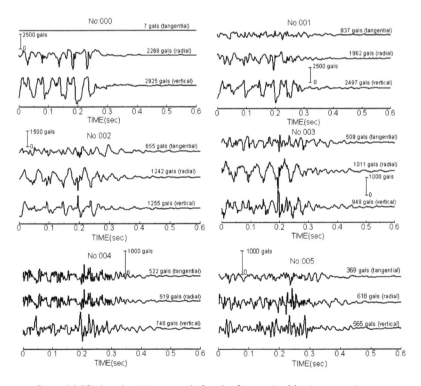

Figure 11.13 Acceleration records for the first series blasting experiments.

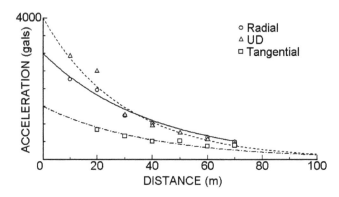

Figure 11.14 Attenuation of maximum ground acceleration with respect to distance.

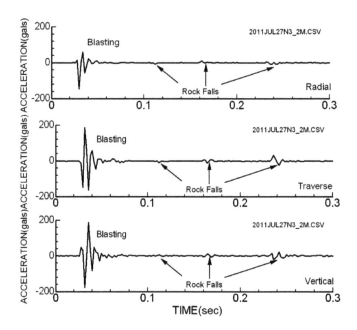

Figure 11.15 Rock-fall induced ground motions triggered by blasting.

Figure 11.14 shows the attenuation of maximum acceleration as a function of distance. As indicated previously, the vertical component is the largest while the tangential component is the smallest. However, the attenuations with distance are different from each other and the tangential component attenuates gradually compared with the rapid attenuation other components. Figure 11.15 shows an acceleration record in which ground motions induced by rock falls seen in Figure 11.12(a) also triggered by blasting.

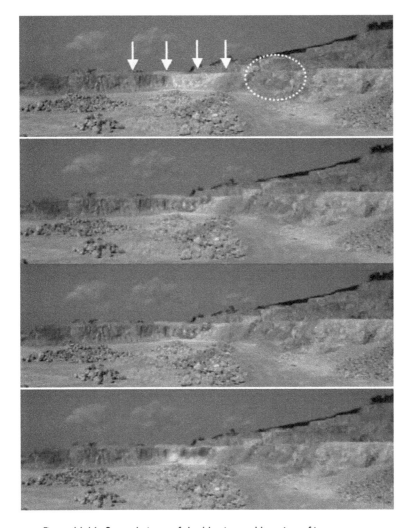

Figure 11.16 Several views of the blasting and location of instruments.

11.3.4 Motobu quarry

A trial measurement was done at Motobu limestone quarry on Okinawa Island, Japan (Figure 11.16). The monitoring of blasting-induced vibrations is done using 4 stand-alone accelerometers with trigger mode, whose locations with respect to blasting are shown in Figure 11.1 also. The first row consists of 6 holes of 13.5 m deep spaced at a distance of 3.5 m while the second row had 7 holes spaced with distance of 3.5 m. The hole was filled with 5 m high ANFO and 8.5 m stemming material. However, one of the accelerometers did not function during the blasting.

Figure 11.17 shows the acceleration records. The axial component was larger than the UD component and they attenuate with distance from the blasting location.

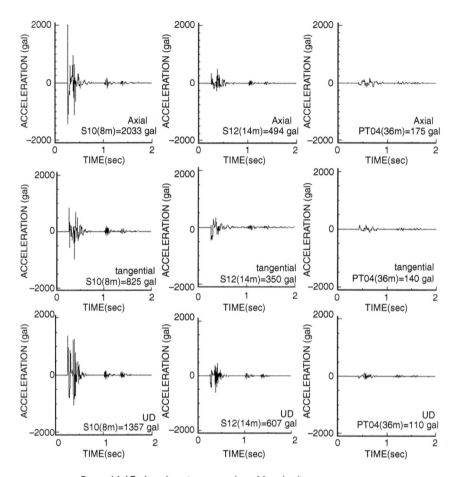

Figure 11.17 Acceleration records at Motobu limestone quarry.

In the same records, some ground motions induced by the falling rock blocks are also noticed.

11.4   MEASUREMENTS AT UNDERGROUND OPENINGS

The author has conducted measurements of blasting-induced motions in several tunnels in Japan and Turkey. The results of measurements conducted by the author are briefly explained in this sub-section.

11.4.1   Kuriko tunnel

The Kuriko Tunnel in Fukushima prefecture is excavated through Mt. Kuriko. The geology of the tunnel consists of granite, rhyolite, tuff, andesitic dykes and intercalated

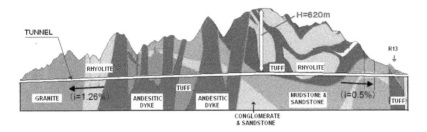

Figure 11.18 Geological cross section of Kuriko evacuation tunnel beneath Mt. Kuriko (modified from Haruyama and Narita, 2009).

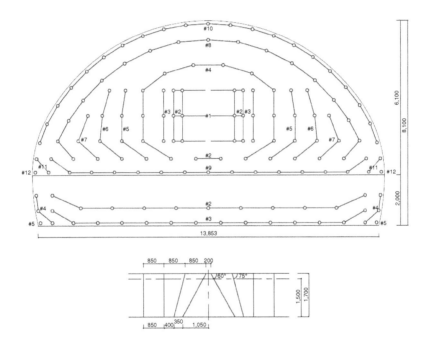

Figure 11.19 A typical layout and plan view of blast-holes used at Kuriko Tunnel.

sedimentary rocks such as sandstone, mudstone and conglomerate. While granite is exposed in Fukushima side (East), folded sedimentary rocks with folding axis aligned in north-south outcrops on Yonezawa side (West). Sedimentary rocks are covered with tuff and rhyolite and intruded with andesitic dykes (Figure 11.18).

The length of blast-hole rounds was 1.5 m and a total of 112.6 kg dynamite was used at the location where measurement is done. Figure 11.19 shows the layout of blast-holes while Figure 11.20 shows the layout of instrumentation and their views with respect to tunnel advance direction. Figure 11.21 shows the acceleration records. The axial and tangential components were larger than the radial component and they attenuate with distance from the blasting location. Furthermore, several shocks are recorded in relation to blasting sequence.

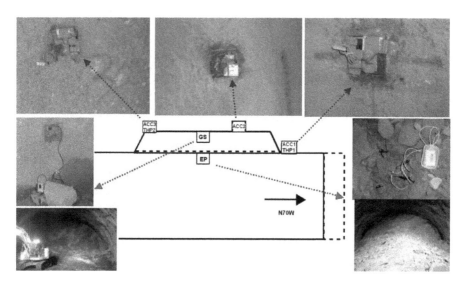

Figure 11.20 Layout of instrumentation at Kuriko Tunnel.

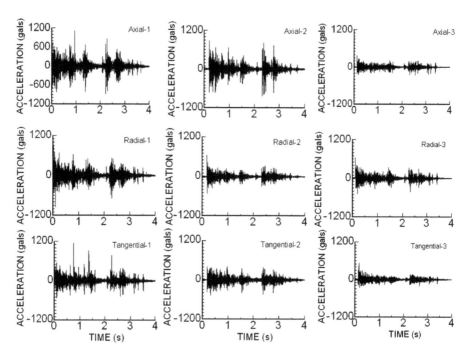

Figure 11.21 Acceleration records for each component.

Figure 11.22 A view of the accelerometer and its fixation in the main tunnel.

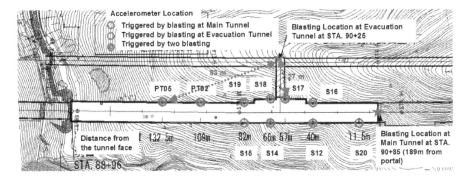

Figure 11.23 Locations of blasting and accelerometers (note that accelerometer S15 had a battery problem during blasting at the main tunnel).

11.4.2 Taru-Toge tunnel

Taru-Toge tunnel is being constructed as a part of expressway project connecting Shin-Tomei Expressway and Chuo Expressway at the boundary of Shizuoka and Yamanashi Prefectures in the Central Japan. The tunnel passes through a series of mudstone, sandstone, conglomerate layers with folding axes aligned north-south.

(a) Instrumentation and installation

The total number of accelerometers was 13 and the accelerometers were fixed to the plates of the rockbolts or steel ribs (Figures 11.22 and 11.23). In addition, three more accelerometers were attached to the plates of rockbolts at the passage tunnel between the main tunnel and the evacuation tunnel for wave velocity measurements during the blasting operation at the main tunnel.

The accelerometers can be synchronized and they can be set to the triggering mode with a capability of recording pre-trigger waves for a period of 1.2 s. The trigger threshold and the period of each record can be set to any level and chosen time as desired. The accelerometer is named QV3-OAM-SYC and it has a storage capacity of 2 GB and it is a stand-alone type. It can operate for two days using its own battery and

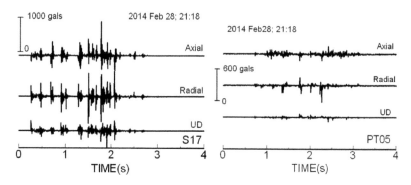

Figure 11.24 Acceleration records at measurements stations S17 (27 m) and PT05 (83 m).

the power source can be solarlight if appropriate equipment is used. In other words, it is an eco-friendly acceleration monitoring system. Figure 11.23 shows the location of accelerometers triggered during each blasting operation. The blasting (Blasting-1) at the evacuation tunnel was done at 21:28 on February 28, 2014. The second blasting (Blasting-2) was carried out at 10:54 on March 1, 2014.

(b) Characteristics of ground vibration during blasting

Blasting-1

The blasting (Blasting-1) at the evacuation tunnel was done at 21:28 on Feb. 28, 2014 and the amount of explosive was 71.2 kgf with 10 rounds with delay of 0.4–0.5 s. The threshold value for triggering was set to 10 gals and the total number of the triggered accelerometers was 9. The most distant accelerometer was PT05 and its distance from the blasting location was 83m. The highest acceleration was recorded at the accelerometer denoted as S17 and its value was about 1000 gals. Figure 11.24 shows the acceleration records at the stations denoted S17 and PT05. As noted from the figure, the amplitude of accelerations decreases with distance as expected. Another interesting observation is that the acceleration records are not symmetric with respect to the time-axis. Furthermore, acceleration wave amplitudes differ depending upon the direction.

　　Figure 11.25 also shows the Fourier spectra of each component of acceleration waves. As noted from the figures the Fourier spectra of the radial and axial components of the accelerometer S17 consist higher amplitude contents and higher frequency content. As for the distant acelerometer PT05, the vice-versa is valid as expected.

　　Figure 11.26 shows the acceleration response spectra of acceleration records taken S17 and PT05 for each respective direction. As noted from the figure, the acceleration response spectra have very short natural periods as expected.

　　Figure 11.27 shows the attenuation of maximum acceleration at all stations. As the wave forms are asymmetric with respect to time axis, peak values are plotted as positive peak (PP) and negative peak (NP) with the consideration of their position in relation to the blasting location. As noted from the figure, the attenuation is quicker on the unblasted side compared with those on the blasted side. The data is somewhat

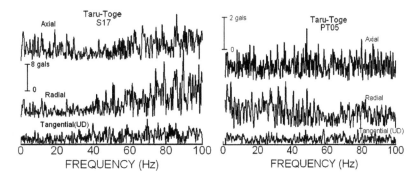

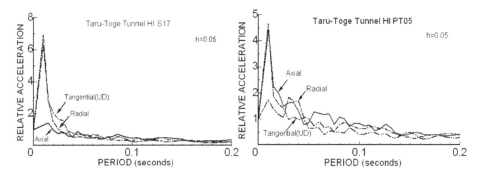

Figure 11.25 Fourier spectra of acceleration records shown in Figure 11.24.

Figure 11.26 Acceleration responses spectra of acceleration records shown in Figure 11.24.

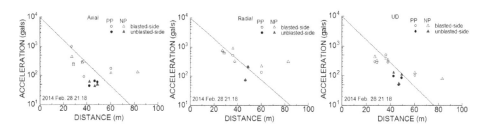

Figure 11.27 Attenuation of maximum ground acceleration with distance for Blasting-1.

scattered and this may be related to the existence of structural weakness zones in the rock mass.

Blasting-2

The second blasting (Blasting-2) at the main tunnel was carried out at 10:54 on March 1, 2014 and the total amount of blasting was 30.4 kgf with a 0.4–0.5 second delays per round. Figure 11.28 shows the tunnel face before and after blasting of the lower

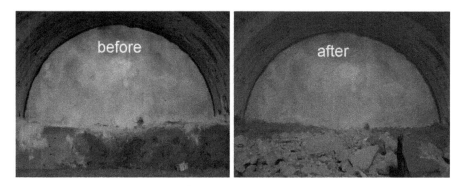

Figure 11.28 Views of tunnel face before and after blasting.

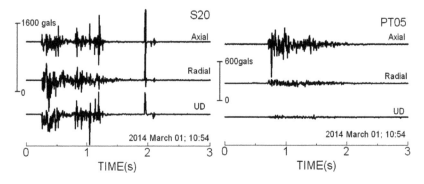

Figure 11.29 Acceleration records at measurements stations S20 (11.5 m) and PT05 (132.5 m).

bench. The threshold value for triggering was set to 10 gals and the total number of the triggered accelerometers was 14. The most distant accelerometer was PT05 and its distance from the blasting location was 132.5 m. The highest acceleration was recorded at the accelerometer denotes as S20 and its value was about 1030 gals.

Figure 11.29 shows the acceleration records at the stations denoted S20 and PT05. As also noted from the figure, the amplitude of accelerations decreases with distance as expected. Another interesting observation is that the acceleration records are not symmetric with respect to the time-axis. Furthermore, acceleration wave amplitude differ depending upon the direction of measurements. The comments for Fourier spectra basically would be the same except those for the tangential component. Lower frequency content waves become dominant for the tangential component as seen in Figure 11.30. Figure 11.31 shows the acceleration response spectra of acceleration records taken by S20 and PT05 for each respective direction. As noted from the figure, the acceleration response spectra has very short natural periods as expected.

Figure 11.32 shows the attenuation of maximum acceleration at all stations triggered. As the wave forms are non-symmetrical with respect to time axis, peak values are plotted as positive peak (PP) and negative peak (NP) with the consideration of their position in relation to the blasting location. As the passage tunnel exists on the west side between the main tunnel and the evacuation tunnel, the maximum accelerations are

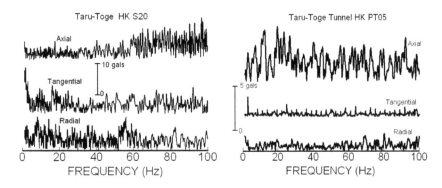

Figure 11.30 Fourier spectra of acceleration records shown in Figure 11.29.

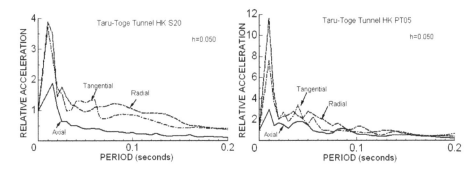

Figure 11.31 Acceleration response spectra of acceleration records shown in Figure 11.29.

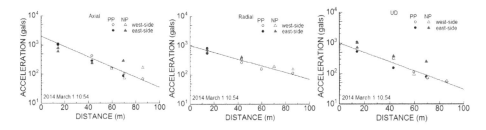

Figure 11.32 Attenuation of maximum ground acceleration with distance for Blasting-2.

somewhat smaller at the station S18. Despite some scattering of measured results, the attenuation of maximum acceleration decreases with the increase of distance exponentially. The data scattering may also be related to the existence of structural weakness zones in the rock mass.

11.4.3 Zonguldak tunnels

Several tunnels in association of rehabilitating the intercity roadways in Turkey have been excavated in Zonguldak and its close vicinity using the drilling and blasting

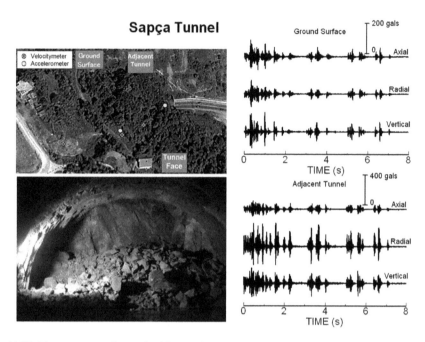

Figure 11.33 The position of tunnels, blasting location and measurement locations and measured responses.

technique. Geniş *et al.* (2013) have been monitoring the blasting-induced vibrations in several adjacent tunnels near Sapça, Üzülmez and Mithatpaşa tunnels. In this subsection, the outcomes of monitoring of vibrations in tunnels and at ground surface are briefly presented for assessing the effects of blasting on the adjacent structures.

Sapça tunnels are double-tube two-lane tunnels. The main purpose of the measurements was to see the effect of blasting at the new tunnel on the adjacent tunnel and ground surface. Figure 11.33 shows the position of tunnels, blasting location and measurement locations. Tunnels were excavated through intercalated sandstone, siltstone and claystone. Figure 11.33 also shows the measurements at ground surface (70 m) and at the tunnel face (32 m) of the adjacent tunnel. The amount of blasting was 6–12 kg for each round. Although the ground motions are less on the ground surface than those at the adjacent tunnel due to the distance from the location of blasting, the ground motions are relatively high regarding the UD component.

Üzülmez are also two-lane double tube tunnels. The pillar distance between two tunnels is about 11 m, three accelerometers were installed in the pillar side of the adjacent tunnel and two accelerometers at the mountainside. One more accelerometer was installed at the tunnel where blasting was carried out. Figure 11.34 shows the measurement results. Despite the distance, the measurements were highest in the tunnel of blasting and the accelerations were high at the pillar side.

Mithatpaşa tunnels fundamentally have similar geometrical features while rock mass is limestone and some karstic caves were encountered during excavation. Figure 11.35 shows the position of tunnels, blasting location and measurement

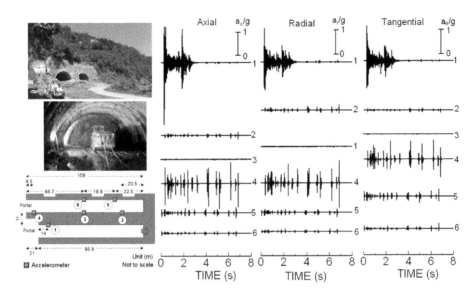

Figure 11.34 The position of tunnels, blasting location and measurement locations and measured responses.

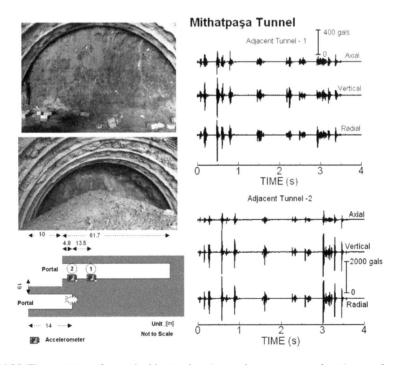

Figure 11.35 The position of tunnels, blasting location and measurement locations and measured responses.

Table 11.2 Values of constants for maximum ground velocity and ground acceleration for different situations.

Coefficient	Inside Tunnels		Adjacent Tunnels		Ground Surface	
	Acceleration (gal)	Velocity (kine)	Acceleration (gal)	Velocity (kine)	Acceleration (gal)	Velocity (kine)
A_o	4000	140	2000	80	6000	200
a	4600	4600	4600	4600	4600	4600
b	240	240	240	240	240	240
c	100	120	120	100	100	120

locations and measured responses. The blasting was very close to the portal of the tunnel. The total amount of the explosives was 192 kg while the amount of explosive for rounds ranged from 10 kg to 25 kg. The acceleration was high for UD and traverse components.

Geniş *et al.* (2013) extended Eq. 11.2 to the estimation of maximum ground velocity in addition to the maximum ground acceleration and determined the constants of Eq. 11.3 for tunnels from the measurements presented in this subsection. They are given in Table 11.2. The coefficients are slightly different from those for measurements at lignite mines. The reason may be the difference of explosives and confinement in underground excavations to those at ground surface.

11.5 MULTI-PARAMETER MONITORING DURING BLASTING

The real time monitoring of the stability of tunnels is of great importance when tunnels are prone to failure during excavation such as rock bursting or squeezing. It is also known that when rock starts to fail, the stored mechanical energy in rock tends to transform itself into different forms of energy. Experimental studies by the author showed that rock indicates distinct variations of multi-parameters during deformation and fracturing processes. These may be used for the real-time assessment of the stability of rock structures. The parameters measured involve electric potential variations, acoustic emissions, rock temperature, temperature and humidity of the tunnel in addition to the measurements of convergence and loads on support members during the face advance.

An application of this approach was done to the third Shizuoka tunnel of the second Tomei Expressway in Japan (Aydan *et al.*, 2005). The parameters measured involve electric potential variations, acoustic emissions, rock temperature, temperature and humidity of the tunnel in addition to the measurements of convergence and loads on support members during the face advance. The tunnel excavation of the tunnel was done through drilling-blasting technique. Each blasting operation caused both dynamic and static stress variations around the tunnel and its close vicinity. Measurements were carried out in two phases. In the first phase, the effect of the face advance of Nagoya-bound tunnel on Tokyo-bound tunnel was investigated. Figure 11.36 shows the layout of instrumentation.

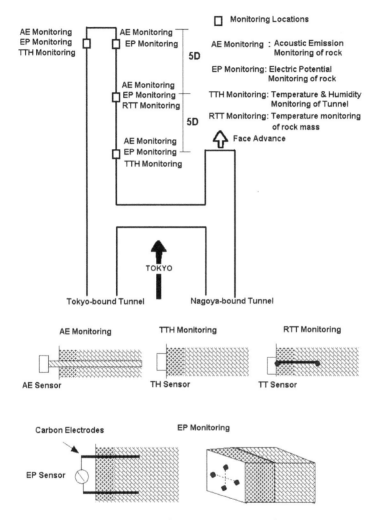

Figure 11.36 The layout of instrumentation and face advance.

Figure 11.37 shows the measured AE and electric potential responses as a function of time. Vertical bars in the same figure indicate the blasting operations. After each blasting operation, distinct AE and EP variations were observed. These variations cease after a certain period of time. The electric potential increases simultaneously and tends to decrease as time goes by. AE response also showed the same type of response. When the tunnel is stable, it is expected that these variations should disappear on the basis of experimental observations and theoretical considerations by the author.

The second phase measurements were planned to see both the effect of face advance on AE and EP responses in the same tunnel as well as that of the adjacent tunnel. Furthermore, two new electric potential measurement devices with higher impedance were used in addition to low impedance electric potential measurement devices. Figure 11.38 shows the layout of instrumentation and face advance schedule.

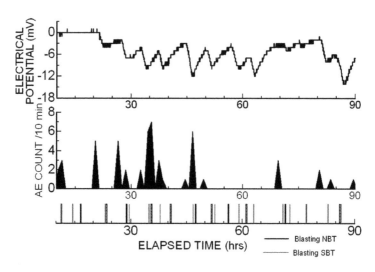

Figure 11.37 AE and EP responses measured resulting from face advance during the period from August 19 to August 26, 2004.

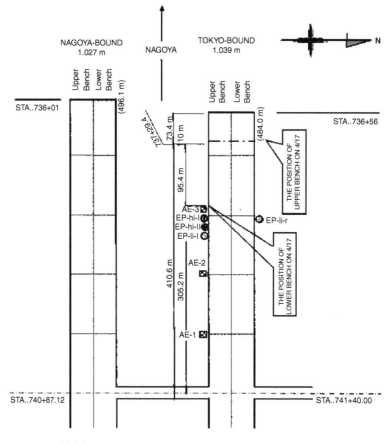

Figure 11.38 The layout of instrumentation and face advance schedule.

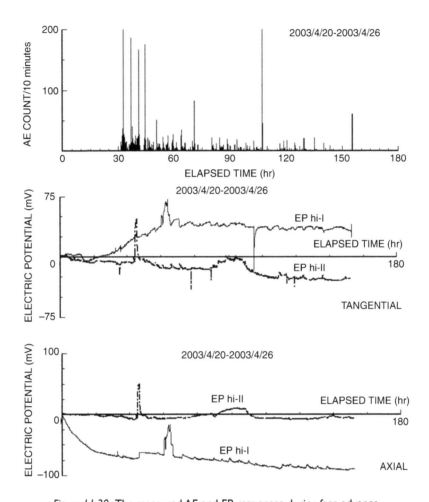

Figure 11.39 The measured AE and EP responses during face advance.

The responses of AE and EP are shown in Figure 11.39 and they were basically similar to those of the first phase. Nevertheless, the amplitude of variations was much larger than those of the previous phase. This was thought to be the closeness of the instruments to the tunnel face where blasting operations were carried out. The electric potential measurements with low impedance and high impedance electric potential devices were almost the same. However, the high impedance electric potential measurement devices are desirable. In addition to that, some problems were noted with the fixation of electrodes into the rock mass and the damage by flying rock fragments during the blasting operation. However, the electric potential variations may sometimes include the far field effects from the deformation of the earth's crust associated with earthquakes. During the measurement of electric potentials an additional device would be necessary outside the tunnel. In this particular case, such a device was installed at a place about 2 km west of the tunnel.

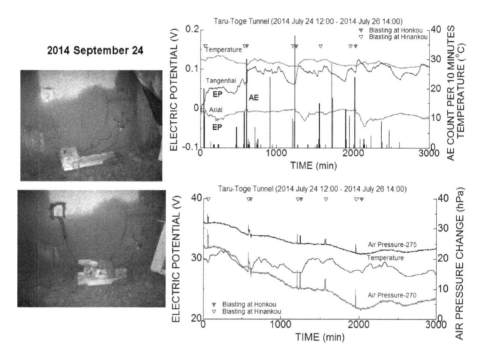

Figure 11.40 AE and EP responses measured resulting from face advance during the period from July 24 to July 26, 2014.

The same instrumentation was repeated at Taru-Toge tunnel. Figure 11.40 shows the variations of acoustic emissions (AE) and electric potential variations in relation to blasting at the main tunnel and evacuation tunnel. The overall responses are similar to those measured at Shizuoka 3rd tunnel.

A trial measurement was carried out on the ground water pressure variation during blasting operation at Demirbilek open-pit lignite mine. Figure 11.41 shows the monitoring results during the blasting operation numbered Demirbilek-07-blasting. The 1500 mm deep borehole with a diameter of 160 mm was drilled and filled with water. A water pressure sensor was installed in the borehole. The water pressure was measured using the same monitoring system. Although very slight time-lag exists between the peaks of ground acceleration and groundwater, the water pressure fluctuation occurs in response to the ground acceleration fluctuation. Furthermore, the ground water level decreases thereafter due to seepage into surrounding rock as well as the increased permeability due to the damage rock by blasting.

11.6 THE POSITIVE AND NEGATIVE EFFECTS OF BLASTING

In this section, the positive and negative effects of blasting and blasting operations are presented and discussed.

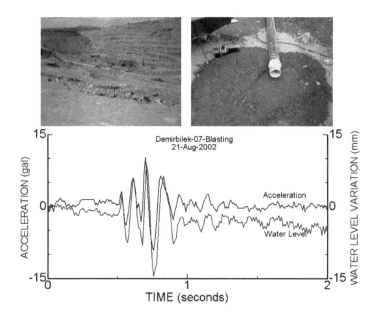

Figure 11.41 Water level fluctuation during blasting at Demirbilek open-pit mine.

11.6.1 In-situ stress inference

Aydan (2013) proposed a method to estimate the stress state from the damage zone around blasted holes. This method was applied to the damage zone around blasted holes and some stress inferences made for the tunnel face at Kuriko Tunnel and Tarutoge Tunnels. The estimations are compared with those from other methods. In this section, these examples are briefly presented.

(a) Kuriko tunnel

The damage around blastholes is shown in Figure 11.42. The fracture zone formations around the blasted holes are almost elliptical with longitudinal axis almost horizontal. These results imply that lateral stress is higher than the vertical stress on the tunnel face plane.

The lateral stress coefficient was taken as 2.2 and the inclination of the maximum principal stress was assumed to be 0 from horizontal by taking into account the actual shape of the damage zone around blastholes. The results are given in Table 11.3 for a blasthole pressure of 150 MPa and inferred plastic zones for different yield criteria and actual plastic zone are shown in Figure 11.43. The estimated damage zones are quite close to those shown in Figure 11.42. The in-situ stress estimations are also similar to the in-situ stress measurements by the AE method, in which the lateral stress coefficient was found to be 1.7.

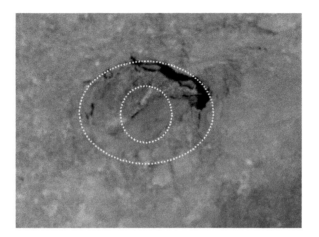

Figure 11.42 Views of damage and estimated yield zones around blastholes.

Table 11.3 Inferred in-situ stress parameters and yield function parameters used in computations

σ_{10} (MPa)	k	σ_c (MPa)	σ_t (MPa)	ϕ (°)	m	S_∞ (MPa)	b_1 (MPa)
9.0	2.2	90	6	60	14.93	360	0.045

(b) Taru-Toge tunnel

Imazu *et al.* (2014) utilized the fault striation method (Aydan, 2000) and blasted hole damage method (BHMD) (Aydan, 2013) at Taru-Toge tunnel. The estimated lateral stress coefficient from the fault striation method at the tunnel cross-section ranges between 1.6 and 1.7. The BHDM was applied to damage zone around blasted holes and some stress inferences were made for the tunnel face. Figure 11.44 shows the damage

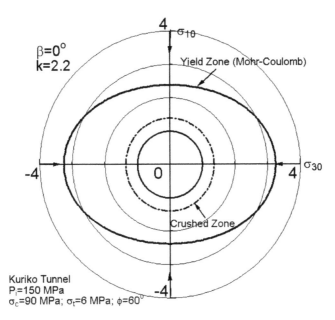

Figure 11.43 Estimated damage zones around blastholes for Kuriko Tunnel.

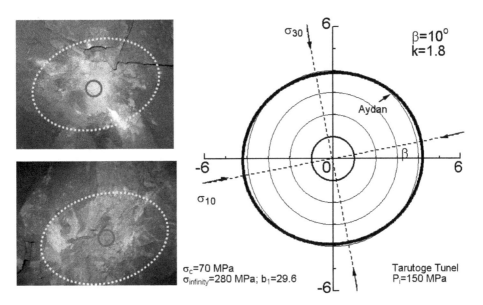

Figure 11.44 Estimated stress state from the damage zone around blasted holes.

zone around two blasted holes and inferred stress state. The estimations indicate that the lateral stress coefficient is about 1.8 and it is inclined with an angle of 10 degrees to the west. It is also interesting to note that this ratio is also very close to that estimated from the fault striation method.

11.6.2 Rock mass property estimation from wave velocity using blasting induced waves

The mechanical properties of rock mass may be estimated using the elastic wave velocity. Direct relations or normalized relations exist in literature (i.e. Ikeda, 1970; Aydan et al., 1993, Sezaki et al., 1990). The uniaxial compressive strength (UCS) of intact rock ranges between 40–70 MPa and its p-wave velocity is in the range of 3 and 4 km/s. The direct relations between the UCS of rock mass and elastic wave velocity proposed by Aydan et al. (1993, 2013) and Sezaki et al. (1990) are as follows respectively:

$$\sigma_{cm} = 5(V_{pm} - 1.4)^{1.43} \tag{11.4}$$

$$\sigma_{cm} = 1.67(V_{pm} - 0.33)^2 \tag{11.5}$$

$$\sigma_{cm} = 0.98 V_{pm}^{2.7} \tag{11.6}$$

where σ_{cm} and V_{pm} are uniaxial compressive strength (UCS) and p-wave velocity of rock mass.

Another relation, which is called "rock mass strength ratio" is as follows

$$\sigma_{cm} = \left(\frac{V_{pm}}{V_{pi}}\right)^2 \sigma_{ci} \tag{11.7}$$

where σ_{ci} and V_{pi} are uniaxial compressive strength (UCS) and p-wave velocity of intact rock.

Aydan et al. (2016) developed a portable system to measure p-wave and s-wave velocity of surrounding rock mass. The measurement system consists of 5 accelerometers connected to each other with wire and operated through "start-stop" switch. This system was first used in March 1, 2014 and September 1–3, 2015 at Tarutoge tunnel. The distance between accelerometers in present system can be up to 6 m (measurement length is about 30 m) and the sampling interval is 50 microseconds. Figure 11.45 shows the installation of the device in Tarutoge tunnel and measured velocity responses are shown in Figure 11.46.

The estimated wave velocity of rock mass in the instrumented zone is estimated to be 1.9–2.6, 2.0–2.7 and 1.8–2.4 km/s during March 1, 2014. The normalized UCS of rock mass is estimated to be about 0.2–0.46 times that of intact rock. If direct relations are used and normalized by that of intact rock (Aydan et al., 2013), the normalized UCS of rock mass would be 0.1–0.2 times that of intact rock.

The measurement during September 1–3, 2015 yielded that the p-wave velocity was about 2.7 km/s. In view of p-wave velocity of intact rock, the results are quite close to those of March 1, 2014 measurements. The normalized UCS of rock mass is estimated to be about 0.46 times that of intact rock.

11.6.3 Instability problems

(a) Evaluation of effects of blasting on bench stability and responses

The width and height of benches ranges between 15 to 45 m and 8 to 10 m, respectively at Demirbilek open-pit lignite mines (Aydan et al., 2013). The slope angle of the

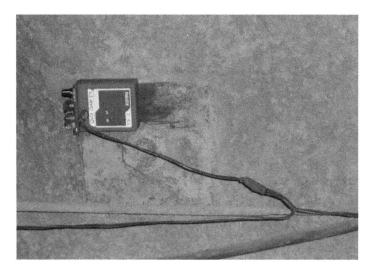

Figure 11.45 Fixation of an accelerometer for wave-velocity measurement.

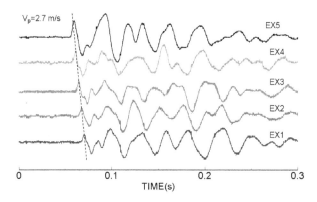

Figure 11.46 Acceleration records during the arrival of P-waves induced by blasting.

benches also varies between 50 and 60 degrees depending upon the mining operations and layouts. The failure of benches was observed in the upper marl unit and they were of planar type on the east slope of the open-pit mine. Figure 11.47 shows the failure of benches at different locations at the east side of the open-pit mine. As seen in the pictures, the planar sliding failure is the dominant failure mode. The engineers of the open-pit mine also reported that the stability issues become very important following the heavy rains. On the other hand, benches on the west slope of the open-pit mine were more stable although some cracking and their opening were observed. The instability problems on the benches of the west slope associated with normal fault. Unless the fault plane cut through the benches, there were no major slope stability problems on the west slope benches.

Figure 11.47 Views of stability problems on the benches of the east slope.

Aydan and his co-workers (Aydan *et al.*, 1996b; Aydan and Ulusay, 2002) proposed a method to estimate the movements of slopes involving sliding failure. This method was further elaborated in subsequent studies (Aydan *et al.*, 2006, 2008, 2009a, 2011; Tokashiki and Aydan, 2011; Aydan and Kumsar 2010). The authors utilize this technique for assessing the response and stability of the slope subjected dynamic forces as well as gravity and pore water pressures. The distance of the blasting location is one of the most important parameters for analyzing the response and stability of bench slopes.

First the effect of vibrations measured by single blasthole experiment with a distance of 100 m and a ANFO explosive of 50 kgf on the benches consisting of the upper marl unit was investigated. The computations indicated that no movement would occur under dry and fully saturated conditions. Then, the number of blastholes was assumed to be 18, which is commonly used during blasting operations in the open-pit mining of lignite mines in Turkey. The amplitude of the accelerations was increased by 4.7 times that of the single blasthole with the consideration of 18 blastholes with an ANFO explosive of 50 kgf. In the computations, both horizontal and vertical accelerations were considered. If rock mass is assumed to be dry, the computations indicated that no relative movement along bedding planes emanating from the toe of the benches would occur. If the ground water coefficient is more than 0.75, the relative movement along the bedding planes occurs. Figure 11.48 shows the computed relative displacement responses in relation to the horizontal base acceleration. We introduce a water force coefficient denoted as r_u to count the effect of ground water in the body subjected to slide. It is defined as the volume of water to the total volume of the body prone to sliding. As noted from the figure, if rock mass is fully saturated ($r_u = 1.0$) or nearly fully saturated ($r_u = 0.8$), some permanent displacement would occur after each blasting

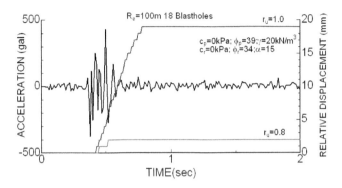

Figure 11.48 Relative displacement responses of the benches of the east slope for each blasting operation.

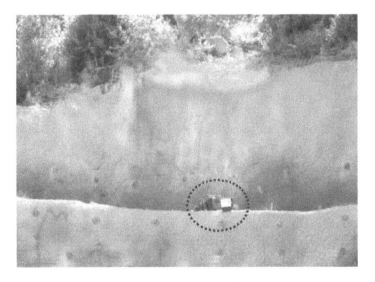

Figure 11.49 Rock fall at the portal of Mithatpaşa tunnel.

operation. The sliding body in benches would gradually be displaced and the separation of the sliding body from the rest of benches would occur at the upper levels and relative offsets at the toe of the slope. When the relative displacement becomes more than the half of the typical block size, blocks from the benches would fall to the lower level benches (Aydan *et al.*, 2009b). This process would repeat itself successively at a given time interval.

It is well known that the blasting may cause the individual blocks to topple or slide. Figure 11.49 shows an example of rockfall at the portal of Mithatpaşa tunnels described in Subsection 11.3.4. Particularly such events may cause casualties as well as property damage. The stability of such blocks can be evaluated using some dynamic limiting equilibrium approaches mentioned before. Similar events can also be found

Figure 11.50 Blasting induced failure at Gökgöl cave.

at karstic caves. Geniş and Çolak (2015) describes such an example at Gökgöl karstic cave in Zonguldak (Figure 11.50). The major collapse hall at the cave was caused by uncontrolled blasting operations in 1960 for enlarging a national highway.

11.6.4 Vibration effects on buildings

Using the acceleration records of accelerometer Acc-1 and accelerations computed from the records of velocity-meter Vel-3, a series of response analysis is carried out. Figure 11.51 shows the normalized acceleration response spectra of each component of acceleration records, respectively with damping coefficient (h) values of 0.000, 0.025, 0.050.

The results indicate that structures having a natural period less than 0.05s could be very much influenced. The effect of blasting should be smaller for structures having natural periods greater than 0.1 s. Figure 11.52 and 1.53 show the responses of a structure with a natural period of less than 0.05 second and damping coefficient of 0.05 (H = 0.059) for each acceleration components from the accelerometer measurements and velocity-meter measurements. As understood from Figures 11.52 and 11.53, the absolute acceleration acting on structures should be less than the accelerations of input motion. It would be safe to assume that the induced accelerations caused by blasting should act on the ground and structure system without any reduction.

11.6.5 Air pressure due to blasting

A typical pressure-time profile for a blast wave in free air is shown in Figure 11.54. It is characterized by an abrupt pressure increase at the shock front, followed by

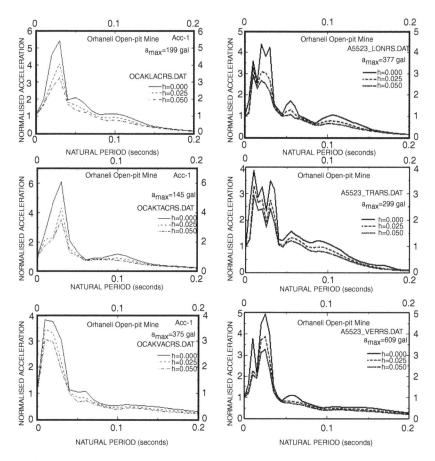

Figure 11.51   Normalized acceleration response spectra of the records of accelerometer Acc-1 and records of velocity-meter Vel-3.

a quasi exponential decay back to ambient pressure p_o and a negative phase in which the pressure is less than ambient. The pressure-time history of a blast wave is often described by exponential functions such as Frielander's equation (Smith and Etherington, 1994)

$$p = p_o + p_s \left(1 - \frac{t}{T_s}\right) e^{-bt/T_s} \tag{11.8}$$

where p_s: peak overpressure; T_s: duration of the positive phase; i_s: specific impulse of the wave which is the area beneath the pressure-time curve from the arrival at time to t_o the end of the positive phase.

It is common to use the scaled distance to evaluate the effects of blasting given as

$$Z = \frac{R}{\sqrt[n]{W}} \tag{11.9}$$

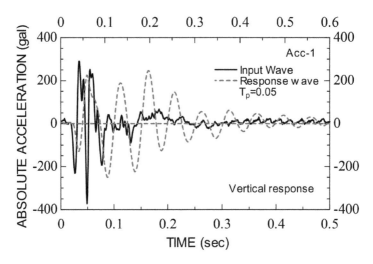

Figure 11.52 Absolute vertical acceleration response of a structure system (Acc-1).

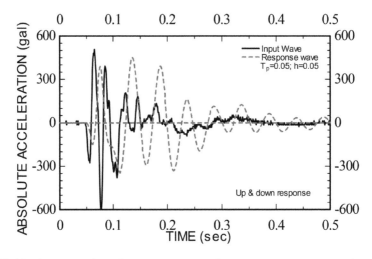

Figure 11.53 Absolute vertical acceleration response of a structure system using the acceleration computed from the velocity records of velocity-meter Vel-3.

where R is distance and W is mass of the explosive. When the value of n is 3, it is called Hopkinson blast scaling. USBM suggests the value of n as 2. The attenuation of air pressure is generally given in the following form

$$p = AZ^{-\alpha} \tag{11.10}$$

If W is given in kg and R in m, the unit of p is kPa and the value of A is generally about 185 and the power α is ranges between 1.2 and 1.5.

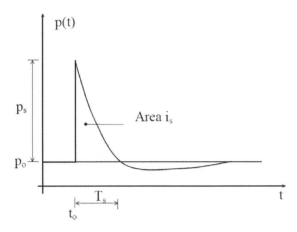

Figure 11.54 Air pressure variation during blasting.

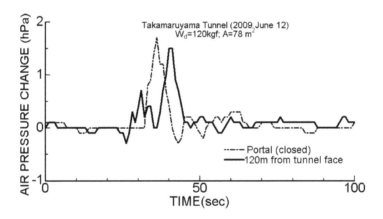

Figure 11.55 Air pressure fluctuations at Takamaruyama tunnel.

Air pressure changes were measured at Takamaruyama, Kuriko and Tarutoge tunnels using TR-73U produced by TANDD corporation. Although the device may not be appropriate for very sensitive air pressure changes induced by blasting, the measurements are quite meaningful as seen in Figures 11.55–11.57. The cross section of the tunnels are almost the same. Takamaruyama and Kuriko tunnel are single tubes while the Tarutoge tunnel has a branch connecting to the evacuation tunnel. The air pressure decreases at the evacuation tunnel when the shock wave passes by the branch and then increases as the air pressure is confronted by the door at the tunnel portal. Furthermore, the amplitude of the waves decreases as the distance increases.

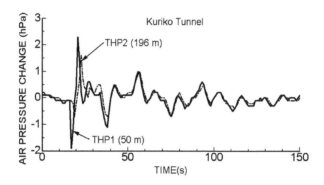

Figure 11.56 Air pressure fluctuations at Kuriko tunnel.

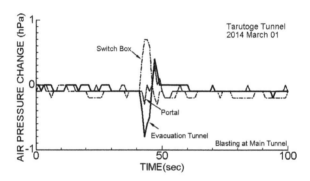

Figure 11.57 Air pressure fluctuations at Tarutoge tunnel.

11.6.6 Flyrock distance

Flyrocks is another major issue for the safety of people and damage to machinery and equipment when blasting is employed (Figure 11.58). The flyrock issue was caused by either inappropriate stemming, blasting sequence, charges or existence of some weak zones in rock mass. It is reported that the flyrock may travel up to a distance of 900 m in worst cases. The ejection velocity may reach up to 300 m/s. Flyrock distance in common practice is less than 50 m and it may reach a distance of 95 m sometimes. There are some empirical relations to estimate the maximum distance of flyrocks from the blasting relations.

One of the empirical relation is given by Lunborg (1981).

$$L_{max} = 260D^{2/3} \tag{11.11}$$

The unit of fly distance is m and D is given in inches.

The fly distance may be obtained from the simple physical laws of an object thrown with an initial angle (β_o) and velocity (v_o) at a given height (h_o). The air resistance may

Figure 11.58 Flyrock during blasting operation at Işıkdere open-pit lignite mine.

be taken into account as a viscous drag F_d. The fundamental equations of the flying object may be written as

$$\frac{d^2x}{dt^2} = -\frac{F_d}{m}; \quad \frac{d^2y}{dt^2} = -g - \frac{F_d}{m} \tag{11.12}$$

where g is gravitational acceleration. If the viscous resistance is given in the following form

$$F_d = \eta m v^n \tag{11.13}$$

The governing equation becomes

$$\frac{d^2x}{dt^2} = -\eta v^n; \quad \frac{d^2y}{dt^2} = -g - \eta v^n \tag{11.14}$$

where η is viscous drag coefficient. When $n=2$, it is called Stoke's law. If $n=1$, it is possible to solve the above differential equation. Otherwise the differential equations

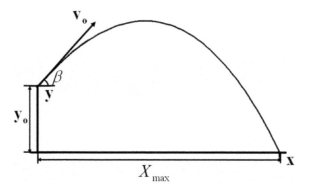

Figure 11.59 Trajectory of a flyrock and its conditions.

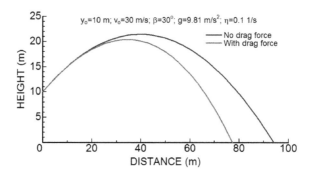

Figure 11.60 Comparison of the effect of viscous drag resistance of air on the fly trajectories of the flyrock.

become non-linear. The solution of Eq. 11.12 without drag force would yield the trajectory of a flyrock with conditions illustrated in Figure 11.59 as

$$x = v_o \cos \beta t \quad \text{and} \quad y = y_o + v_o \sin \beta t - \frac{g}{2}t^2 \qquad (11.15)$$

The maximum fly distance of flyrock in air can be estimated from Eq. (11.15) with condition of $y = 0$ as

$$X_{max} = \frac{v_o \cos \beta}{g} \left(v_o \sin \beta + \sqrt{(v_o \sin \beta)^2 + 2y_o g} \right) \qquad (11.16)$$

When $y_o = 0$, then, the travel distance of a flyrock takes the following form

$$X_{max} = \frac{(v_o)^2 \sin 2\beta}{2g} \qquad (11.17)$$

The maximum fly distance would be obtained when $\beta = 45°$. If $\beta = 0°$, the travel distance of a flyrock ejected from a given height (y_o) can be obtained from Eq. 11.16 as

$$X_{max} = v_o \sqrt{\frac{2y_o}{g}}$$ (11.18)

As noted from these relations, the initial conditions are very important for the fly distance of flyrocks.

If drag forces are taken into account, the solution of Eq. 11.12 yields the trajectory of the flyrock as

$$x = \frac{v_o \cos \beta}{\eta} \left(1 - e^{-\eta t}\right) \quad \text{and} \quad y = y_o + \frac{1}{\eta} \left(v_o \sin \beta + \frac{g}{\eta}\right) \left(1 - e^{-\eta t}\right) - \frac{g}{\eta} t$$ (11.19)

The solution of Eq. 11.12 for $n > 1$ requires the utilization of numerical techniques as it is difficult to obtain the closed-form solutions. Figure 11.60 compares the effect of viscous drag resistance of air on the fly trajectory of the flyrock.

Chapter 12

Dynamics of rockburst and possible countermeasures

12.1 MECHANICS OF ROCKBURSTS

Rock bursting is generally associated with the violent failure of brittle hard rocks such as igneous rocks, gneiss, quartzite and siliceous sandstone (i.e. Jaeger and Cook, 1979; Stacey 1981; Ortlepp and Stacey, 1994; Aydan *et al.*, 2001, 2004). It has been a well known phenomenon of instability in mining for a long time. When hard rocks are tested under uniaxial loading conditions, the fragments of rocks can be thrown a considerable distance once the peak strength of rock is exceeded. The failure surface is mostly associated with an extensional straining and with fine powder. Rock bursting in underground excavations is quite similar to that under laboratory conditions. When rock bursting occurs in underground openings, rock fragments detach from surrounding rock and are thrown into an opening in a violent manner like bombshells. A less severe form of rock bursting is observed as spalling.

It is known that rock bursting occurs in hard rocks having high deformation modulus while squeezing is observed in weak rocks having a uniaxial strength less than 20–25 MPa (Aydan *et al.*, 2004). Figure 12.1 shows typical stress-strain responses for both bursting and squeezing rocks. Bursting rocks are characterized with their high strength, higher deformation modulus and brittle post-peak behaviour. On the other hand, squeezing rocks are characterized with low strength, smaller deformation modulus and ductile post-peak behaviour.

The violent detachment of rock fragments during rock bursting is associated with how the stored mechanical energy is dissipated during the entire deformation process. As shown in Figure 12.2, if the intrinsic stress-strain response of rocks could not be achieved through its surrounding system in laboratory tests or underground openings, a certain part of stored mechanical energy would be transformed to kinetic energy. This kinetic energy results in the detachment of rock fragments, which may be thrown into the opening with a certain velocity depending upon the overall stiffness of the surrounding system and deformation characteristics of the bursting material (Jaeger and Cook, 1979). The first author observed this phenomenon even in granular crushed quartz samples confined in acrylic cells and dry initially sheared Fuji clay. It is observed that wrapping samples with highly deformable rubber-like strings greatly reduced the violent detachment of fragments. Although such materials could not delay or increase the overall confinement, they act as dampers to reduce the velocity and acceleration of detaching rock fragments, which may be one of very important observations in dealing with the rockburst problem in underground excavations.

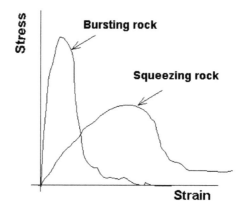

Figure 12.1 Typical stress-strain responses of squeezing and bursting rocks (from Aydan et al., 2004).

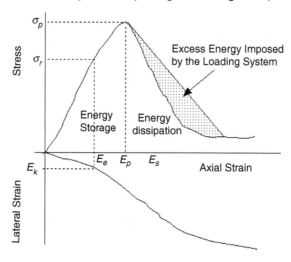

Figure 12.2 A simple illustration of the mechanical cause of rock bursting phenomenon (from Aydan et al., 2004).

For rocks exhibiting bursting phenomenon, the following identity must hold (Aydan *et al.*, 2011):

$$E_S = E_K + E_T + E_P + E_O \qquad (12.1)$$

where E_S, E_K, E_T, E_P and E_O stand for stored mechanical energy, kinetic energy, thermal energy, plastic work done and other energy forms, respectively. For a very simple case of one-dimensional loading of a block prone to bursting, one can easily derive the following equation for the velocity of the rock fragment if one assumes that the stored mechanical energy is totally transformed into kinetic energy (i.e. Arıoğlu *et al.*, 1999; Kaiser *et al.*, 1996):

$$v = \frac{\sigma_c}{\sqrt{\rho E}} \qquad (12.2)$$

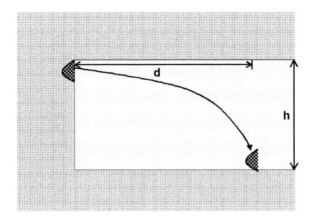

Figure 12.3 An illustration of fragment ejection from tunnel face.

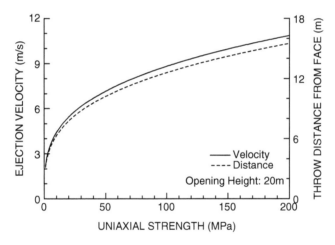

Figure 12.4 Ejection velocity and throw distance of rock fragment for an opening height of 20 m as a function of uniaxial strength of surrounding rock.

where σ_c, ρ and E are uniaxial strength, density and elastic modulus of the detaching rock block. Furthermore, the maximum ejection distance d of the block for a given height h of the opening and horizontal ejection angle can be easily obtained from the physics as follows:

$$d = v \sqrt{\frac{h}{g}} \qquad (12.3)$$

where g is the gravitational acceleration. Figure 12.4 shows the ejection velocity and throw distance of block as a function uniaxial strength of surrounding rock mass.

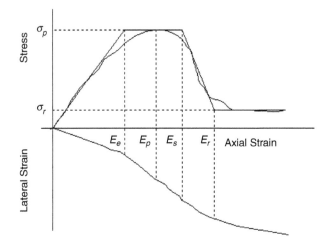

Figure 12.5 Illustration of strain limits for different states of rock under compression tests.

12.2 STRESS CHANGES IN THE VICINITY OF TUNNEL FACE

12.2.1 Static stress changes

In this section, an unsupported circular tunnel subjected to an axi-symmetric initial stress state is considered and the variation of displacement and stresses along the tunnel axis were computed using the elastic finite element method (Figure 12.6). The stresses and displacement vary with distance from the tunnel face. This is particularly an important aspect for assessing the tunnel stability as well as constructing the final tunnel lining (Aydan, 2011).

The first result is concerned with the radial displacement of the tunnel surface. The radial displacement is normalized by the largest displacement and is shown in Figure 12.7. As seen from the figure, the radial displacement takes place in front of the tunnel face. The displacement is about 28–30% of the final displacement. Its variation terminates when the face advance is about +2D. Almost 80% of the total displacement takes place when the tunnel face is about +1D. The effect of the initial axial stress on the radial displacement is almost negligible.

The next result is concerned with the variation of radial, tangential and axial stress around the tunnel at a depth of 0.125R as shown in Figure 12.8. As noted from the figure, the tangential stress gradually increases as the distance increases from the tunnel face. The effect of the initial axial stress on the tangential stress is almost negligible. The radial stress rapidly decreases in the close vicinity of the tunnel face and the effect of the initial axial stress on the radial stress is also negligible. The most interesting variation is associated with the axial stress distribution. The axial stress increases as the face approaches, and then it gradually decreases to its initial value as the face effect disappears. This variation is limited to a length of 1R(0.5D) from the tunnel face. It is also interesting to note that if the initial axial stress is nil, some tensile axial stresses may occur in the vicinity of the tunnel face.

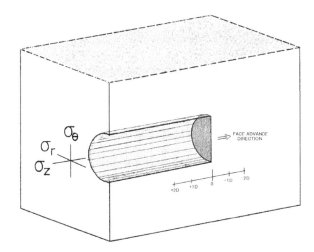

Figure 12.6 Computational model for elastic finite element analysis.

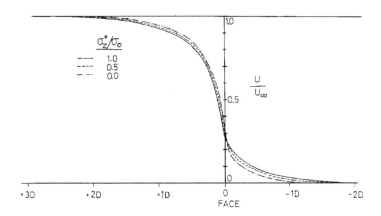

Figure 12.7 Normalized radial displacement of the tunnel surface.

The final example shows the stress distributions along r-axis of the tunnel at various distances from the face for the initial axial stress equal to initial radial and tangential stresses (Figure 12.9). As noted from the figure the maximum tangential stress is 1.5 times the initial hydrostatic stress and it doubles as the distance from the tunnel face is +5R, which is almost equal to theoretical estimations for tunnels subjected to hydrostatic initial stress state. The stress state near the tunnel face is also close to that of spherical opening subjected to hydrostatic stress state. The stress state seems to change from spherical state to the cylindrical state. It should be noted that it would be almost impossible to simulate the displacement and stress changes simultaneously in the vicinity of tunnels in 2D simulations using the stress-release approach irrespective of constitutive law of surrounding rock as a function of distance from tunnel.

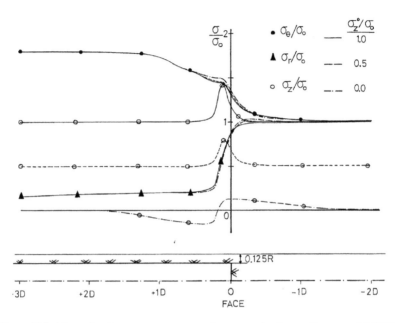

Figure 12.8 Normalized stress components along tunnel axis at a distance of 0.125R.

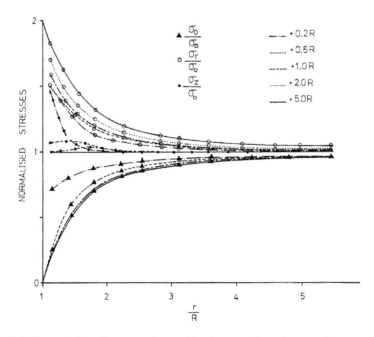

Figure 12.9 The variation of stresses along r-direction at various distances from tunnel face.

12.2.2 Dynamic stress changes

Excavation of tunnels is done through drilling-blasting or mechanically such as TBM and/or boom header excavators. The most critical situation on stress state is due to the drilling-blasting type excavation since the excavation force is applied almost impulsively. However, the drilling-blasting type excavation may cause a damage zone around the opening so that high-stress concentrations are moved into surrounding rock from the excavation perimeter. The effect of impulsive application of excavation force is evaluated for an axisymmetric cylindrical tunnel under initial hydrostatic stress is evaluated by a dynamic visco-elastic finite element method and discussed in Section 10.3.2.

The sudden application of the excavation force results 1.6 times the static ground displacement at the tunnel perimeter and shaking disappears almost within 2 seconds. Another important observation was that the tangential stress was greater than that under static condition. Furthermore, very high radial stress of tensile character occurs nearby the tunnel perimeter. Although the tunnel may be subjected to transient stress state just after excavation, stresses converge to their static equivalents provided that the surrounding rock behaves elastically. In other words, the surrounding rock may become plastic even though the static condition may imply otherwise.

12.3 EXAMPLES OF ROCKBURSTS

Rockburst problems have been reported in mining engineering and civil engineering literature (i.e. Jaeger and Cook, 1979; Stacey, 1981; Kaiser *et al.*, 1993, 1996; Shimokawa *et al.*, 1977; Panet, 1969) and it could be particularly a very severe problem during the excavation as it involves detachment of rock fragments with high velocity. Mont Blanc tunnel in France, Gotthard tunnel in Switzerland, Dai-Shimizu tunnel and Kanetsu tunnel in Japan are some of the well-known examples of rock bursting in tunneling (i.e. Panet 1969; Amberg, 1983). Rockburst problems are also one of the common instability modes in deep mining in hard rocks and numerous examples are reported from South Africa and Canada (i.e. Bosman and Malan, 2000; Kaiser *et al.*, 1993, 1996).

Most rockburst examples are associated with mining and there are few examples in civil engineering works. Some major civil engineering examples are described in chronological order using a recent study by Aydan and Geniş (2010) in the following while the examples from the mining field are briefly described as there is an excellent compilation of case history data on rockburst in the mining engineering field by Kaiser *et al.* (1996).

12.3.1 Major rockburst examples in civil engineering

(a) Mont Blanc Tunnel

Mont Blanc Tunnel was built between Italy and France (Panet, 1969). The construction was started in 1957 and completed in 1965. The tunnel is 11.6 km long and 8.6 m wide. The main body of the tunnel passes through granite with mylonitized fracture zone. Rockburst problems started to be observed when the overburden was more than 1000 m. The uniaxial strength of rock ranged between 100 and 140 MPa and the maximum in-situ principal stress ranged between 59 and 80 MPa. The rockburst was

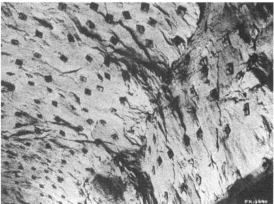

Figure 12.10 Views of slabbing in Mont Blanc tunnel (arranged from Panet, 1969).

observed as popping and slabbing (Figure 12.10). Slabbing was observed when rock mass was jointed. Rockburst occurred within 3 to 4 days and the face distance was 0.3–0.6 times the tunnel diameter. Rockbolting with steel straps were used to stabilize and to prevent violent detachment of rock fragments from sidewalls.

(b) Furka railway tunnel (Switzerland)

Furka railway tunnel is 15.5 km long and maximum overburden is 1520 m (Amberg, 1983). The cross section of this single track tunnel varied from 26 to 42 m². It is remarkable that the permanent support only consisted of rock anchors and a shotcrete lining. In sections with high rock pressures it was necessary to excavate an elliptical or circular profile using steel ribs in addition to anchors and the shotcrete with wire-mesh. Rock mass consisted of mainly granite or gneiss. While spalling was observed in granite, buckling was the main mode of failure in gneiss. Uniaxial strengths of gneiss and granite were 95 and 110 MPa, respectively while the maximum initial stress was 37–38 MPa. Popping and spalling were observed when the tunnel face was at a distance of 0.5 to 1.5D.

(c) Kaunertal pressure tunnel (Austria)

The Kaunertal pressure tunnel was 12 km long and had a diameter of 8 m. It was excavated using the bench method (Detzlhofer, 1970). Rock mass was gneiss and the maximum overburden was 1000 m. The uniaxial strength and elastic modulus of rock was 100 MPa and 10 GPa, respectively. The tunnel experienced rockbursting in the form of slabbing-buckling when the overburden was greater than 600–700 m.

(d) Lærdal tunnel (Norway)

The maximum overburden of the 24.5 km long Lærdal Tunnel is 1450 m, which corresponds to a vertical stress of the magnitude 39 MPa. 20 km of the total length (24.5 km) has an overburden more than 800 m (Grimstad, 1999). Despite the large overburden, the major principle stress is sub horizontal in most of the tunnel. Stress induced fracturing was observed to 3.5 m depth from the tunnel periphery. The magnitude of σ_1 was 31.5 MPa. The intermediate principal stress, σ_2 (=28.7 MPa) was dipping 36° in a direction parallel to the valley side and the tunnel axis. The minor principal stress, σ_3 (=21.4 MPa) was dipping 47° almost perpendicular to the valley side. The uniaxial strength of rock was ranging between 120–190 MPa. Rockburst was observed as popping or spalling. Spalling occurred at sidewalls and crown. Myrvang *et al.* (1984) also reported that Hoyanger roadway tunnel (overburden: 1450 m; σ_1: 34; σ_c: 60 MPa), Lanefjord roadway tunnel (overburden: 700 m; σ_1: 40; σ_c: 60 MPa), and Heggura roadway tunnel (overburden: 700 m; σ_1: 25; σ_c: 100 MPa) experienced rockbursting problems.

(e) Lötschberg and Gotthard tunnels (Switzerland)

The new rail link through the Alps involves two new rail tunnels, one through the Lötschberg, the Lötschberg Base Tunnel with a length of 34.6 km, and the other one through the St. Gotthard, the Gotthard Base Tunnel with a length of 57 km (Vuillemeur *et al.*, 1997; Aeschbach, 2002; Henke, 2005). The Lötschberg Base Tunnel runs from Frutigen in the Kandertal valley to Raron in the Valais. The Lötschberg Base Tunnel is designed as a tunnel system with two separate single-track tubes. The distance between the two tubes varies from 40 to 60 m, depending on the quality of the massif. Transverse tunnels connect the tubes every 300 m. The so-called transit function requires the European clearance profile EBV4 with 4 m headroom height. The cross-sectional diameter for excavation with a tunnel-boring machine (TBM) is 9.43 m. Excavation with the traditional blasting method yields a surface area between 62 and 78 m².

The southern half of the Lötschberg tunnel consists of granodiorite and gneiss belonging to Aar massif while the northern half consists of sedimentary formations of limestone, phyllite and sandstone. The depth of the Aar massif granite and granodiorite rock cover along the Lötschberg Base Tunnel route reaches up to 2000 meters in two places. The depth is greater than 1500 meters for a total of about 9.3 km. The risk of rockburst occurring is therefore relatively high. The risk of rockburst is categorized into four classes according to expected tangential stress on tunnel perimeter:

- Class A: Very high rockburst risk: 75–100% $\sigma_\theta > 130$ MPa
- Class B: High rockburst risk: 50–75%   120 MPa $< \sigma_\theta < 130$ MPa
- Class C: Moderate rockburst risk: 25–50%   110 MPa $< \sigma_\theta < 120$ MPa
- Class D: Low rockburst risk: 0–25%   100 MPa $< \sigma_\theta < 110$ MPa.

On the basis of this classification, about 4.1 km of the Lötschberg Base Tunnel were classified as class A, about 1.4 km as class B, about 1.4 km as class C and about 300 m as class D. Although uniaxial compressive strength of hard rocks ranges between 75 and 205, the average uniaxial strength is about 180 MPa.

For the TBM-excavation two distinct failure phenomena could be identified: a) Block formation in front of the TBM head: The blocks have no particular shape; they are not slab-shaped like fragments typically encountered in spalling and slabbing processes (Figure 12.11). When these instabilities appear, instead of showing a flat aspect with clear marks from the cutters, the tunnel face is quite irregular. The marks from the cutters are then only visible in a smaller area of the face. In some cases this

(a) spalling	(b) popping

Figure 12.11   Rockbursting problems in Lötschberg Base Tunnel (arranged from Vuillemeur *et al.*, 1997).

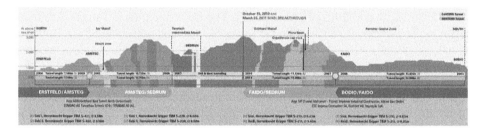

Figure 12.12   Final geological longitudinal section of Gotthard tunnel (from http://www.plurimedia. eu/2016/06/04/gothard-base-tunel).

block-formation seems to be assisted by the presence of weak failure planes filled with chlorite. The blocks reduce the TBM utilization time and penetration rate. b) Onion-skinning initiated at the level of the TBM shield: About 0 to 4 m behind the tunnel face, scales of low thickness peel off the walls in the excavation. When high overburden depth is encountered, deep notches up to about 1 m may appear, typically in a symmetrical pattern. In some cases, the notches prevented the grippers from making good contact with the rock mass, which caused significant output losses. The project geologist believes the difficulties encountered in this particular case are to a large part attributable to the increased fracturing of the rock. Most often, these phenomena were noticed in zones of strong massive rocks. Some strong acoustic phenomena have also been reported, but almost no violent projections of blocks have so far been observed. Apparently, most energy releases occur within 4 meters from the face within the TBM shield.

Gotthard tunnel, which is a 57 km long twin tube single track railway tunnel, has been recently completed (Figure 12.12(a)). However, the tunnel at the Piora section was anticipated to experience heavy squeezing problem where rock mass is called sugarized dolomite due to alpine tectonism. In this particular location, heavy water inflow was also anticipated. However, the dolomitic rock at the tunnel level was not sugarized and TBM excavated the tunnel without any major problem. The geology of the central

Figure 12.13 Rockburst in the vicinity of Faido Multifunction Station (from Hagedorn *et al.*, 2005).

Alpine zone, where the tunnel is situated, consists of three major gneiss zones: the Aar massif in the north, the Gotthard massif in the middle and the Pennine gneiss zone in the south. These units consist mainly of very strong igneous and metamorphic rock with a high initial strength. In general these zones should not cause major geotechnical difficulties during construction. Nevertheless, some dangers and risks due to rockburst were anticipated at sections where the overburden is very high and it is higher than 2000 m and reaches up to 2300 m for 5 km.

Rockburst was observed in Faido Multifunction Station where a major fault zone crosses the tunnel (Figure 12.13) (Hagedorn *et al.*, 2008). This fault strikes at an average angle of about 15° to the tunnel axis and dips at about 80°. The rockburst events took place over a total length of 400 m in the EON tube near the fault and rock mass was mainly gneiss. During 2004 rockbursts took place mainly at the face of the EON tube to the north of the cross cavern. The p-wave velocity of rock mass is 5.33 km/s.

In addition, there is a recent report of rockburst in an exploration tunnel of Gotthard tunnel (Aydan and Geniş, 2010). The tunnel shape is horse-shoe type and popping starts at the bottom of the tunnel and later tunnel face starts to fail in the form of slabbing.

(f) Underground Research Laboratory (URL) 420 m Level Research Adit (Canada)

The URL is located within the Canadian Shield in the Lac du Bonnet granite batholith. The batholith is one of a number of similar post-tectonic and post-metamorphic batholiths within the Bird River and Winnipeg River sub-provinces of the Canadian Shield. The batholith has an areal extent of 1400 km² on surface and extends in depth to between 6 and 25 km. The granite at the URL is approximately 2,600,000,000 years old. The site has interesting and varied geology and is crosscut by two low-dipping thrust faults, or fracture zones with a deeper third thrust fault that appears to die out before passing below the URL excavations. The blocks between the thrust faults define different structural domains that can be distinguished by the presence of intrusions and segregations and by the pattern and frequency of sub-vertical fracturing, as well as by differing in situ stress regimes. Shaft collar excavation and construction of the surface facilities took place during 1982 and 1983.

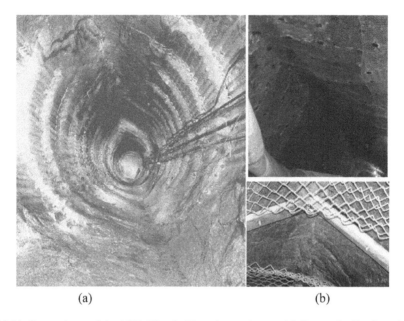

(a)   (b)

Figure 12.14 Some views of the URL Mine-by Tunnel experiment. (a) Picture by Read *et al.* (1998);
(b) Pictures by Aydan in 2005.

Excavation of the URL shaft to a depth of 255 m began on 12 May 1984 and
continued for the remainder of the year. The loop of horizontal excavations on the
240 Level (240 m below surface) and the raise-bored ventilation shaft were completed
by 1987. The main shaft was extended to a depth of 443 m in 1988, followed by
the excavation of the 420 Level and the ventilation shaft over the following three
years. A cylindrical test tunnel with a length of approximately 46 m and a diameter of
3.5 m was excavated at a depth of 420 m at the Underground Research Laboratory. The
tunnel was excavated parallel to the intermediate principal stress so that the differential
stress was very high – promoting the maximum development of excavation induced
damage (Read *et al.*, 1998).

Excavation sequences in which the face was advanced by 0.5 m to 1 m steps took
place between October 1991 and July 1992 within a rock mass monitored by arrays
of accelerometers, extensometers, strain cells and thermistors and convergence arrays
measuring the changing shape of the tunnel. The granite has average isotropic P- and
shear-wave velocities of 5763 m/s and 3376 m/s, respectively, giving a Vp/Vs ratio of
1.707.

As expected, breakout notches occurred along the entire length of the Mine-by
tunnel as the excavation progressed (except for the last meter before the tunnel face
immediately following each excavation sequence) (Figure 12.14). Breakout notches
are a common linear feature of boreholes and underground tunnels. They generally
form parallel to the direction of weakest stress and perpendicular to the length of the
cavity (Bell and Gough, 1979). The breakout notches in the Mine-by tunnel generally

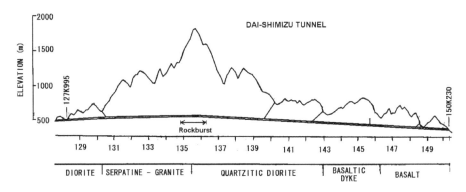

Figure 12.15 A longitudinal geological section of Dai-Shimizu tunnel (Shimokawa et al., 1977).

occurred between 10° and 20° to the vertical, towards the southeast. This broadly agrees with the measured direction of minimum stress. Stress-induced fracturing around the tunnel initiated at approximately 0.3 to $0.5\sigma_{ci}$. Damage initiation under uniaxial compression occurred at 0.3 to 0.5 of the peak strength and the in-situ observation was quite similar to that observed under laboratory conditions.

(g) Dai-Shimizu Tunnel (Japan)

Dai-Shimizu tunnel is a 22.2 km long double track railway tunnel through which the Joetsu-Shinkansen line passes. Rockmass consists of mainly quartzitic diorite and its construction was completed in 1982 (Figure 12.15). Overburden varies between 470 to 1150 m. The maximum in-situ stress is about 24–26 MPa while the uniaxial compressive strength of rock is about 150 MPa. Rockburst was observed when the overburden exceeded 700 m. Rockbursting occurred just after blasting or during drilling for next blasting round at the face as well as at the crown. Sometimes it occurred at the sidewall. The section was supported by rockbolts and steel ribs together with wire-mesh (Shimokawa et al., 1977).

(h) Kanetsu tunnel (Japan)

Kanestu Tunnel is built on the Kanetsu expressway and it is close to Dai-Shimizu Tunnel. It is about 11 km long. The maximum overburden is 1200 m and rockburst was observed when the overburden was greater than 600–700 m (Figure 12.16). It occurred as popping from the tunnel face and soon after blasting. Spalling occurred at sidewalls after face advance. 4 m rockbolts and shotcrete with wire mesh was used to deal with rockbursting section. Rock mass is diorite where rockburst was observed and its uniaxial compressive strength ranges between 126 and 282 MPa (Inoma, 1981).

(i) Kawaore powerhouse access tunnel (Japan)

A large underground opening for a pumped-storage scheme is planned at a depth of 500 m in the Central Japan. The in-situ stress measurement showed that unusually

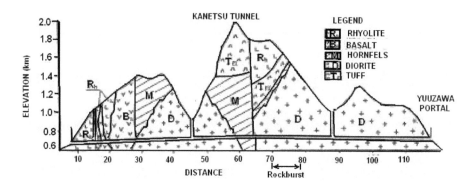

Figure 12.16 A longitudinal geological section of Kanetsu tunnel (Inoma, 1981).

Table 12.1 Physical and mechanical properties of intact rock.

Class	γ (kN/m³)	E_i (GPa)	ν	σ_{ci} (MPa)	σ_{ti} (MPa)	ϕ_i (°)	V_{pi} (km/s)
B	26	25–46	0.16–0.22	218–217	7.4–11.2	59–62	4.6–5.5
CH	26	5–21.4	0.18–0.27	110–153	4.7–9.2	53–58	3.9–5.1

high in-situ stresses exist at the site, which was not experienced at other underground power house constructions in Japan. Rock is granite and it belongs to the geological era of cretaceous (Ishiguro et al., 1997). Rock masses at the site were classified as B and CH on DENKEN classification system of Japan. B Class rock mass has less discontinuity sets with large spacing. Furthermore, rock is fresh and staining or weathering of discontinuity walls does not exist. On the other hand, CH Class rock mass has more discontinuities and its discontinuity spacing becomes smaller. Furthermore, staining and/or weathering of discontinuity walls exist. The staining of discontinuity walls in this site is due to thermal alteration resulted from volcanic dykes, which took place after the placement of granitic rock. Laboratory and in-situ tests are carried out to obtain physical and mechanical properties of rock mass. Tests involve physical properties such unit weight, porosity and wave velocity and mechanical properties such as elastic modulus, Poisson's ratio, uniaxial and triaxial compression tests, Brazilian tests and permeability tests. Cores are obtained from the sites, which are classified as B Class rock and CH Class rock and they are tested in the laboratory. Table 12.1 gives the physical and mechanical properties of intact rock for each rock class.

Various in-situ stress measurement techniques were used to infer the stress state and the horizontal stress was almost 4 times the overburden stress, which is quite an unusual situation in Japan (Ishiguro et al., 1997). The longitudinal axis of the cavern was changed so that it would be aligned parallel to the largest horizontal stress. Although the construction of the underground powerhouse was suspended for an indefinite period due to changes in energy demand, the access tunnel was constructed.

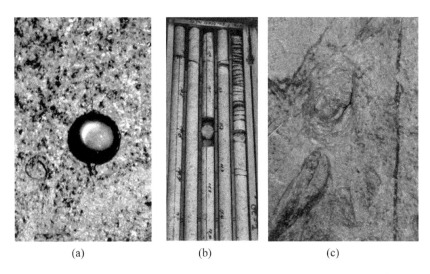

(a)   (b)   (c)

Figure 12.17 Views of yielding around boreholes in exploration adits and core-disking: (a) Tensile fracture around borehole in an exploration adit; (b) core disking; (c) yielding around borehole in access tunnel.

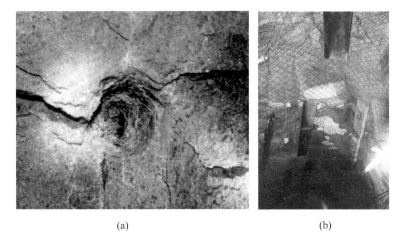

(a)   (b)

Figure 12.18 Views of tensile fracture and dog-ear yielding in a borehole and (b) Rockbolting, straps and wire mesh at an exploration adit support.

Rockburst problem was observed when the axis of the access tunnel was perpendicular to the maximum horizontal stress. Sounds of rock breakage and small scale spalling were observed at the tunnel face soon after the blasting.

Boreholes drilled into surrounding rock for different purposes in exploration adits and access tunnel were carefully investigated. As seen in Figure 12.17 and Figure 12.18, various forms of stress-induced fractures and yielding were observed around these boreholes.

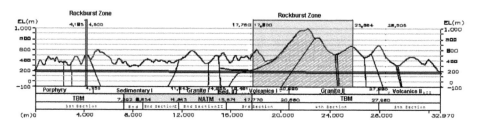

Figure 12.19 A longitudinal geological section of Andon-Yongchun waterway tunnel.

(j) Andong-Yongchun waterway tunnel (South Korea)

The Andong-Yongchun waterway tunnel passes through high mountains with an over-burden ranging from approximately 200 m to 800 m (Figure 12.18). It is approximately 32.97 km in length and is the longest waterway tunnel at present in Korea (Lee *et al.*, 2004). The tunnel was constructed through three different rock types, specifically, intrusive rock, sedimentary rock, and volcanic rock. Each of these rocks had faults, dykes, and two to four sets of joints. The tunnel was excavated by drilling-blasting with NATM concepts and TBM. The NATM section has widths of 3.8 m to 4.0 m and the TBM has a diameter of 3.5 m. Tunnel lengths of NATM and TBM are 10.6 km and 22.4 km respectively. The uniaxial strength of rock ranged between 78 to 162 MPa. From their construction of the tunnel, Lee *et al.* (2004) concluded that

1   There is a high potential for rockburst if tunnel depth is greater than 400 m and excavated by TBM in the granite or volcanic rock (Figure 12.19).
2   Fast tunnel advancing speed with little damage on rock mass plays a role in rock-burst at lower depths, less than 400 m. It is important to note that there are rockbursts even though tunnel depth is less than 400 m in TBM tunnel.
3   However, there are no rockbursts in the NATM tunnel if the tunnel depth is less than 400 m no matter what the host rock is.

(k) Er-Tan underground power house (China)

The 3,492 MW Ertan hydropower plant comprises a 240 m concrete arch dam and Asia's largest underground powerhouse, which is 280 m long by 25.5 m wide and 65 m high (Li *et al.*, 1998). The site is located near Panzhihua on the Yalong River (a tributary of the Yangtze) in Sichuan province. Sichuan is a mountainous inland province in the south west of China with high hydro potential. The underground complex includes a powerhouse cavern (281 m × 26 m to 31 m × 66 m) with six 550 MW units, a trans-former cavern (215 m × 19 m × 25 m) and a surge chamber (201 m × 19 m × 69 m). The project also has the two large diversion tunnels (1,000 m × 20 m × 23 m), two spillway tunnels (each 850 m × 13 m × 13.5 m). On the basis of a number of stress measurements, the region of the river valley is divided into three different in-situ stress zones. A rather high horizontal stress of up to 30 MPa was measured and it was even over 60 MPa at the bottom of the river bed. Rock mass is syenite and basalt and the

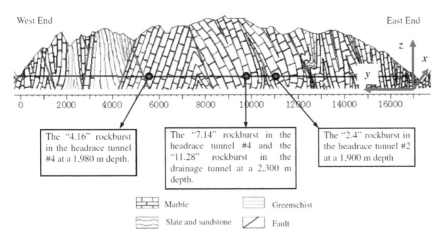

West End

East End

| The "4.16" rockburst in the headrace tunnel #4 at a 1,980 m depth. | The "7.14" rockburst in the headrace tunnel #4 and the "11.28" rockburst in the drainage tunnel at a 2,300 m depth. | The "2.4" rockburst in the headrace tunnel #2 at a 1,900 m depth |

Marble   Greenschist

Slate and sandstone   Fault

Figure 12.20 Major intense rockburst event in Jiping II Hydropower tunnels (from Yang *et al.*, 2012).

cavern was located at a depth of 250 m below ground surface. The diversion tunnel on the left-bank experienced rockburst. The strength of intact rock ranged between 40–80 MPa (Li *et al.*, 1998).

(l) Jingping tunnel (China)

The Jinping II Hydropower Station, which traverses Jinping Mountain, has seven parallel tunnels, each 17 km long with a maximum depth of 2,525 m. A number of extremely intense rockbursts occurred during the excavations of the tunnels, killing several construction workers, injuring many more, and damaging several sets of equipment (Yang *et al.*, 2012). Rockburst intensity in the auxiliary tunnels was classified into four grades: light class (grade I), moderate class (grade II), intense class (grade III), and extremely intense class (grade IV). A number of extremely intense rockbursts occurred during the excavation of these seven tunnels. Four typical ones among them with distinctive characteristics were named as: the "11.28" event in the drainage tunnel, the "7.14" event in the headrace tunnel #4, the "4.16" event in the headrace tunnel #3, and the "2.4" event in the headrace tunnel #2 (Figure 12.20). Even the TBM was destroyed by the rockburst events. The rockburst took place in the close vicinity of cutter head and shield. The rock in the vicinity of the rockburst event consisted of thick, gray and off-white medium-fine-grained crystalline massive marbles, including fresh and hard calcite, pinstripe biotite, and other minerals. The marble is characterized by a brittle behaviour and high strength with a uniaxial compressive strength of about 100 MPa and a tensile strength of 3–6 MPa.

12.3.2 Major rockburst examples in mining

Rockburst problem is well known for a long-time in mining engineering field and it is generally associated with deep mining. First reported examples of rockburst come from South Africa, which are followed by USA, Canada, and other countries (Jaeger and Cook, 1979; Stacey, 1981; Ortlepp and Stacey, 1994; Ortlepp, 2000; Kaiser *et al.*, 1993, 1996). Mining is now reaching depths greater than 3500 m, which at least implies the overburden pressure of 65 MPa. Since hard rocks have strength ranging between 100 and 350 MPa, there is a high possibility of yielding of intact rock due to undesirable geometrical shape of mine stopes. Furthermore, the scale of mines is quite large so that the existing faults may also be activated due to mining operations. Therefore, there is a high possibility of rockburst as a result of the activation of faults in addition to stress variations due to geometrical changes under high in-situ stresses. However, it should be noted that civil engineering applications (even large underground power houses) are less likely to activate faults as a result of underground excavations. As Kaiser *et al.* (1996) compiled and analyzed various major case histories of rockbursts in mining field in a recent publication, the readers are recommended to refer the publication of Kaiser *et al.* (1996). A detailed summary of rockbursts in mining could be also found in the special issue on the rockburst phenomenon and its counter measures by Aydan and Geniş (2010).

12.4 LABORATORY TESTS ON ROCKBURST PHENOMENON

12.4.1 Sandstone block from Shizuoka Third Tunnel

Aydan *et al.* (2004, 2011) described a laboratory test on large rock samples of silicious sandstone obtained from the Shizuoka Third Tunnel having a circular hole with a diameter of 58 mm. The sample had preexisting discontinuities healed by calcite filling. The height and width of sample were 300 mm and 200 mm, respectively. Figure 12.21

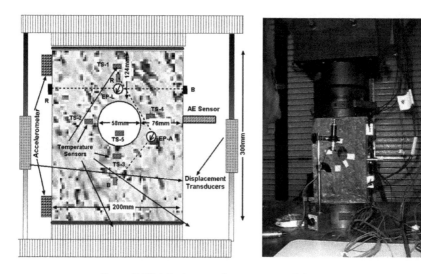

Figure 12.21 The layout of instruments of the sample.

shows the instrumentation layout and a view of the sample. Figure 12.22 shows the acceleration responses measured on sandstone sample ($300 \times 200 \times 138$ mm) having a circular hole with a diameter of 58 mm. The overall acceleration responses are quite similar to those observed in uniaxial compression experiments. However, multiple acceleration responses with growing amplitudes occurred before that during the final rupture state. This phenomenon may be related to the ejection of small fragments from the perimeter of the model tunnel before the final failure of the sample. The observations of the experiments on the model tunnel in a brittle hard rock resemble the rockburst phenomenon experienced in rock engineering. Figures 12.23 and 12.24 show a view of ejection of the rock fragments from the perimeter of the model tunnel and multi-parameter responses.

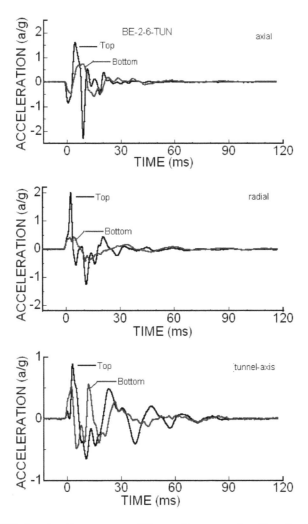

Figure 12.22 Acceleration response of the sandstone sample with a circular hole.

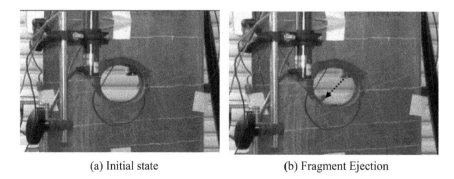

(a) Initial state   (b) Fragment Ejection

Figure 12.23 Views of the sample with a model tunnel during the experiment.

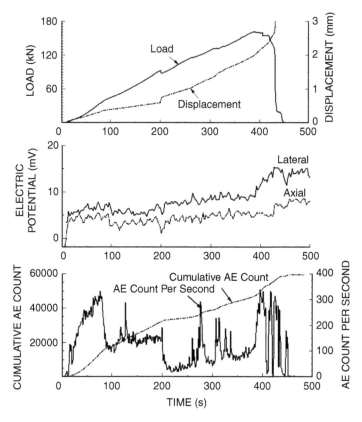

Figure 12.24 Multi-parameter responses of model tunnel sample Shiztun-2 during deformation and fracturing process.

Table 12.2 Mechanical properties of intact rock samples.

Sample	UW (kN/m³)	UCS(peak) (MPa)	Elastic Modulus (GPa)	UCS(residual) (MPa)	Vp(peak) (km/s)	Vp(residual) (km/s)
TGC-1	24.8	40.5	5.05	23.53	4.03	3.57
TGC-3	25.3	25.9	3.16	5.33	3.61	2.71
TGC-2	25.0	111.4	12.96	4.76	5.17	3.37

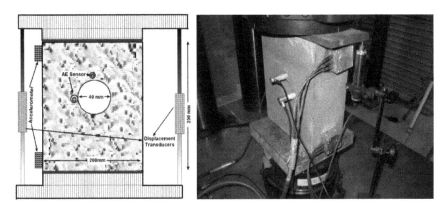

Figure 12.25   Illustration of instrumentation of rock block sample and its view.

12.4.2   Sandstone sample from Tarutoge Tunnel

An experimental set-up consisting of a compression testing device with a capacity of 2000 kN was used and three prismatic rock blocks with a size of $200 \times 250 \times 90$ mm having a circular opening with a diameter of 40 mm subjected to uniaxial compression. Rock samples were obtained from Tarutoge Tunnel, which is being constructed using drilling-blasting technique as a part of an expressway project connecting Shin-Tomei Expressway and Chuo Expressway at the boundary of Shizuoka and Yamanashi Prefectures in the Central Japan (Imazu *et al.*, 2014). The cores obtained from the rock blocks were tests for obtaining their uniaxial strength. The uniaxial strength of rocks ranged from 26 MPa to 110 MPa, implying that the rock strength could be quite anisotropic with respect to bedding plane orientation (BP) (Table 12.2). During experiments, the rock blocks were attached with accelerometers and acoustic emission sensors in addition to displacement transducers and load cell to measure the dynamic response of rock blocks during deformation and fracturing processes (Figure 12.25).

Infra-red Camera X8400sc/X6500sc was used to observe infra-red thermography. Frame rate was 100/s during observations (Figure 12.26). Three blocks numbered TGT1-1, TGT2-1 and TGT2-2 were tested. Figure 12.27 and Figure 12.28 show the multi-parameter responses of rock blocks and associated infrared images together with visual images at several time steps for Block TGT2-1. The temperature of infrared thermography images ranged between 30 and 33 Celsius in all experiments. Although the appearance of tensile fractures at roof, which was expected from the stress state

Figure 12.26 Views of infrared camera used in experiments and a typical experiment.

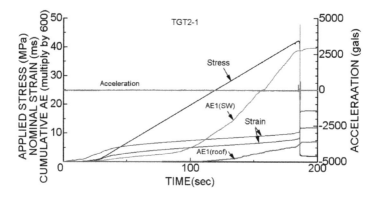

Figure 12.27 Multi-parameter responses of TGT2-1 rock sample with circular opening.

induced by the applied loading and boundary conditions, was not clearly observed, spalling started to occur at side walls as expected. While visible spalling was observed at the 80–85 percent of the total load level, AE occurrences started at the 35–40 percent of the total load level, which roughly corresponds to the anticipated yielding stress level at the sidewalls.

Before the macroscopic failure of samples, the ejection of fragments from the side wall of the openings induced vibrations. These vibrations were clearly noted in acceleration responses. When rock blocks fail, the amplitude of accelerations reached the level of 3 to 4.5 times the gravitational acceleration. The acceleration response was also not symmetric with respect to time axis as noted previously by Aydan (2003), Ohta and Aydan (2010) and Aydan et al. (2011). It should be noted that the accelerations were measured at the top and bottom of the sample near loading platens. In other words, accelerations at the points where the ejection of fragments occurred should be much greater than those measured at the top and bottom of the sample.

As noted from infrared thermography images shown in Figure 12.28, high temperature bands were observed at locations where plastic straining and cracks occurred. These high temperature spots grow and the ejected fragments appear as high temperature spots. Once macroscopic cracks develop, then those high temperature bands cool

TGT2-1

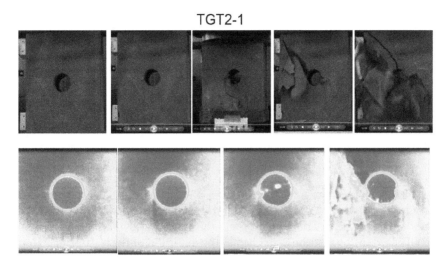

Figure 12.28 Visual and infrared thermography images of rock block sample TGT2-1.

down. These observations clearly show that infrared thermography technique could be a great tool for engineers to identify high temperature zones as an indicator of likely locations of rock failure in the vicinity of excavation surfaces for assessing the real-time safety of underground excavations.

12.5 PREDICTION OF ROCKBURST POTENTIAL

Several methods are proposed to assess the susceptibility of rockbursting in underground excavations. Ortlepp and Stacey (1994) recently presented a detailed review of these methods. These methods can be broadly classified as energy method, elastic-brittle plastic method, and extensional strain method. Nevertheless, none of these methods were validated for assessing the susceptibility of rockbursting in tunneling and its intensity in practice. In this section, a method developed by Aydan *et al.* (2001, 2004) for the assessment of susceptibility of rockbursting in tunneling in hard rocks is discussed and this method is essentially a slight extension of their method proposed for tunnels in squeezing rocks. After presenting the fundamentals of the method, it is compared with other methods to check its merits and de-merits. Furthermore, its several applications to a tunnel with great overburden are given and discussed.

12.5.1 Energy method

The energy method has been used in mining for a long time and it is based on the linear behaviour of materials. When the material behaviour becomes non-linear, it becomes difficult how to define the energy.

Table 12.3 Rockburst susceptibility assessment by Q-system (from Barton *et al.*, 1974).

Rockburst Level	σ_c/σ_1	σ_t/σ_1
Light-popping	10–5	0.66–0.33
Mild rockburst	5–2.5	0.33–0.16
Heavy rockburst	<2.5	0.16

12.5.2 Extensional strain method

Stacey (1981) proposed the extensional strain method for assessing the stability of underground openings in hard rocks. He stated that it was possible to estimate the spalling of underground cavities in hard rocks through the use of his extensional strain criterion. The extensional strain is defined as the deviation of the least principal strain from linear behaviour. This definition actually corresponds to the definition of initial yielding in the conventional theory of plasticity. This initial yielding is generally observed, at the 40–60 percent of the deviatoric strength of materials.

12.5.3 Elasto-plastic method

In elasto-plastic method, there are several models to model the strength of rock. The simple one is the brittle plastic model, in which the strength is reduced from the peak strength to its residual value abruptly. Aydan *et al.* (2001) recently combined both squeezing and rockbursting phenomena and more generally the strength reduction model is proposed as a function of strain level. This model, at least, treats both phenomena in a unified manner.

12.5.4 Empirical methods

One of the empirical methods involves the rock classifications. Among the rock classifications, the Q-system of NGI probably is the one, which directly addresses the rockburst phenomenon on the basis of the ratio of uniaxial strength of rock σ_c and tangential stress σ_t to the maximum initial rock stress σ_1. The criterion is summarized as shown in Table 12.3.

The energy criteria on the basis of case studies compiled by Kaiser *et al.* (1996) is also proposed. Their proposal is re-arranged and given in Table 12.4.

12.5.5 A unified method by Aydan *et al.* (2001, 2004)

The stability assessment of structures against rockburst and its prediction should be carried out according to the following procedure:

i)   Estimating the possibility of rockburst occurrence according to various methods described above (12.5.1 to 12.5.2).

Table 12.4 Rockburst susceptibility assessment by Kaiser *et al.* (1996).

Rockburst Level	σ_t/σ_c	Energy (kJ/m^2)
None or little	0.35	–
Minor	0.35–0.45	<5
Moderate	0.45–0.55	5 to 10
Major	0.55–0.70	10 to 25
Heavy	>0.70	25 to 50
Extreme	>0.70	>50

ii)   Taking appropriate measures against, prevention of or reducing rockburst occurrence.

iii) Real-time monitoring of the various responses of surrounding rockmass and predicting the rockburst.

The first and second items against rockburst problems are used for assessing the rockburst susceptibility. However, the prediction of rockburst events is not possible using the conventional deformation measurements. As Aydan and his co-workers (Aydan *et al.*, 2001, 2002, 2003) showed that the sensitivity of deformation measurements even under the laboratory conditions is not sufficient to infer the possibility of rock failure. However, they experimentally showed that there may be some parameters, which may act as precursors of the failure of rocks. These parameters are acoustic emission (AE) events, variations of electric potential, electrical resistivity, magnetic field and temperature, which are products of the energy transformation of the stored mechanical energy during its dissipation. The real-time multi-parameter monitoring system for predicting rockbursts and also other forms of failure should utilize these parameters in addition to conventional deformation measurements. Nevertheless, the utilization of all parameters may not be always possible due to the limitations of the measurement devices and the excavation procedure.

12.5.6   Application of rockburst prediction methods to tunnels under hydrostatic stress condition

(a) Extension of Aydan's method for tunnels in squeezing rocks to rockbursting

Bosman and Malan (2000) recently reported that the overall behaviour of hard rocks could be very similar to that of squeezing rocks. The fundamental difference between squeezing and bursting is probably the strain levels associated with different states as illustrated in Figure 12.5. As noted from Figure 12.5, the strain levels for bursting rocks are much smaller than those for squeezing rocks. Figure 12.29 shows a plot of normalized strain levels by the elastic strain limit defined in Figure 12.5 for a uniaxial compressive strength range between 1 and 100 MPa. This figure is an extension of the earlier plot for squeezing rocks together with new data. The horizontal axis of the figure is the uniaxial strength of surrounding rock. It should be noted that intact rock strength and rock mass strength are not differentiated on the basis of experiences and

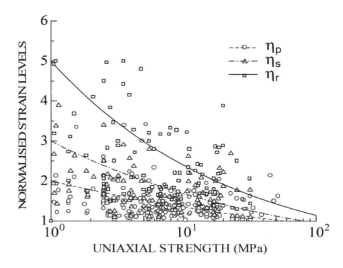

Figure 12.29 Comparison of normalized strain levels and empirical relations for squeezing and bursting rocks.

databases for rock masses and intact rocks. In other words, if the uniaxial strength values of intact rocks and rock masses are similar, their mechanical behaviour should be quite similar to each other.

The relations proposed for squeezing rocks by Aydan *et al.* (1993, 1996) are also applicable to hard rocks with bursting potential. Furthermore, the empirical relations for other mechanical properties, which are required for analyses, can be applicable for rocks with bursting potential. For circular tunnels under hydrostatic initial stress state as shown in Figure 12.30, the strain levels and plastic zone radii can be obtained as follows (Aydan *et al.*, 1993, 1996):

Elastic state

$$\xi = \frac{\varepsilon_\theta^a}{\varepsilon_\theta^e} = 2\left(\frac{1-\beta}{\alpha}\right) \leq 1 \tag{12.4}$$

Elastic perfectly plastic state

$$\xi = \frac{\varepsilon_\theta^a}{\varepsilon_\theta^e} = \left\{\frac{2[(q-1)+\alpha]}{(1+q)\,[(q-1)\beta+\alpha]}\right\}^{\frac{f+1}{(q-1)}} \tag{12.5}$$

$$\frac{R_{pp}}{a} = \left\{\frac{2[(q-1)+\alpha]}{(1+q)\,[(q-1)\beta+\alpha]}\right\}^{\frac{1}{(q-1)}} \tag{12.6}$$

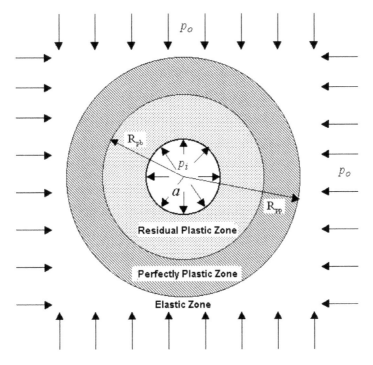

Figure 12.30 States around a circular tunnel and notations.

Elastic perfectly plastic and brittle plastic state

$$\xi = \frac{\varepsilon_\theta^a}{\varepsilon_\theta^e} = \eta_{sf} \left\{ \frac{\dfrac{2\left[(q-1)+\alpha\right]}{(1+q)\,(q-1)}(\eta_{sf})^{\frac{(1-q)}{f+1}} - \dfrac{\alpha}{q-1} + \dfrac{\alpha^*}{q^*-1}}{\beta + \dfrac{\alpha^*}{q^*-1}} \right\}^{\frac{f^*+1}{(q^*-1)}} \tag{12.7}$$

$$\frac{R_{pp}}{a} = \left\{ \frac{\dfrac{2\left[(q-1)+\alpha\right]}{(1+q)(q-1)}(\eta_{sf})^{\frac{(1-q)}{f+1}} - \dfrac{\alpha}{q-1} + \dfrac{\alpha^*}{q^*-1}}{\beta + \dfrac{\alpha^*}{q^*-1}} \right\}^{\frac{1}{(q^*-1)}} \tag{12.8}$$

where $\beta = \dfrac{p_i}{p_0}$; $\alpha = \dfrac{\sigma_c}{p_0}$; $\alpha^* = \dfrac{\sigma_c^*}{p_0}$; $q = \dfrac{1+\sin\phi}{1-\sin\phi}$; $q^* = \dfrac{1+\sin\phi^*}{1-\sin\phi^*}$.

(b) Energy method

The energy method has been used in mining for a long time and it is based on the linear behaviour of materials. When the material behaviour becomes non-linear, it becomes difficult how to define the energy. The overstressed radius of rock around a circular tunnel under hydrostatic stress state can be obtained with the use of Mohr-Coulomb yield criterion and elastic stress components as

$$\frac{R_p}{a} = \left\{ \frac{(q+1)(1-\beta)}{[(q-1)+\alpha]} \right\}^{\frac{1}{2}} \tag{12.9}$$

The total energy per unit area in the overstressed zone is then obtained as follows

$$w_{el} = \frac{1+\upsilon}{E}(p_o - p_i)^2 a \left[1 - \left(\frac{a}{R_p} \right)^2 \right] \tag{12.10}$$

(c) Extensional strain method

Stacey (1981) proposed the extensional strain method for assessing the stability of underground openings in hard rocks. He stated that it was possible to estimate the spalling of underground cavities in hard rocks through the use of his extensional strain criterion. The extensional strain is defined as the deviation of the least principal strain from linear behaviour. This definition actually corresponds to the definition of initial yielding in the theory of plasticity. This initial yielding is generally observed, at the 40–60 percent of the deviatoric strength of materials. If this criterion is applied to circular tunnels under hydrostatic initial stress state, the radius R_o, at which the extensional strain is exceeded, can be shown to be

$$\frac{R_o}{a} = \left(\frac{\left[\frac{1+\upsilon}{E}(p_o - p_i) \right]}{\varepsilon_c} \right)^{\frac{1}{2}} \tag{12.11}$$

Since

$$\varepsilon_c = \upsilon\varepsilon_e \quad \text{and} \quad \sigma_c = E\varepsilon_e \tag{12.12}$$

One can easily obtain the following:

$$\frac{R_o}{a} = \left(\frac{1+\upsilon}{E} \left(\frac{1-\beta}{\alpha} \right) \right)^{\frac{1}{2}} \tag{12.13}$$

(d) Elastic-brittle plastic method

In elastic-brittle plastic method, the strength of rock is reduced from the peak strength to its residual value abruptly. If this concept is applied to circular tunnels under hydrostatic initial stress state, the radius R_p of plastic region can be obtained as follows:

$$\frac{R_p}{a} = \left\{ \frac{\frac{2-\alpha}{(1+q)} + \frac{\alpha^*}{q^*-1}}{\beta + \frac{\alpha^*}{q^*-1}} \right\}^{\frac{1}{(q^*-1)}} \tag{12.14}$$

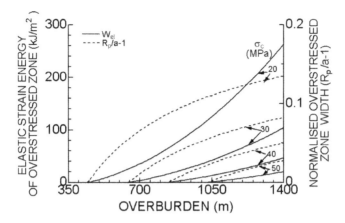

Figure 12.31 Relation between overburden and elastic energy and radius of overstressed zone.

The approach presented above can also be extended to situations, which involve complex excavation geometry and initial stress states. However, the use of numerical methods are necessary under such circumstances as described by the authors previously (Aydan et al., 1995, 2004).

(e) Comparisons

Figure 12.31 shows the relation between overburden and elastic energy and radius of overstressed zone different uniaxial compressive strength of rock. The potential of bursting is quite high if the strength of rock is low. The methods described in the previous sub-sections are compared with each other by considering a circular tunnel under hydrostatic initial stress state. The uniaxial strength of rock mass is assumed to be 20 MPa and the internal pressure was set to 0 MPa. The parameters required for analysis are obtained from empirical relations proposed by Aydan et al. (1993, 1996). In the computations, overburden is varied and the radius of plastic zone or overstressed zone is computed. Figure 12.32 compares the computed radius of plastic zone or overstressed zone. As expected from theoretical relation (12.13) of the extensional strain method, the overstressed zone must appear at shallower depths as compared with predictions of the other methods. The other three methods predict the yielding at the same depth. This difference is due to the value of yielding stress level associated with the extensional strain criterion. The radius of the plastic zone, estimated from the elastic-brittle plastic method, becomes quite large, and it even exceeds the one estimated from extensional strain method. The estimations from the method of Aydan et al. (2004) and the energy method are quite close to each other. They are also more reasonable as compared with estimations from other methods.

The method of Aydan et al. (2004) was applied to a tunneling project. The tunneling project is associated with an expressway construction and passes beneath high mountains in Central Japan. Rock mass properties for this tunnel, which is 10 km long and 12 m in diameter, were re-estimated from the empirical relations developed

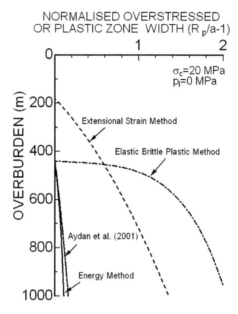

Figure 12.32 Variations of radius of plastic zone or overstressed zone with overburden, estimated from different methods.

for RMQR classification by Aydan *et al.* (2013). Figure 12.33 shows the variations of overburden RMQR and estimated level of bursting or squeezing and tunnel wall deformation along the tunnel alignment.

The final application is concerned with the comparison of the deformation behaviour of a circular tunnel in bursting and squeezing rock masses (Figure 12.34). In the computations, the value of the competency factor for both situations was chosen as 1. As expected from the behaviour of squeezing and bursting rocks shown in Figure 12.2, the tunnel wall strains become larger for tunnels in squeezing rock as compared with that in bursting rock mass. In addition, the radius of plastic zone in squeezing rock is larger than that in bursting rock.

12.5.7   Application of rockburst prediction methods to tunnels under non-hydrostatic stress condition

The analytical solutions for displacement, strain and stress field around cavities exhibiting non-linear behaviour under non-hydrostatic conditions are generally difficult to obtain. However, some analytical solutions were obtained by Kirsch (1889), for circular cavities, by Ingliss (1913) for elliptical cavities and Mindlin for circular cavities in gravitating media when the surrounding medium behaves elastically. Mushelivski (1953) devised a general method based on complex variable functions for arbitrary shape cavities.

Gerçek (1986, 1997) proposed a semi-numerical technique to obtain the integration constants stress functions based on Mushleviski's method. Galin (1955, see

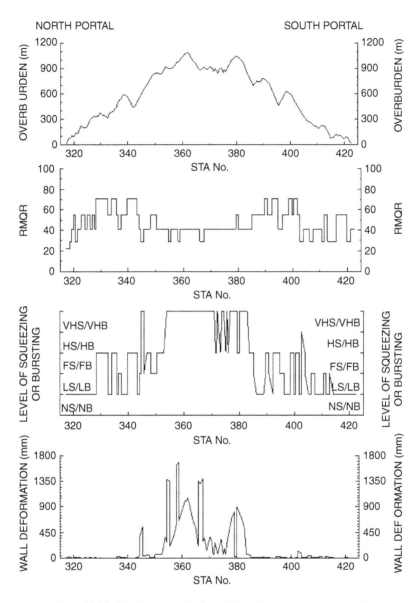

Figure 12.33 Predicted results for a 10 km long expressway tunnel.

Savin, 1961 for English version) was first to develop analytical solutions for circular holes in Tresca-type material under non-hydrostatic initial stress state. His solution was extended to Mohr-Coulomb materials by Detournay (1983). He further discussed problem of non-enveloping yield zone around the circular hole. Kastner (1962) proposed a method for estimating the approximate yield zone around circular cavities under non-hydrostatic stress condition using Kirsch's solutions. This method is also employed by Zoback *et al.* (1980) to estimate the shape of borehole breakouts, which

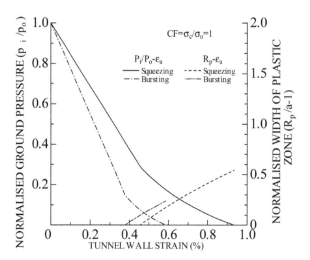

Figure 12.34 Comparison of computed tunnel wall strain and plastic zone of circular tunnels in squeezing rock and bursting rock.

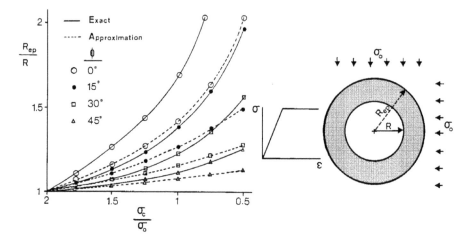

Figure 12.35 Comparison of approximately estimated and exact yield zone (from Aydan, 1987).

was used to infer the in-situ stress state. Gerçek (1993) and Geniş (2002) also used the same concept for arbitrary shaped cavities to estimate the extent of possible yield zone. Although this method estimates the extent of yield zone smaller than the actual one as shown by the first author (Aydan, 1987) for circular cavities under hydrostatic stress state as shown in Figure 12.35, it yielded the estimated shape of yield zone similar to that by exact solutions. Furthermore, it may also provide some rough guidelines for anticipated zone for reinforcement by rockbolts. Aydan and Geniş (2010) extended the same concept to estimate overstressed zones about cavities of arbitrary shape based on stress state computation method proposed by Gerçek (1993) and using

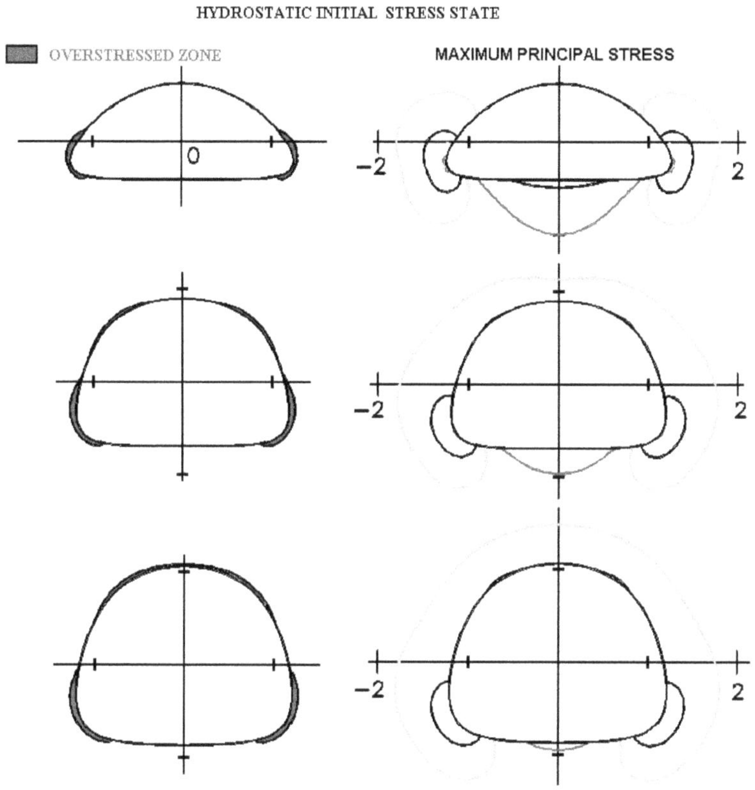

Figure 12.36 The overstressed zones around a tunnel subjected to hydrostatic initial stress state at different stages of excavations and the contours of maximum principal stress.

strain energy, distortion energy, extension strain and no-tension criteria in addition to Mohr-Coulomb yield criteria.

Figure 12.36 shows the overstressed zones around a tunnel subjected to hydrostatic initial stress state at different stages of excavations and the contours of maximum principal stress. The most critical stress state is during the excavation of top heading and the stress state becomes more uniform as the excavation approaches the circular shape. Furthermore, the extent of the tensile stress zone gradually decreases in size as the excavation progress.

An interesting yield zone developed around a circular opening excavated in a grano-dioritic hard rock at a level of 420 m in URL in Winnipeg, Canada as shown in Figure 12.14. Figure 12.37 shows the prediction of overstressed zone around the circular opening at URL. Except shearing and tension yielding model, all methods estimated the most-likely location of the yield zone. In addition to the estimations by the approximate approach, FEM analyses incorporating the yield criteria adopted in the approximate method were performed (Figure 12.38). Material properties used

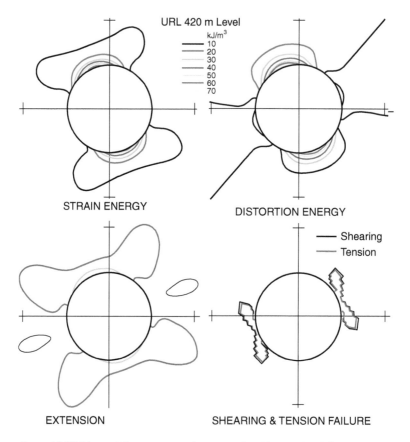

URL 420 m Level

kJ/m³
— 10
— 20
   30
— 40
   50
— 60
   70

STRAIN ENERGY

DISTORTION ENERGY

— Shearing
— Tension

EXTENSION

SHEARING & TENSION FAILURE

Figure 12.37 View of the opening and estimated yield zones by different methods.

are given in Table 12.5. As noted from the figure, it seems that the distortion energy concept yields results close to the observations as shown in Figure 12.14.

12.6 MONITORING OF ROCKBURST

The motivation was actually associated with the possibility of rockburst phenomenon at the Hida Tunnel, which is the longest roadway tunnel in Japan, under an overburden greater than 1000 m (Aydan *et al.*, 2005, 2016). The Third Shizuoka Tunnel of the Second (recently renamed as New) Tomei Expressway was selected as a preliminary application site for the multi-parameter real-time monitoring system (Aydan *et al.*, 2005, 2016; Imazu *et al.*, 2013, 2014). This system was expanded to include the infrared monitoring system at the Tarutoge Tunnel the connection expressway between the Chuo Expressway and Tomei Expressways. In this section, the application of the monitoring systems to Tarutoge Tunnel is explained.

Taru-Toge tunnel is being constructed as a part of the expressway project connecting Shin-Tomei Expressway and Chuo Expressway at the boundary of Shizuoka

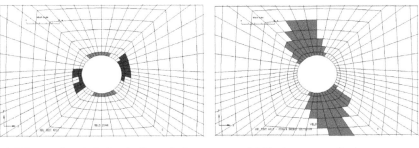

(a) Conventional elasto-plastic analysis   (b) Strain energy criterion

(Blue: tension; red: compression)

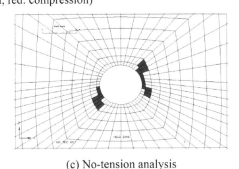

(c) No-tension analysis

Figure 12.38 Yield zones estimated from non-linear FEM analyses.

Table 12.5 Material properties used in FEM analyses.

Elastic Modulus (GPa)	Poisson's ratio ν	Drucker-Prager friction coefficient (α)	Drucker-Prager strength (k)
3.00E+01	2.00E−01	2.71E−01	1.64E+01

Unit weight ρ (gf/cm^3)	Friction angle ϕ (deg)	Cohesion (MPa)	Tensile strength σ_t (MPa)
2.63E+00	4.50E+01	1.50E+01	3.00E+01

and Yamanashi Prefectures in the Central Japan (Figure 12.39(a)). The tunnel passes through a series of mudstone, sandstone and conglomerate layers with folding axes aligned north-south (Figure 12.39(b)). Furthermore it is bounded by Fujikawa Fault in the east and Itoikawa-Shizuoka Tectonic Line. Mt. Fuji is to the east of the tunnel. Figure 12.40 shows the local geology of the southern part of the tunnel where blasting operations are carried. In the section, six major faults (F1–F6) exist.

This multi-parameter real-time monitoring system was used for assessing the real-time stability of the Taru-Toge tunnel (Imazu *et al.*, 2015; Aydan *et al.*, 2016), which is still under construction. The infrared monitoring system was used twice during the construction of the tunnel. The multi-parameter measurement system was basically

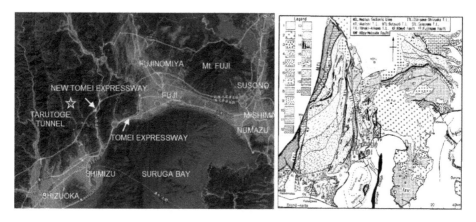

(a) Location of Taru-Toge Tunnel   (b) Regional Geology

Figure 12.39 Location of Taru-Toge tunnel and regional geology.

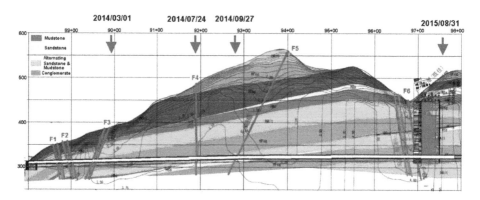

Figure 12.40 Local geology of at the monitoring locations of Taru-Toge tunnel (south portal side).

similar to that used in the Third Shizuoka Tunnel and it involved electric potential (EP) variations, acoustic emissions (AE), temperature (T), humidity (H) and air pressure (P) of the tunnel in addition to the measurements of convergence and loads on support members.

The electric potential measurements were done using a specially developed high impedance electric potential devices. Although the devices can measure electric potential variations of DC type with a sampling rate of 10 Hz, they were set to a sampling interval of 120 seconds. Acoustic emissions (AE) were measured using a compact acoustic emission system, which consists of AE sensor, amplifier and pulse-counting logger (Tano *et al.*, 2005). The system operates using batteries. TR-73Ui produced by TANDD is used to measure temperature, humidity and air pressure and it is denoted as THP device. The device can also measure the blasting induced air-pressure variations.

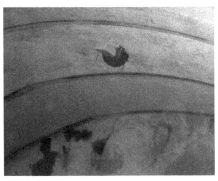 

(a) Instruments installed at the crown   (b) Instruments installed at the sidewall

Figure 12.41 Views of the instruments installed at the crown and sidewall of the tunnel.

The protection of instrumentation devices in the very close vicinity of tunnel face during blasting operation was an extremely difficult issue. In the previous studies, the devices were installed in larger holes at the sidewall and covered by some protection sheaths in the Third Shizuoka Tunnel (Aydan *et al.*, 2005). When measurements were carried out at the very close vicinity of the tunnel face in the Tarutoge tunnel, the instruments were put in an aluminium box attached to rockbolt head and protected by a semi-circular steel cover fixed to the tunnel surface by bolts. Fig. 12.41 illustrates the installation of the instruments in the close vicinity of the tunnel face. Multi-parameter measurements were taken in July and September 2014 and August-September, 2015. These measurements are presented herein.

12.6.1 Multi-parameter monitoring results during July 20–26, 2014

Two locations were selected for multi-parameter response measurements. Because of some malfunctioning of some devices at one of the monitoring stations, we present the results of the station where all devices worked properly. It should be noted that the devices were set at locations as close as possible to the tunnel face. The variations of air-pressure, temperature, electric potential and acoustic emissions are shown in Figures 12.42 and 12.43 as a function of blasting operations. It is interesting to note that rapid air-pressure variations are directly related to blasting events. The temperature of the main tunnel (Honkou) increased soon after each blasting event, and it decreased as a function of ventilation.

Acoustic emissions occurred during each blasting operation and ceased soon after the blasting, which implied that the surrounding rock mass was stable following the stress re-adjustment. The acoustic emission count was higher when the blasting event was carried out at the main tunnel while it was lower when the blasting was carried out at evacuation tunnel (Hinankou) due to distance.

The electrical potential variations occurred mainly soon after the blasting operation at the main tunnel. The blasting operations at the evacuation tunnel had less

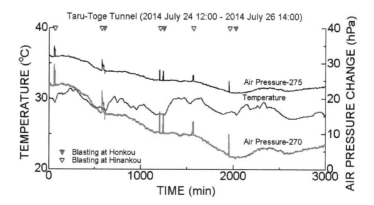

Figure 12.42 Air pressure and temperature variations at the main tunnel.

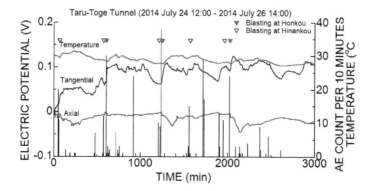

Figure 12.43 Electrical potential and acoustic emission variations at the main tunnel.

influence on the electric potential variations than those on acoustic emissions. It was also interesting to note that the variations disappeared after a certain period implying the tunnel behaviour was stable. The electrical potential in tangential direction increased while the electric potential decreased along the axial direction following the blasting operation. These results were quite consistent with the previous observations by Aydan et al. (2005) carried out on the third Shizuoka Tunnel as explained in Chapter 11.5.

12.6.2   Multi-parameter monitoring results during September 20–26, 2014

Two locations were, again, selected for multi-parameter responses (Aydan et al., 2015). Station 1 and Station 2 are approximately 45 and 25 m away from the face, respectively. The system consisted of AE, EP and climate monitoring devices. The major problem with measurements was how to protect devices during blasting. Figure 12.44 shows

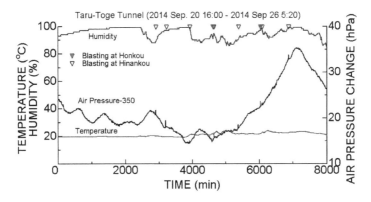

Figure 12.44 Air pressure, humidity and temperature variations at the main tunnel.

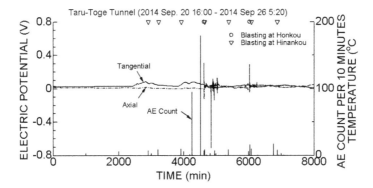

Figure 12.45 Electrical potential and acoustic emission variations at Station 1 (St1) in the main tunnel.

the variations of climatic temperature, humidity and air pressure in the main tunnel during September 20–26, 2014. The blasting times at the main tunnel (Honkou) and evacuation tunnel (Hinankou) are shown in the figure and they are closely associated with rapid air pressure changes. Figure 12.45 shows the variations of acoustic emissions and electric potential variations at Station 1 (St1). The tunnel was approaching a major fracture zone and the seepage from the face was increasing during the blasting operations. The acoustic emission counts were higher after each blasting operation although the counts decreased after a certain period of time.

The electric potential variations showed very high-spike-like responses during blasting and disappeared after a certain period time, implying the tunnel response is stable. Nevertheless, the high variations may be associated with sudden increase of pore-pressure and seepage conditions in the vicinity of the tunnel face. The sudden fluctuations of seepage conditions induced electric potential variations as noted experimentally by Mizutani et al. (1976) and Aydan and Daido (2002).

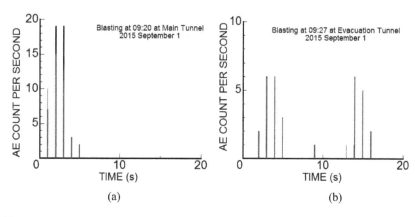

Figure 12.46 Acoustic emissions responses (a) blasting at 9:20 in the main tunnel; (b) blasting at 9:27 in the evacuation tunnel.

12.6.3   Acoustic emission responses at the tunnel face

Aydan *et al.* (2016) attempted to measure acoustic emissions and vibrations in the immediate vicinity of the main tunnel during August 31 and September 1, 2015. The blasting induced vibrations terminate within 2–2.5 s after each blasting as noted in Chapter 10. Figure 12.46 shows the acoustic emissions counts obtained from the acoustic emission devices with a sampling rate of 1s in this particular monitoring. The blasting with about 150 kgf explosives was carried out at 9:20 in the main tunnel. As noted from Figure 12.46(a) AE counts terminate within 5 seconds. The ventilation system was started after about 60–70 seconds in order to eliminate noises from the operation of construction and ventilation equipment. The blasting at evacuation tunnel with a similar amount of explosives was also carried out about 7 minutes after that in the main tunnel. The response was quite similar as seen in Figure 12.46(b). Nevertheless, some further acoustic emission events occurred in the main tunnel. When the tunnel was stable, the acoustic emission counts decreased soon after the excavation. This is in accordance with previous observations by Aydan *et al.* (2005), Imazu *et al.* (2014) as well as those presented in the previous section.

12.6.4   Infrared monitoring system

As an application of the infrared thermography technique to real underground excavations, Aydan *et al.* (2016) made some observations at Taru-Toge Tunnel, from which rock blocks were obtained and presented in the previous section. An actual tunnel excavation using drilling and blasting technique involves the blasting, mucking, shotcreting, installation of rockbolts and steel ribs and drilling of holes for the next blasting round. Forced ventilation is imposed soon after the blasting operation to clear the dust and cloud from the blasted tunnel face. Figure 12.47 and Figure 12.48 show naked eye and infrared thermography images of the evacuation tunnel just before and soon after the blasting operation. An interesting observation is that the infrared thermography image clearly illustrates the tunnel face condition while the naked-eye image is quite blurred.

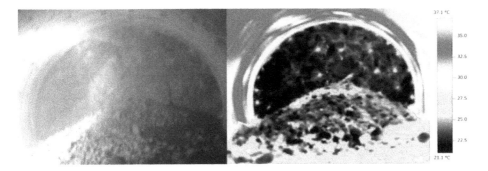

Figure 12.47 Visual and infrared images of the evacuation tunnel soon after blasting.

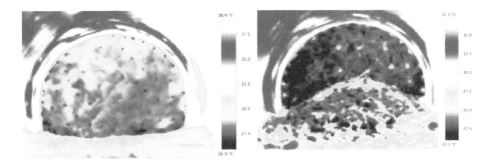

Figure 12.48 Infrared images of the evacuation tunnel before and after blasting.

This is a very important feature of infrared thermography imaging technique, which is not affected by the dust cloud soon after the blasting. The remains of blasted holes are clearly observed in infrared images as hot spots (Figure 12.47–12.48). The cooler areas are likely to correspond to de-stressed zones at the tunnel face and the ground-water seepage locations. In addition, high temperature zone next to the blasted tunnel face is the shotcrete layer, which was undergoing hydration process. This observation may also be of great value to assess the quality of shotcrete and its hardening process (Figure 12.48).

12.7 COUNTERMEASURES AGAINST ROCKBURST

12.7.1 Allowing rockburst to occur

The simplest approach would be to allow rockburst to occur. The practical experience together with some computational findings indicate that the rockburst is likely to take place within a distance of 1D from the tunnel face as explained in Sections 12.2 and 12.3. If the blasting technique is employed, there should be some time interval between blasting and the start of works at the tunnel face. When the vertical stress

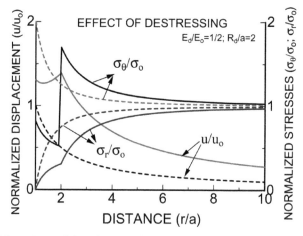

(a) deformation modulus of distressed zone is the half of the original rock mass

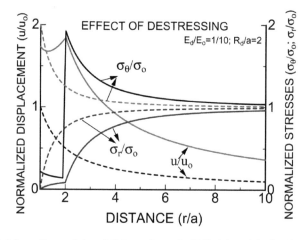

(b) deformation modulus of distressed zone is 1/10f of the original rock mass

Figure 12.49 Computational examples on the effect of de-stressing on stress re-distribution.

is the dominant principal field stress, the most critical situation appears during the excavation of upper benches if a bench type excavation procedure is utilized. However, if the horizontal stress is the dominant field stress, then the crown would be a quite critical location when the excavation height increases. In the application of this technique, it is required to monitor the micro-seismicity and other variations nearby the opening as described in monitoring section.

12.7.2   De-stressing or pre-conditioning

The mining engineering community developed this technique to deal with rockburst problem (i.e. Blake, 1972; Mitri *et al.*, 1988; Toper *et al.*, 2000). The fundamental

Figure 12.50 Post-failure views of dry initially sheared Fuji clay samples unwrapped and wrapped with rubber strings.

concept is to create a broken zone with low deformation modulus and certain load bearing capacity in surrounding rock around the tunnel perimeter through blasting. However, the development of some practical and theoretical techniques for quantitative design of such de-stressed zones is necessary. If a TBM type excavation technique is employed, this operation may become quite cumbersome. Figure 12.49 shows an example of computation to evaluate the effect of de-stressing on the stress re-distribution for an underground opening subjected to hydrostatic stress state with reduced deformation modulus (1/2 and 1/10). These two examples also clearly show why rockburst risk is reduced theoretically.

12.7.3 Flexible and deformable support

It is experimentally observed that wrapping samples with highly deformable rubber-like strings in laboratory tests greatly reduced the violent detachment of fragments as seen in Figure 12.50. Although such materials could not delay or increase the overall confinement, they act as dampers to reduce the velocity and acceleration of detaching rock fragments, which may be one of very important observations in dealing with rockburst problem in underground excavations. The actual implementation of flexible and deformable support may be achieved through ductile rockbolts (such as swellex, SA yielding rockbolts, cable rockbolts) together with wire mesh. Furthermore, the application of steel-fiber reinforced shotcrete could be effective. Nevertheless, such shotcreting operation without rockbolts cannot be sufficient and efficient. The actual examples described in Section 12.2 clearly indicate that ductility of the support system is the key concept in dealing with rockburst. However, the most difficult aspect in this procedure is actual implementation of this support system for protection of workers against any unexpected detachment of rock fragments. The real-time monitoring would be quite useful to incorporate during the implementation of this technique.

12.8 CONCLUSIONS

In this chapter, the state of art on rockburst problems in underground excavations is presented. Although the problem has been well known for a long time and many techniques have been developed for this problem, there is still some room for further developments on the cycle of prediction-monitoring-preventing/controlling of rockburst. The estimations of rockburst potential by various available theoretical and empirical methods are reasonable. However, it should be noted that the rockburst is governed by the strength of intact rock and the concept of rock mass strength should not be employed. It seems that relations proposed by Aydan *et al.* (1993, 1996) previously can also be used for rocks prone to bursting with some confidence. The energy method yields similar results to those estimated from the proposed method. Furthermore, the dynamic aspect of the problem needs further study. Especially the estimation of ejection velocity of rock fragments is quite important, as it is closely associated with the safety of workers during excavations.

The multi-parameter monitoring systems involving AE, geo-electro-magnetic signals may be quite useful. Some specific conclusions are:

1   The real-time stability of the tunnel face may be assessed using AE counting system and geo-electric potential (GEP) system during the excavation.
2   When the tunnel face is stable, both AE and GEP activities nearby the tunnel face cease within a short period of time.
3   The AE and GEP activities after initial phase may be indicators that the rock mass surrounding the tunnels is stable.
4   The AE and GEP activities have a maximum value when the tunnel face distance is about 0.5D.
5   De-stressing and the use of flexible support system could be quite useful in rock masses prone to rockburst.
6   TBM is not a suitable excavation method in rock masses prone to rockbursts.

Chapter 13

Dynamics of rockbolts and rock anchors and their non-destructive testing

Rockbolts and rock anchors are commonly used as principal support members in underground and surface excavations. They are generally made of steel bar or cables, which are resistant against corrosion. These support members may be subjected to earthquake loading, vibrations induced by turbines, vehicle traffic and long-term corrosion. Figure 13.1 shows the ruptured rock anchors used at a rock slope by the 2008 Iwata-Miyagi earthquake. The acceleration response of underground cavern has been already presented in Chapter 8 in relation to the 2009 L'Aquila earthquake. Figure 13.2 shows the state of rock anchors and rockbolts at an underground power house. This chapter presents some theoretical, numerical and experimental studies on rockbolts and rock anchors under shaking and their non-destructive testing for soundness evaluation.

13.1 TURBINE-INDUCED VIBRATIONS IN AN UNDERGROUND POWER HOUSE

Aydan *et al.* (2008, 2012) performed some investigations and measurements at a pumped-storage scheme consisting of two reservoirs and an underground powerhouse and was constructed about 40 years ago. The underground powerhouse is 55 m long, 22 m wide and 39 m high and it has two turbines. The maximum water level variation may reach 45 m at the full capacity. Vibrations induced by turbines in an underground

(a) Overall view   (b) Close-up view

Figure 13.1 Rock anchors ruptured by the 2008 Iwate-Miyagi earthquake.

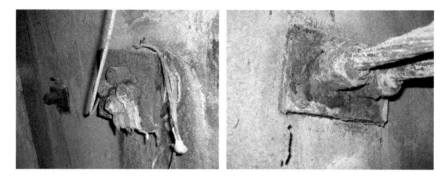

Figure 13.2 The state of rock anchors and rockbolts at an underground power house.

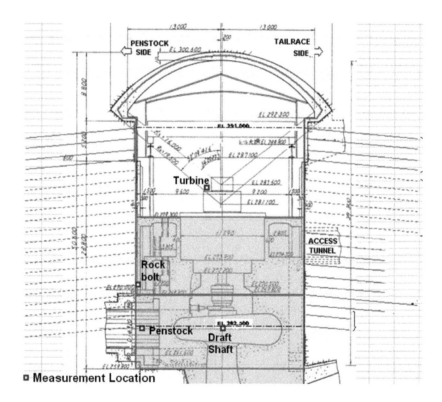

Figure 13.3 Locations of vibration measurements.

power house were measured at several locations in relation to the assessment of the behaviour of rockbolts and rock anchors during operation. The measurements were carried out at penstock, draft shaft, turbine top, a rockbolt above the penstock (Figures 13.3 and 13.4). Figure 13.5–13.8 shows the records of measured vibrations at the draft-shaft, turbine top and penstock during operations. The maximum accelerations range between 120–250 gals. It should be noted that these vibrations are transmitted

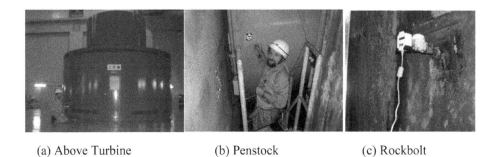

(a) Above Turbine   (b) Penstock   (c) Rockbolt

Figure 13.4 Views of locations of vibration measurements.

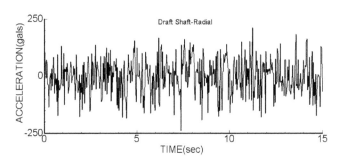

Figure 13.5 Radial acceleration records at the drat-shaft.

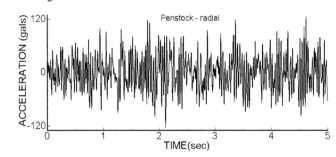

Figure 13.6 Radial acceleration records at the penstock.

to the surrounding rock mass. Furthermore, the vibrations are higher when the water flow is stopped or started in relation to the operation of the powerhouse turbines of this pumped-storage scheme.

13.2 DYNAMIC BEHAVIOUR OF ROCKBOLTS AND ROCK ANCHORS SUBJECTED TO SHAKING

Owada and Aydan (2005) and Owada *et al.* (2004) carried out some model experiments on the development of axial forces in rock anchors and grouted rockbolts stabilizing

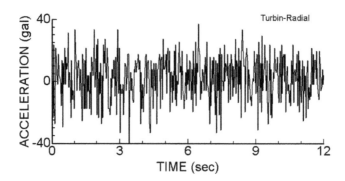

Figure 13.7 Radial acceleration records at the turbine.

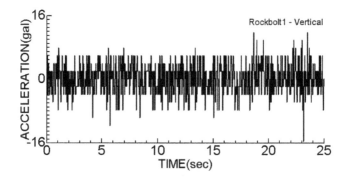

Figure 13.8 Radial acceleration records at the rockbolt above the penstock.

the potentially unstable blocks in sidewalls of the underground openings using shaking tables by considering the situation illustrated in Figure 13.9. The model rockbolts tested in the tests were rock anchors and fully grouted rockbolts. The experimental results are described in the following two sub-sections.

13.2.1   Model tests on rock anchors restraining potentially unstable rock-block

The physical model and instrumentation used in experiments are illustrated in Figure 13.10. Two laser displacement sensors produced by KEYENCE (LB-01) together with a data-acquisition and logger system were used to measure the displacement of the model and also the motion of the shaking table. The displacement of shaking table was derived to obtain the base acceleration. If the sampling interval is 10 ms, the acceleration obtained by derivation of displacement response is almost the same as the acceleration response measured directly using the accelerometers (AR-10TF) produced by TOKYO-SOKKI. In addition to displacement response measurements, acoustic emissions were also measured by AE-Tester (Human Data Co. & NF Sensor) nearby the potential sliding plane. A small shaking table, which is capable of imposing acceleration up

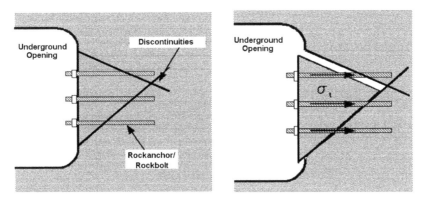

Figure 13.9 Possible situation of potentially unstable rock blocks at side walls of underground openings.

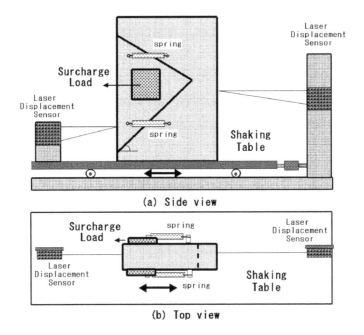

Figure 13.10 The experimental set-up for model tests.

to 650 gal, is used. The frequency of imposed acceleration waves can be adjusted as desired.

Blocks were used for model tests. The block was 274 mm high, 137 mm wide and 37 mm thick. A plane of discontinuity with an inclination of 58° was introduced into the block. The friction angle of the saw-cut surface was measured by tilting test and its value ranged between 24–26°. Two springs with a length of 30 mm as the model rock anchors and rock anchors were fixed at both sides of the block so that the springs cross the potential sliding surface as shown in Figure 13.10.

The displacement-load relation of springs was obtained by an additional loading test. The following relation between displacement (mm) and load (gf) of the spring was established:

$$T = 57\delta^{0.5} \qquad (13.1)$$

The total weight of the potentially unstable block was 242.4 gf and the block was unstable without springs due to the high inclination angle of the sliding plane. From the limit equilibrium analyses, the horizontal pull-out force (T) at the time of sliding under the dead weight and surcharge load can be obtained as:

$$\frac{T}{W_t} = \tan(\alpha \pm \phi) \qquad (13.2)$$

where α and ϕ are inclination and friction angle of discontinuity plane, respectively. Plus sign is for upward motion and minus sign is for downward motion. Eq. (13.2) implies that to move the block upward, the force will be almost 10 times the load for downward motion. In other words, the base acceleration, which may cause the downward sliding of the block, may be incapable of causing the upward sliding and the block will remain stationary under such loading situations. If the deformation of the sliding block itself is negligible, the axial force on the spring may be estimated from the rigid motion of the block. The extension of the spring as a function of the block motion can be shown to be

$$\delta_s = \ell - \ell_o = \sqrt{(\ell_o + \delta_h)^2 + \delta_v^2} - \ell_o \qquad (13.3)$$

where ℓ_o, δ_h and δ_v are the initial length of springs and horizontal and vertical movement of the sliding block. The force in the spring can then be obtained by inserting the spring extension value from Eq. (13.3) into Eq. (13.1).

Four shaking experiments were carried out on the models described above. The responses measured in an experiment numbered 4 are shown in Figure 13.11. The maximum base acceleration was 590 gals. As noted from the figure, the anchor force increases after each slip event in a step-like fashion and becomes asymptotic to a certain value thereafter. Furthermore, the block does not return to its original position. This implies that the axial force becomes higher than the applied pre-stress level following each slip event. It was interesting to note that the axial force on the rock anchor at the end of shaking was more than twice that of initial stage. Finally, the bolt-force becomes sufficient to stop the sliding of the block, if it does not fail. The experiments clearly demonstrate that the bolts and anchors used as a part of the support system may experience greater load levels than that at the time of their installation after each passage of dynamic loads. This outcome has very important implications that support systems may fail during their service lives not due to their deterioration but also dynamic loads resulting from different causes as mentioned in the introduction.

Some additional model experiments were carried out on rock blocks in the roof, supported by rock anchors using the same shaking table together with a similar instrumentation and modeling of rock anchors. Although the experimental results are not shown here, it is observed that the shaking induces dynamic forces in the rock anchors

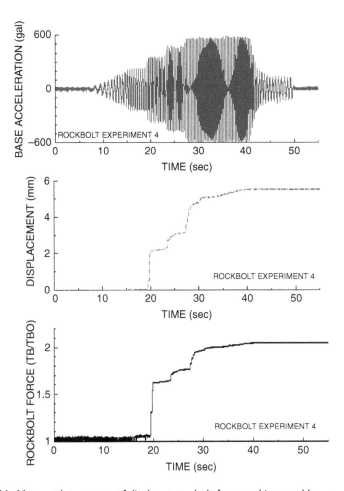

Figure 13.11 Measured responses of displacement, bolt-force and imposed base acceleration.

and the suspended rock block returns to its original position so that no residual forces occur in the rock anchors. This is a quite important conclusion for the evaluation of behaviour of rockbolts and rock anchors in the roof of openings under dynamic loads.

13.2.2 Model tests on fully grouted rockbolts restraining a potentially unstable rock-block against sliding

Owada and Aydan (2005) and Owada *et al.* (2004) also carried out some model experiments on the development of axial forces grouted rockbolts stabilizing the potentially unstable blocks in sidewalls of the underground openings or rock slopes subjected to planar sliding as shown in Figure 13.12 using shaking tables. An acrylic bolt equipped with three strain gauges shown in Figure 13.13 was used and the inclination of the bolt was varied as 45, 90 and 135 degrees with respect to sliding plane, which is inclined

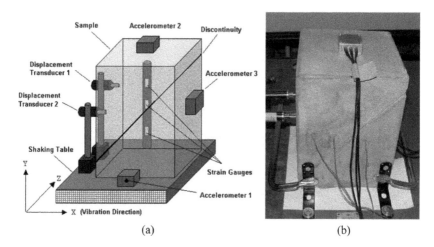

(a) (b)

Figure 13.12 An illustration of experimental set-up (a) and its view (b).

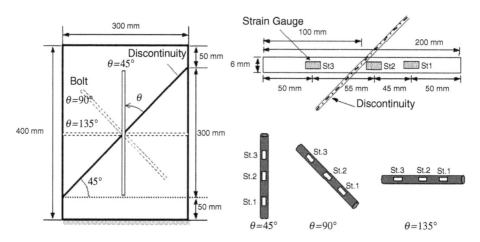

Figure 13.13 A drawing of an instrumented sample and position of strain gauges.

at an angle of 45 degrees to the horizontal. Strain gauge numbered 2 (St2) is set next to the discontinuity plane. The dynamic response of the acrylic bolt under impulsive gravity loading has been already shown in Figure 10.2. The samples with different orientation of model rockbolt were also tested under static loading condition and the experimental results are shown in Figure 13.14 and Figure 13.15.

Figure 13.16 to Figure 13.19 show the applied base acceleration and strains in the bolt and displacement of the potentially unstable block. As noted from the figures, some residual strains occur in the bolts following the termination of shaking. Similarly, the block is displaced permanently. Strain at Gauge 2 is always largest as expected. Although the measured strain levels are small as compared with their yield strain level, the permanent straining results from the permanent displacement of the potentially unstable block.

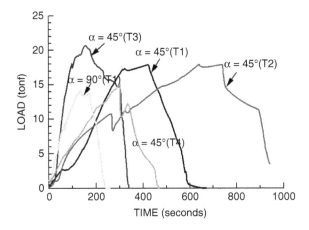

Figure 13.14 The load response of samples with different bolt orientations.

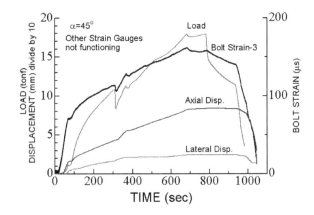

Figure 13.15 Load, strain and displacement responses of a sample with 45 degrees rockbolt.

As shown in Figure 13.19, the amplitude of acceleration is increased step wise up to 450 gals. When the acceleration was less than 200 gals, the straining of the bolt and displacement of the potentially unstable block was small. However, strains and permanent displacement become larger after each acceleration level increment.

13.2.3   A theoretical approach for evaluating axial forces in rock anchors subjected to shaking and its applications to model tests

A theoretical approach was developed by Owada and Aydan (2004) for evaluating the axial forces in rock anchors in their experiments presented in the previous subsections. This approach is described in this subsection and applied to experiments.

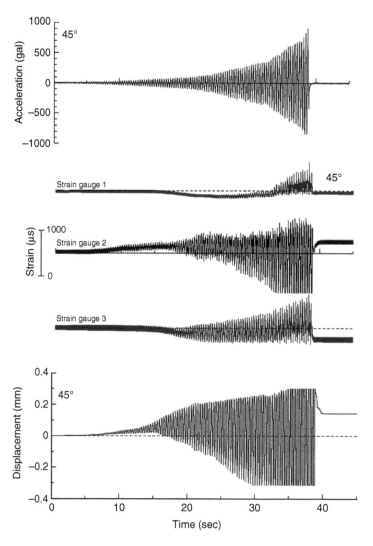

Figure 13.16 The responses of bolt strains and displacement of unstable block in relation to applied base acceleration (bolt angle 45°).

The following limiting equilibrium equations of the potentially unstable block for s and n directions can be written as follows (Figure 13.20):

$$\sum F_s = W_t \sin \alpha + E \cos \alpha - T \cos(\alpha - \beta) - S = m \frac{d^2 s}{dt^2} \qquad (13.4a)$$

$$\sum F_n = W_t \cos \alpha + E \sin \alpha - T \sin(\alpha - \beta) - N = m \frac{d^2 n}{dt^2} \qquad (13.4b)$$

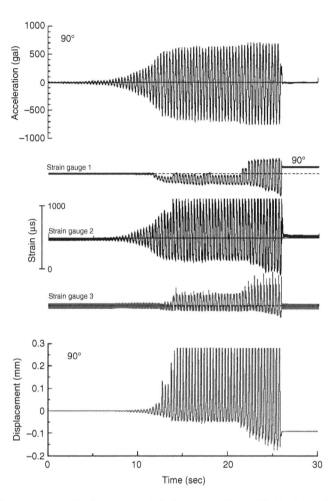

Figure 13.17 The responses of bolt strains and displacement of unstable block in relation to applied base acceleration (bolt angle 90°).

Let us assume that the inertia force for *n*-direction during sliding is negligible and the resistance of the failure plane is purely frictional as given below

$$\left|\frac{S}{N}\right| = \tan(\phi)$$ (13.5)

One can easily obtain the following equation for the rigid body motion of the sliding rockmass body

$$m\frac{d^2s}{dt^2} = A$$ (13.6)

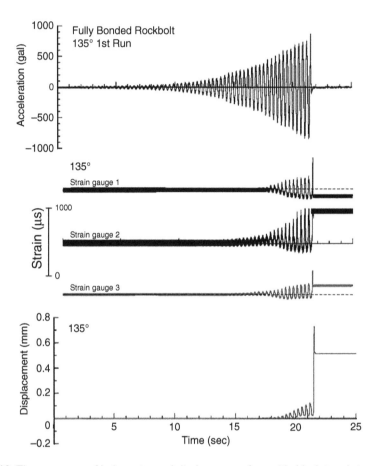

Figure 13.18 The responses of bolt strains and displacement of unstable block in relation to applied base acceleration (bolt angle 45°).

where $A = W_t(\sin \alpha - \cos \alpha \tan \phi) + E(\cos \alpha + \sin \alpha \tan \phi) - T(\cos(\alpha - \beta) + \sin(\alpha - \beta) \cdot \tan \phi)$.

Since the shaking force E will be proportional to the mass of the sliding body, it can be related to ground shaking in the following form:

$$E = \frac{a_g(t)}{g} W_t \qquad (13.7)$$

where a_g and g are ground acceleration and gravitational acceleration, respectively.

The theoretical approach is applied to model tests shown in Figure 13.11 by selecting that the friction angle is 26° on the basis of friction tests using tilting machine. The computed results are shown in Figure 13.21. As noted from the figure, the computed results are quite similar to experimental results both quantitatively and qualitatively. However, the computations indicate that the yielding should start earlier than the measured results. The discrepancy may result from the complexity of actual frictional

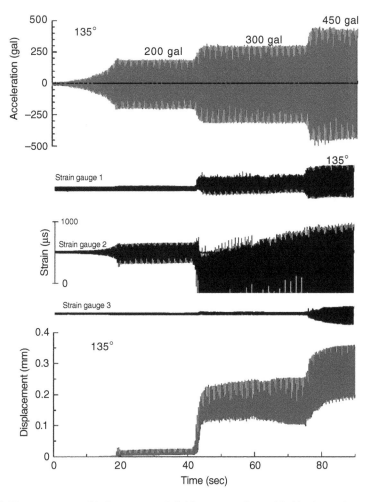

Figure 13.19 The responses of bolt strains and displacement of unstable block in relation to applied base acceleration (bolt angle 135°).

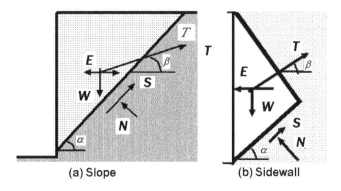

Figure 13.20 Mechanical model.

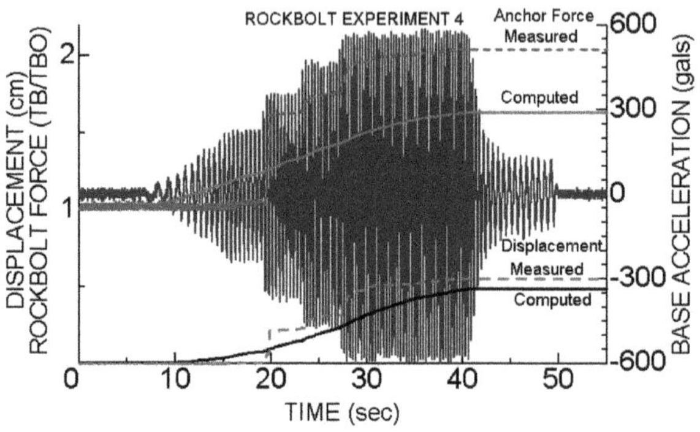

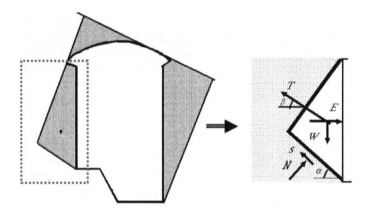

Figure 13.21 Comparison of computed responses with measured responses.

Figure 13.22 Identified unstable blocks and force equilibrium conditions.

behaviour of sliding surface. Nevertheless, the theoretical model is capable of modeling the dynamic response of the support system.

13.2.4 Application of the theoretical approach to rock anchors of an underground power house subjected to turbine-induced shaking

Rock mass generally contains geological discontinuities and these discontinuities play a major role on the local instabilities in underground openings (i.e. Aydan, 1989; Kawamoto *et al.*, 1991). The potentially unstable blocks were identified around an underground cavern shown in Figure 13.3 and Figure 13.4 on the basis of in-situ investigations as well as geological investigations during the construction phase as shown in Figure 13.22 (Aydan, 2016; Aydan *et al.*, 2012).

The axial force T of rock anchors is also computed from a relation similar to Eq. 13.1, given below:

$$T = C\delta^b \tag{13.8}$$

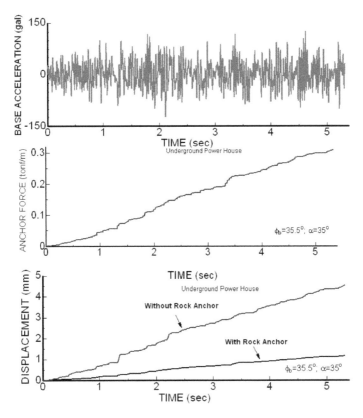

Figure 13.23 Computed displacement and axial force responses.

The displacement of rock anchors induced by the relative displacement during each relative motion due to sliding was computed from geometrical considerations. It should be also noted that there is no sliding if the rock anchor force attains a certain level. Furthermore, the oscillations of anchor force due to visco-elastic response during each increment of axial force were neglected in computations.

Figure 13.23 shows the computational results for the block shown in Figure 13.22 for two different situations using the induced radial vibration record at the penstock. In the first case, no rock anchors were considered while rock anchors were assumed to be installed in the second case. The inclination of discontinuity was set to 35° and its friction coefficient was determined from tilting tests as 35.5°. When no rock anchors are installed, the block tends to slide downward when vibrations induced are sufficient to cause sliding. However, the relative sliding movements of the block are restricted and the amount of sliding becomes less and the anchor force tends to become asymptotic to a certain level for the given amplitude of vibrations. This computational result also implied that the axial forces may increase during their service life when forces resulting from vibrations in the vicinity of cavern and/or earthquakes from time to time. These increases may also lead to the rupture of rock anchors in long-term besides the reduction of cross sections of rock anchors due to corrosion.

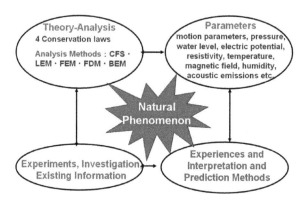

Figure 13.24 The main concept for developing non-destructive assessment of soundness of rockbolts and rock anchors.

13.3 NON-DESTRUCTIVE TESTING FOR SOUNDNESS EVALUATION

The geo-environmental conditions imposed on rockbolts and rock anchors may be very adverse and they may cause some corrosion in a short period of time. While rockbolts are generally made of steel bars, rock anchors may be either high-strength steel bars or cables consisting of several wires. As the surface area of cable rock anchors is much larger they are much more prone to corrosion as compared with steel bars.

The maintenance of existing rock engineering structures such as underground pow-erhouses, dams and slopes requires the re-assessment of support members. Although in-situ pullout tests and lift-off tests would be the best option to assess the soundness and acting stress of the rockbolts and rock anchors, the cost is quite high and it is labour-intensive to perform such tests. Furthermore, they may require re-installation of rockbolts and rock anchors if they are pulled-out or broken during the tests. There-fore, the use of non-destructive tests to assess their soundness and/or acting stresses is desirable with a main concept as illustrated in Figure 13.24. Despite there having been several attempts to assess the state of support members, it is very difficult to say that such techniques are satisfactory for practical purposes. Furthermore, the struc-tural part of rockbolts and rock anchors used during the non-destructive tests is quite limited as compared with the overall size of the support system.

The soundness investigation involves the state of corrosion of steel bars, cracks in tendon and grouting material, debonding of interfaces and acting stresses. There have been several attempts to develop such systems for non-destructive evaluations of the soundness of rock anchors and rockbolts (i.e. Tannant *et al.*, 1995; Aydan *et al.*, 2005, 2008; Ivanovic and Neilson, 2008; Beard and Lowe, 2003). Due to very limited exposure of rockbolts and rock anchors embedded in rock masses, it is necessary to utilize the numerical techniques to capture the fundamental features of the expected responses, which may be used for the interpretation of measured responses from in-situ non-destructive tests. With this in mind, some theoretical models are developed for the axial and traverse dynamic tests and then numerical simulations are carried out based on these fundamental equations by considering possible conditions in-situ. Specifically, the bonding quality, the existence of corroded parts or couplers and pre-stress are simulated through numerical experiments. The results of the numerical simulations can be used for the interpretation of the dynamic responses to be measured in non-destructive dynamic tests in-situ. Several applications of this procedure to practice are given together with measurements obtained from some non-destructive tests under laboratory and in-situ conditions.

13.3.1 Impact waves for non-destructive testing of rockbolts and rock anchors

Aydan (Aydan, 1989; Aydan et al., 1985, 1986, 1987, 1988; Kawamoto et al., 1994) studied the static response of rockbolts and rock anchors and developed some theoretical and numerical models. However, the dynamic responses to be used in non-destructive tests require dynamic equilibrium equations for axial and traverse responses and their numerical representations and solutions. These dynamic equilibrium equations and their numerical representations are given in next sub-sections.

13.3.1.1 Mechanical models

By modification of the static equilibrium equations developed for rockbolts and rock anchors by Aydan (1989), the equation of motion for the axial responses of rockbolts and rock anchors together with the consideration of inertia component including mass proportional damping can be written in the following form (Figure 13.25)

$$\rho \frac{\partial^2 u_b}{\partial t^2} + h_a^* \frac{\partial u_b}{\partial t} = \frac{\partial \sigma}{\partial x} + \frac{2}{r_b} \tau_b \tag{13.9}$$

where ρ, u_b, σ, r_b, h_a^* and τ_b are density, axial displacement, axial stress and radius of tendon, axial damping ratio and shear stress along the tendon-grout interface, respectively. Constitutive relations for axial stress and shear stress may be given in the following form, respectively:

$$\sigma = E_b \varepsilon + \eta \dot{\varepsilon}; \quad \varepsilon = \frac{\partial u_b}{\partial x}; \quad \dot{\varepsilon} = \frac{\partial \varepsilon}{\partial t} = \frac{\partial}{\partial t}\left(\frac{\partial u_b}{\partial x}\right) \tag{13.10a}$$

$$\tau_b = K_g u_b \tag{13.10b}$$

The effect of surrounding grouting annulus and rock mass is taken into account through shear stiffness K_g, which was originally derived by Aydan et al. (1985, 1986, 1987, 1988) and Aydan (1989), and it is specifically given as

$$K_g = \frac{G_g}{r_b \ln(r_b/r_b)} \cdot \left(\frac{X-1}{X}\right) \quad \text{and} \quad X = \frac{G_r}{G_G} \frac{\ln(r_b/r_b)}{\ln(r_0/r_b)} + 1 \tag{13.11}$$

where G_g, G_r, r_b and r_o are shear modulus of grouting material and surrounding rock, radius of hole and radius of influence.

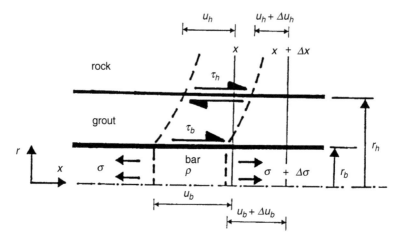

Figure 13.25 Modelling of Dynamic Axial Response of tendons.

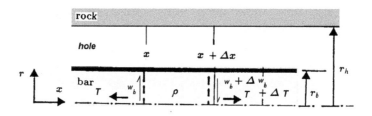

Figure 13.26 Modelling of traverse response of tendons.

As for traverse response of the free part of the tendon with the assumption of the existence of the axial tensile force, the following equation may be written (Figure 13.26):

$$\rho\frac{\partial^2 w_b}{\partial t^2} + h_t^* \frac{\partial w_b}{\partial t} = \sigma_p \frac{\partial^2 w_b}{\partial x^2} \tag{13.12}$$

where w_b, σ_p and h_t^* are traverse displacement, applied prestress and traverse damping ratio of tendon.

13.3.1.2 Analytical solutions

The analytical solutions for Eq. (13.9) and Eq. (13.12) are extremely difficult for given constitutive laws, boundary and initial conditions. However, it is possible to obtain solutions for simple cases, which may be useful for the interpretation of results of site-investigations. Eq. (13.9) may be reduced to the following form by omitting the effect of damping and interaction with surrounding rock as

$$\frac{\partial^2 u_b}{\partial t^2} = V_p^2 \frac{\partial^2 u_b}{\partial x^2} \tag{13.13}$$

where

$$V_p = \sqrt{\frac{E_b}{\rho}}$$

The general solution of partial differential equation (13.13) may be given as

$$u_b = h(x - V_p t) + H(x + V_p t) \tag{13.14}$$

For a very simple situation, the solution may be given as follows:

$$u_b = A \sin \frac{2\pi}{L}(x \pm V_p t) \tag{13.15}$$

where L is tendon length. Thus, the Eigen values of tendon may be obtained as follows

$$f_p = n\frac{1}{2L}V_p, \quad n = 1, 2, 3, \tag{13.16}$$

Similarly, the Eigen values of traverse vibration of the tendon under a given pre-stress may be obtained as follows:

$$f_T = n\frac{1}{2L}V_T, \quad n = 1, 2, 3, \tag{13.17}$$

where

$$V_T = \sqrt{\frac{\sigma_o}{\rho}}$$

13.3.1.3   Finite element formulation

(a) Weak form formulation
The integral form of Eq. (13.9) may be written as follows

$$\int \delta u_b \left(\rho \frac{\partial^2 u_b}{\partial t^2} + h_a^* \frac{\partial u_b}{\partial t} \right) dx = \int \delta u_b \frac{\partial \sigma}{\partial x} dx + \int \delta u_b \frac{2}{r_b} \tau_b dx \tag{13.18}$$

Introducing the following identity into Eq. (13.18)

$$\delta u_b \frac{\partial \sigma}{\partial x} = \frac{\partial}{\partial x}(\delta u_b \sigma) - \frac{\partial \delta u_b}{\partial x}\sigma \tag{13.19}$$

one can easily obtain the weak form of Eq. (13.9) as follows

$$\int \delta u_b \left(\rho \frac{\partial^2 u_b}{\partial t^2} + h_a^* \frac{\partial u_b}{\partial t} \right) dx + \int \frac{\partial \delta u_b}{\partial x}\sigma dx - \int \delta u_b \frac{2}{r_b}\tau_b dx = \int \delta u_b \sigma_n \Big|_{x=a}^{x=b} \tag{13.20}$$

(b) Finite element formulation

Displacement field of rockbolts and rock anchors is discretized in space in a classical finite element form as

$$u_b = [N]\{U_b\} \tag{13.21}$$

With the use of Eq. (13.21), one can write the following

$$a_b = \frac{\partial^2 u_b}{\partial t^2} = [N]\,(\{\ddot{u}_b\}); \quad v_b = \frac{\partial u_b}{\partial t} = [N]\{\dot{u}_b\} \tag{13.22}$$

$$\varepsilon_b = \frac{\partial u_b}{\partial x} = \frac{\partial}{\partial x}\,([N(x)]\{u_b\}) = [B]\{u_b\}; \quad \varepsilon_b = \frac{\partial \dot{u}_b}{\partial x} = \frac{\partial}{\partial x}\,([N(x)]\{\dot{u}_b\}) = [B]\{\dot{u}_b\} \tag{13.23}$$

If Eqs. (13.21) to Eq. (13.23) are inserted into Eq. (13.20), one can easily obtain the following expression

$$[M_e]\{\ddot{u}_b\} + [C_e]\{\dot{u}_b\} + [K_e]\{u_b\} = \{f_e\} \tag{13.24}$$

where

$$[M_e] = \int \rho [N]^T [N] dx; \quad [C_e] = \int h_a^* [B]^T [B] dx + \int \eta [B]^T [B] dx h;$$

$$[K_e] = \int E[B]^T [B] dx - \int \frac{2Kg}{r_b} [N]^T [N] dx$$

Similarly, the finite element form of Eq. (13.12) may be written as follows:

$$[M_e^t]\{\ddot{w}_b\} + [C_e^t]\{\dot{w}_b\} + [T_e^t]\{w_b\} = \{t_e^t\} \tag{13.25}$$

where

$$[M_e^t] = \int \rho [N]^T [N] dx; \quad [C_e^t] = \int h_t^* [N]^T [N] dx; \quad [T_e] = \int \sigma_p [B]^T [B] dx$$

13.3.1.4 Properties of rockbolts/rock anchors, grouting material and interfaces

(a) Properties of rockbolts/rock anchors

Aydan and his group have been performing some destructive and non-destructive tests on non-corroded and corroded iron and steel bars (Aydan, 1989; Aydan, et al., 2005, 2012, 2016 and some unpublished reports). Various relevant parameters are given in Table 13.1. This table also includes the parameters of steel bars sampled from 70 year old reinforced concrete structures. Figure 13.27 shows the strain-stress responses of widely used steel bars and cable anchors under tension. In this figure, steel bar (WC) corresponds to a steel bar with a section of areal reduction to represent the effect of corrosion.

Table 13.1 Physical and mechanical parameters of various steel bars and cables.

Material	Unit Weight (kN/m³)	P-wave velocity (km/s)	S-wave velocity (km/s)	Elastic Modulus (GPa)	Tensile Strength (MPa)
PC Steel Bar	75.6	6.05	3.23	200–210	
PC Wire Cable	69.6	4.80	–	165–170	
Deformed Bar	70–74	5.6–5.9	3.0–3.2	200	
Steel	77–80	5.95	3.23	200	
Iron	76–78	5.91	3.21	200	
Ikejima-Corroded smooth bar		4.83			
Shimizu Bridge, Smooth Bar (70 years)	75.4–78.5			196–213	425–458
Iron bolt (Moyeuvre Mine)	–	5.26	–	–	–
Theoretical estimations	75.6	5.97	3.19	200	

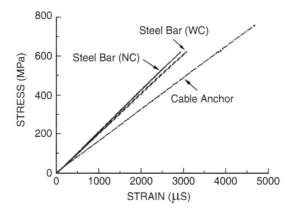

Figure 13.27   Strain-stress response of steel bars and cable anchors.

Table 13.2 Physico-mechanical properties of grouting material (28 days).

Grout Type	Unit Weight (kN/m³)	Elastic Modulus (GPa)	Poisson's ratio	UCS (MPa)	Tensile Strength (MPa)
M-T	20.0–20.4	14.7–18.4	0.17–0.25	19–21.9	1.5–1.8
M-M	19.1	11.6	–	53.1	5.54
M-R	20.2–20.6	13.7–17.3	0.17–0.25	19–22	2.3–2.7

(b) Properties of grouting material and interfaces

The overall behaviour of rockbolts/rock anchors is also influenced by the characteristics of grouting material. Depending upon the composition, material properties of grout may differ. The properties of some grouting material tested by the author are given in Table 13.2.

The load bearing capacity of rockbolts and rock anchors are strongly influenced by the shear strength characteristics of interfaces. These interfaces are tendon-grout

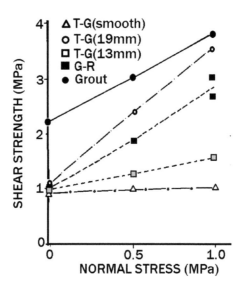

Figure 13.28 Comparison of shear strength of interfaces with that of grouting material (from Aydan, 1989).

(T-G), grout-protection sheath (G-S) and grout-rock (G-R). The short-term, creep and cyclic characteristics of interfaces have been studied by Aydan (1989) and Aydan *et al.* (1985–2016) and some experimental results are shown in Section 3.2 and Figure 13.28 compares the shear strength of various interfaces with that of grouting material. As noted from the figure, the surface morphology of interfaces greatly influences the shear strength of interfaces and it is much less than that of the grouting material.

13.3.1.5   Evaluation of corrosion of rockbolts and rock anchors

The evaluation of corrosion from the measured responses is one of the most important items in the interpretation of non-destructive test investigations. This definitely requires some mechanical models for interpretation. The mechanical properties of corroded part of the steel are much less than those of steel. Furthermore, the corrosion may be limited to a certain zone where the anti-corrosive protection may be damaged due to either relative motions at rock discontinuities or chemical attacks of corrosive elements in the ground water. If the corrosion is assumed to be taking place uniformly around the steel bar for a certain length as illustrated in Figure 13.29, the equivalent elastic modulus (E_b^*), shear modulus (G_b^*) and density (ρ_b^*) of the tendon may be obtained using the micro-structure theory (Aydan *et al.*, 1996, 2005, 2012) as follows:

$$\frac{E_b^*}{E_b} = \frac{(1-\alpha)+\alpha E_c/E_b}{\lambda+(1-\lambda)((1-\alpha)+\alpha E_c/E_b)},$$

$$\frac{G_b^*}{G_b} = \frac{(1-\alpha)+\alpha G_c/G_b}{\lambda+(1-\lambda)((1-\alpha)+\alpha G_c/G_b)},$$

$$\frac{\rho_b^*}{\rho_b} = (1-\lambda)+\lambda((1-\alpha)+\alpha\frac{\rho_c}{\rho_b}, \quad \lambda=\frac{l_c}{L}, \quad \alpha=\frac{A_c}{A_b}$$

(13.26a)

(13.26b)

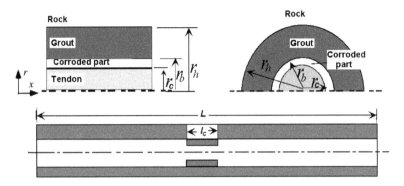

Figure 13.29 Geometrical illustration for the evaluation of effect of corroded part.

where E_c, G_c and ρ_c are properties of corroded part. A_c is corrosion area. If the properties of the corroded part are negligible, then the above equations take the following form

$$\frac{E_b^*}{E_b} = \frac{(1-\alpha)}{\lambda + (1-\lambda)(1-\alpha)}, \quad \frac{G_b^*}{G_b} = \frac{(1-\alpha)}{\lambda + (1-\lambda)(1-\alpha)}. \tag{13.27}$$

If corrosion is uniformly distributed over the total length of the tendon, then one may write the following equation

$$\alpha = 1 - \left(\frac{v_p^*}{v_p^o}\right)^2 \tag{13.28}$$

The dynamic parameters to be obtained from in-situ non-destructive tests may be used to obtain the dimensions of corrosion using the models presented above.

13.3.1.6 Numerical analyses

Some numerical analyses were already reported by Aydan *et al.* (2008) for 10 m long rock anchors. Numerical experiments presented in this section are performed to clarify the dynamic responses of rock anchors and rockbolts under impact waves for various conditions to be encountered in actual situations and in relation to some experiments carried out in the laboratory. Specifically the following conditions are considered together with the consideration of damping effects.

Case 1: Unbonded and non-corroded bar
Case 2: Bonded and non-corroded bar
Case 3: Unbonded bar with corrosion
Case 4: Bonded bar with corrosion
Case 5: Unbonded and non-corroded bar under prestress
Case 6: Unbonded and non-corroded bar under variable prestress
Case 7: Unbonded bar with corrosion under prestress

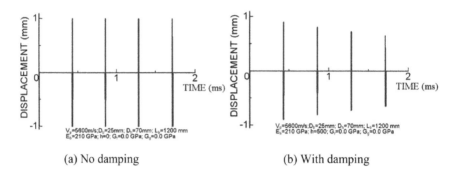

(a) No damping   (b) With damping

Figure 13.30 Axial dynamic displacement responses of unbonded rock anchors with and without damping (Case 1).

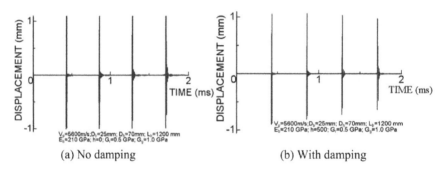

(a) No damping   (b) With damping

Figure 13.31 Axial dynamic displacement responses of bonded rock anchors with and without damping (Case 2).

Rock anchors were assumed to be 1200 mm long with a 26 mm diameter and they are elastic. When rock anchors are bonded, it is bonded along its entire length. The shear modulus of grouting material and rock are assumed to be 2 and 0.5 GPa respectively. The effect of the bonding is taken into account according to Aydan's model for rockbolts. Figures 13.30 and 13.31 shows the axial dynamic response of rock anchors for Case 1 and Case 2 with and without damping. As noted from the figure, the wave travels according to p-wave velocity and the bonding has little influence on the arrival time response of the reflected wave. Nevertheless, some noise-like responses are noted following the fundamental wave and the amplitude of the reflected waves differs. The amplitude of waves decreases with time if damping is introduced, which was also discussed in Chapter 10.

Next two numerical experiments were concerned with the effect of corrosion on the dynamic response of unbonded and bonded rock anchors (Case 3 and Case 4). Figure 13.32 shows the computed dynamic displacement responses for corroded tendon at the middle with a cross sectional reduction of 20%. Compared with the responses in previous cases, there are some reflected waves before the arrival of the reflected main shock. Furthermore, the amplitude of the reflected main shocks is no longer the

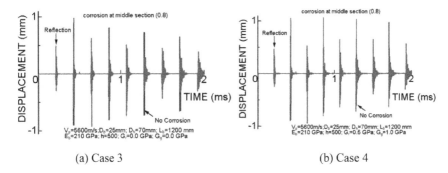

(a) Case 3

(b) Case 4

Figure 13.32 Axial dynamic displacement responses of unbonded and bonded rock anchors with corrosion.

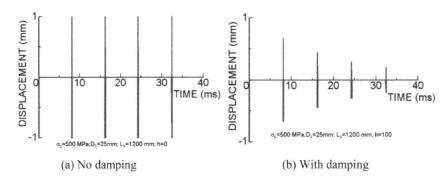

(a) No damping

(b) With damping

Figure 13.33 The effect of pre-stress on dynamic traverse responses of pre-stressed tendons (Case 5).

same and the noise-like waves are noted following the main shock. The amplitude and duration of these noise-like responses become larger as time passes. However, the results from these numerical analyses indicate that it is possible to locate the corrosion location. The amplitude of the reflection from the corroded part depends upon the geometrical and mechanical characteristics of the corroded part.

The effect of the pre-stress in rock anchors on the traverse response of rock anchors is investigated (Case 5). In the numerical tests, the prestress value are varied. Figure 13.33 shows the dynamic displacement responses for a prestress of 500 MPa under undamped and damped conditions. It should be noted that the travel time of the waves entirely depends upon the prestress value. Figure 13.34 shows the dynamic traverse response of the bar under different values of prestress. As also reported by Aydan *et al.* (2008), the travel time of the reflected traverse wave becomes shorter as the value of prestress increases. These results indicate that the traverse wave responses should provide valuable information on the prestress state of actual rock anchors.

The final example is concerned with the effect of the prestressed tendon with a corroded part in the middle section (Case 7). Figure 13.35 shows the dynamic displacement responses for non-corroded and corroded tendons subjected to the same prestress. The corroded part is assumed to have the 80% of the original cross section. When there is no corrosion, the travel time of the reflected displacement wave

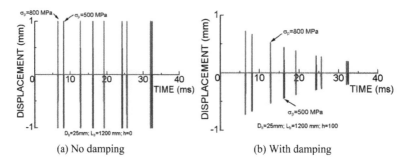

(a) No damping

(b) With damping

Figure 13.34 The effect of pre-stress variation on dynamic traverse responses of pre-stressed tendons (Case 6).

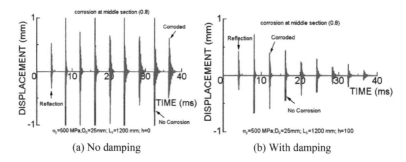

(a) No damping

(b) With damping

Figure 13.35 The effect of corrosion on axial wave responses of pre-stressed tendons (Case 7).

is 8 ms. Since the behaviour of the tendon is assumed to be elastic the reflected wave arrives at each 8 ms interval. However, when it is corroded at the middle section, it is noted that a wave is reflected before the arrival of the main wave. Furthermore, the amplitude of the reflected main shocks is no longer the same and the noise-like waves are noted following the main shock. The amplitude and duration of these noise-like responses become larger as time passes. However, the results from these numerical analyses indicate that it is possible to locate the corrosion location. The amplitude of the reflection from the corroded part is associated with the geometrical and mechanical characteristics of the corroded part.

13.3.1.7 Identification of reflected waves from records

As the application and observation of rockbolts and rock anchors for non-destructive testing is very small, the utilization of reflected waves is the essential item in the evaluation of soundness of the rockbolts and rock anchors. There are different techniques to identify the arrival time of reflected waves for the first fundamental mode. These may be categorized as

a)   Manual pick-up,
b)   Structural function and
c)   Auto-correlation function.

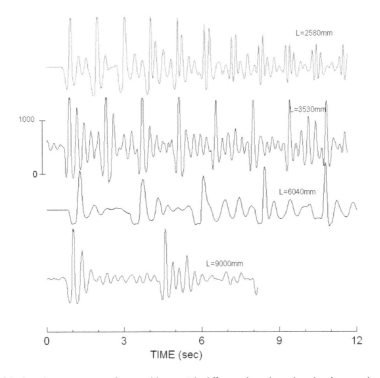

Figure 13.36 Acceleration waves for steel bars with different length under shock wave (unit is gals).

To illustrate these approaches, non-destructive tests are carried out on steel bars having different lengths (2580, 3530, 6040, 9000 mm). The actual records for steel bars for four different lengths under shock waves are shown in Figure 13.36.

The arrival time of reflected waves can be manually picked-up from the records if they distinctly appear in the records. For this purpose, "*peak to peak*" approach should be appropriate. The structural function (SF) used in wave processing is given in the following form

$$SF = \frac{1}{T} \int_{t=0}^{t=T} (\phi(t) - \phi(t+\tau))^2 \, dt \tag{13.29}$$

Similarly, auto-correlation function (ACF) is given by

$$ACF = \frac{1}{T} \int_{t=0}^{t=T} \phi(t)\phi(t+\tau) dt \tag{13.30}$$

where ϕ, t and τ are observation function, time and time lag, respectively. For a periodic function with given constant amplitude, when the time lag coincides with the periodicity of the wave, the values of *SF* and *ACF* would be minimum and maximum. The applications of Eq. (13.29) and Eq. (13.30) to the records shown in Figure 13.34

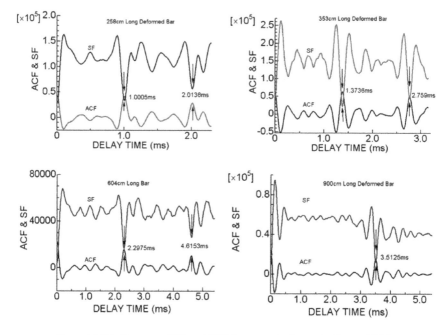

Figure 13.37 Computed SF and ACF for records shown in Figure 13.36.

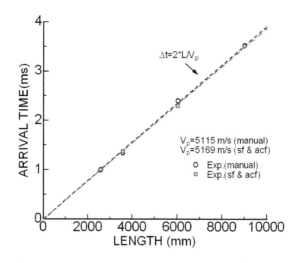

Figure 13.38 The relation between bar length and arrival time of reflected waves determined from three different approaches.

are shown in Figure 13.37. It is interesting to note that both SF and ACF approaches yield exactly the same results.

The results of arrival time of reflected waves obtained from three different approaches are plotted in Figure 13.38. The manually picked-up arrival times yielded that the velocity of the bar is about 5115 m/s while the SF and ACF approaches yielded that the velocity of bars is about 5169 m/s. Despite a slight difference, the results are close to each other. As the bars were exposed to the atmosphere for several years, some rusting of bars was noted. The approach described in Sub-section 13.3.1.5 was applied to the computed results by assuming that corrosion took place along the entire length of the bars and it was found that the ratio of the corrosion was estimated to be 14–25% depending upon the assumed original wave velocity of the bars. If it is assumed to be 5900 m/s, the amount of corrosion would be about 25%.

The next approach would be utilization of frequency content of the recorded waves. As noted from Eq. (13.16), the frequency content to be obtained from Fourier spectra analysis would also yield the wave velocity of the tendon. Fourier transformation is generally used to simulate real wave forms by numerical approximate functions. Let us consider the actual acceleration form is given by the following function.

$$a = a(t) \tag{13.31}$$

Fourier transform of Function (13.31) is replaced by the following function

$$a(t) = \sum_{k=0}^{\infty} \left[A_k \cos(kt) + B_k \sin(kt) \right] \tag{13.32}$$

If Eq. (13.32) is approximated by a finite number (N) of data with a time interval (Δt), Eq. (13.32) can be re-written as

$$\tilde{a}(t) = \frac{A_o}{2} + \sum_{k=1}^{N/2-1} \left[A_k \cos(2\pi f_k t) + B_k \sin(2\pi f_k t) \right] + \frac{A_{N/2}}{2} \cos 2\pi f_{N/2} t \tag{13.33}$$

where

$$f_k = \frac{k}{N \Delta t} \tag{13.34}$$

If time (t) is represented by $m\Delta t$, coefficients A_k and B_k in Eq. (13.32) would be expressed as

$$A_k = \frac{2}{N} \sum_{m=0}^{N-1} a_m \cos \frac{2\pi km}{N} \quad k = 0, 1, 2, \ldots, \frac{N}{2} - 1, \frac{N}{2} \tag{13.35a}$$

$$B_k = \frac{2}{N} \sum_{m=0}^{N-1} a_m \sin \frac{2\pi km}{N} \quad k = 1, 2, \ldots, \frac{N}{2} - 1 \tag{13.35b}$$

For kth frequency, the maximum amplitude, phase angle and power would be obtained as

$$C_k = \sqrt{A_k^2 + B_k^2}; \quad \phi_k = \tan^{-1}\left(-\frac{B_k}{A_k} \right); \quad P_k = \frac{C_k^2}{2} \tag{13.36}$$

Figure 13.39 Samples with bar-type and cable type tendons embedded in rock and a typical experimental set-up.

It should be noted that, the amplitude of Fourier coefficient C_k is multiplied by the half of the period of the record ($T/2$) in Fast Fourier Transformation (FFT). Therefore, Fourier spectra explained above differs from the FFT spectra in the value of amplitudes and its unit.

13.3.1.8 Applications to actual measurements under laboratory conditions

Several bar type tendons (1200 mm long, 36 mm in diameter) and cable type tendons (1300 mm long and 6 wires with a diameter of 6 mm) with/without artificial corrosion under bonded and unbonded conditions have been prepared (Figure 13.39 and Figure 13.40) and the responses of the tendons under single impact waves induced by an impact hammer or special Schmidt hammer like device (ponchi) were measured. Three different sensors are used and two of them had a center hole for inducing impact waves on tendons (Figure 13.41). The waves can be recorded as displacement, velocity or acceleration and the device for monitoring and recording consists of an amplifier and a small hand-held type computer (Figure 13.42). The measurement can be done through a single person and the Fourier spectra of recorded data can be stored in the small hand-held computer and they can also be visualized in the measurement spot. In the following, results obtained would only be given without any reference to the sensor or hammer unless it is mentioned. Most of the results and details of experiments

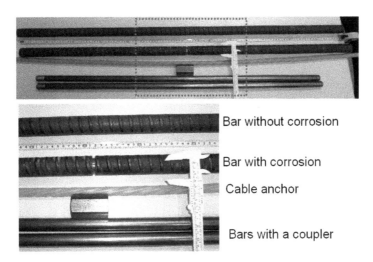

Bar without corrosion

Bar with corrosion

Cable anchor

Bars with a coupler

Figure 13.40 Bar-type and cable type tendons with/without artificial corrosion used in experiments.

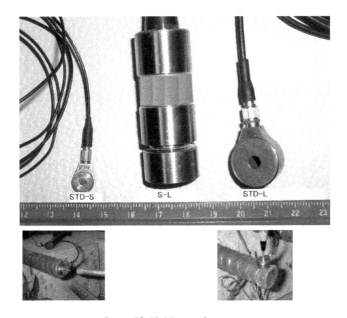

Figure 13.41 Views of sensors.

described in this section can be found in an invited lecture by Aydan (2012) presented in EITAC Annual Convention.

Figure 13.43 shows the wave responses of a 1200 mm steel bar induced by an impact hammer and its numerical simulation using the numerical method described previously. The wave velocity inferred from acceleration records directly was about 5520 m/s. The Fourier spectra of wave records induced by the impact hammer and a special Schmidt-hammer like device (ponchi) are shown in Figure 13.44. The frequency

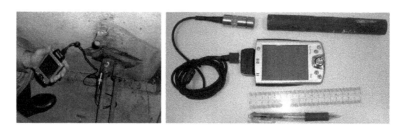

Figure 13.42 Views of PC-pocket type sampling and recording device.

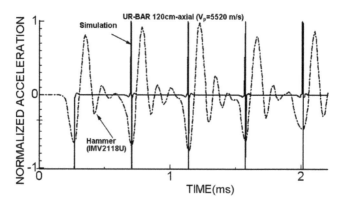

Figure 13.43 Impact hammer induced wave response and its numerical simulation for a 1200 mm long steel bar.

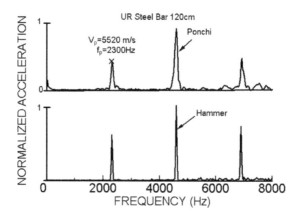

Figure 13.44 Normalized Fourier spectra of recorded acceleration records induced by impact hammer and a special Schmidt-hammer like device.

content interval is about 2300 Hz and the inferred velocity of the steel bar was 5520 m/s. These results are consistent with each other.

Figure 13.45 and Figure 13.46 show the effect of coupler on the steel bar with a length of 1000 mm and the acceleration response of single 1000 mm bar also shown in

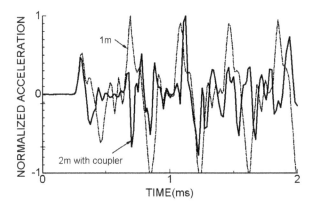

Figure 13.45 Impact hammer induced wave responses of single and two 1 m long two bars connected to each other with a coupler.

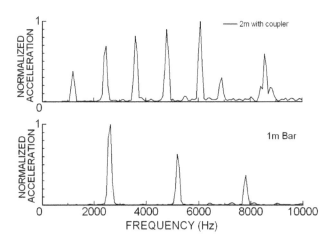

Figure 13.46 Normalized Fourier spectra of recorded acceleration records of single and two 1 m long two bars connected to each other with a coupler.

the figures. As noted from the figure, two reflections occur in the 2 m long coupled bar. The main reflection is due to the other end of the coupled bar and secondary-reflection is due to the coupler.

Figure 13.47 shows the Fourier spectra of 1200 mm bar with and without artificially induced area reduction to simulate corrosion. As expected from the numerical analysis high frequency content would be generated by the partially reflected waves from the artificial corrosion zone. This feature clearly observed in the computed Fourier spectra. Figure 13.48 shows the effect of bonding of the bar to the surrounding rock. Although the frequency of the bonded bar is slightly smaller than that of the unbonded rock anchor, the Fourier spectra for the first mode are quite close to each other. Nevertheless, the frequency content starts to change after the second or higher modes.

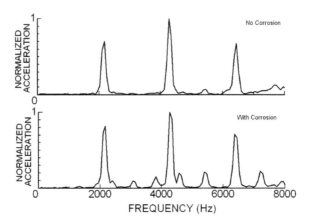

Figure 13.47 Normalized Fourier spectra of recorded acceleration records for bars with or without corrosion.

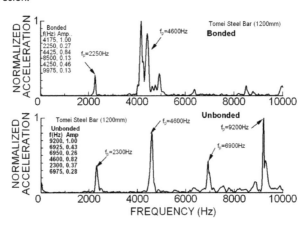

Figure 13.48 Normalized Fourier spectra of recorded acceleration records of bonded and unbonded bars.

13.3.1.9 Some applications to rockbolts and rock anchors in-situ

The monitoring and recording system described in the previous sub-section is utilized for non-destructive testing of rockbolts, bar-type rock anchors and cable-type rock anchors installed in various rock masses (Figure 13.49). Some of these measurements are shown in Figure 13.50 to Figure 13.53 together with numerical simulations for the evaluation of in-situ measurements. As noted from the figures, the existence of grouting causes the leakage of waves into the surrounding medium so that the amplitude of reflected waves drastically reduced.

Figure 13.54 shows the traverse acceleration responses of a cable type rock anchor with a 30 m length during a lift-off test. The numbers in the figure corresponds to axial forces (60, 164, 268, 372, 476 and 580 kN) applied to the rock anchor. The arrival times during the increase of axial are expected to be shorter. This feature is well observed when the axial load was over 268 kN. There are different causes for

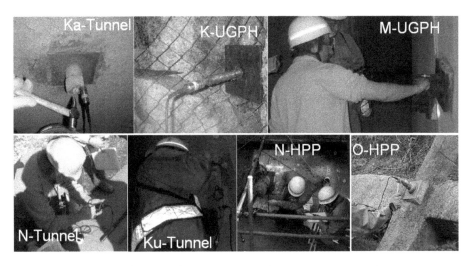

Figure 13.49 Views of some in-situ non-destructive tests examples.

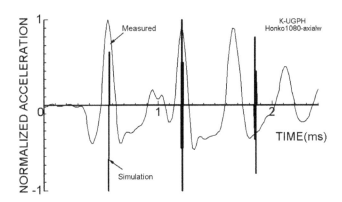

Figure 13.50 Measured and computed acceleration responses of a rockbolt at K-UGPH.

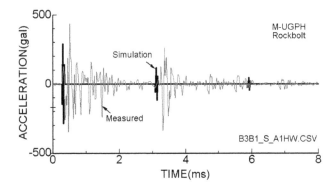

Figure 13.51 Measured and computed acceleration responses of a rockbolt at M-UGPH.

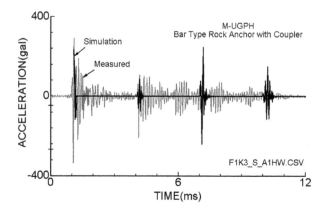

Figure 13.52 Measured and computed acceleration responses of a bar-type rock anchor with a coupler at M-UGPH.

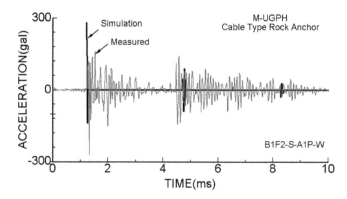

Figure 13.53 Measured and computed acceleration responses of a cable type rock anchor at M-UGPH.

explaining some of the discrepancies between measured and anticipated responses. The most difficult aspect associated with cable type anchors is that the induction of impact wave and measurement location differs due to the inherent structure of the cable anchors.

13.3.1.10 The utilization of wavelet data processing technique and some issues

As described in previous sections, the processing of measured responses requires tremendous efforts for data interpretation. The discrimination of arrival times of reflected waves and their frequency content are essential items for data interpretation. When impact waves are utilized, it is desirable to evaluate the wave packets and the utilization of discrete Fourier spectra is preferable to the continuous Fourier spectra analyses. The wave-let technique actually has the ability to do such analyses, which can evaluate the data for each arrival time of wave packets reflected from the end as well as some structural weaknesses in the rock anchor systems. Figure 13.55 show an

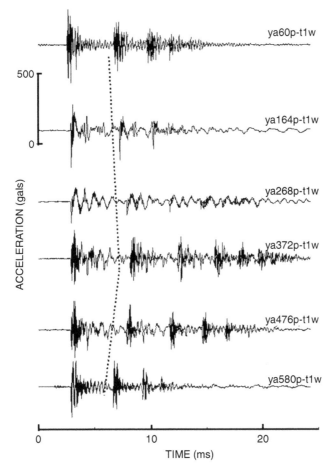

Figure 13.54 Measured traverse acceleration responses of a cable type rock anchor during a lift-off test at N-HPP.

example of the utilization of wavelet technique for one of the experiments reported in sub-section 13.3.1.6. However, it should be also noted that the wavelet data process-ing technique results in a huge amount of processed data, which may sometimes be confusing during data interpretation.

As pointed out previously, the application of impact waves on rockbolts and rock anchor and monitoring responses of reflected waves are carried out at the rockbolt/rock anchor head, which has a very limited physical space. Therefore, it is very desirable to utilize a sensor with a central hole to induce impact waves and easy to mount to the head of rockbolts/rock anchors in real applications.

13.3.2   Guided ultrasonic wave method

The guided ultrasonic wave method is an NDT method using the vibration charac-teristics or propagation velocity characteristics of the ultrasonic wave varying with

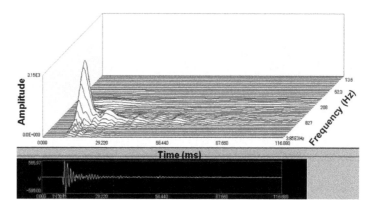

Figure 13.55 A wavelet processing of acceleration response.

the stress of the pre-stressing tendon (i.e. Beard and Lowe, 2003; Chaki and Bourse, 2009; Buys *et al.*, 2009). A short duration Gaussian windowed sine burst is used to excite a guided wave in the bolt from the free end. The wave is then reflected from the other end and from any major defects. From the reflection arrival time and knowledge of the wave velocity dispersion curves, the positions of the defects or the bolt length can be calculated. Although this method is widely adopted for the inspection of the cracks or corrosion of pipelines, it presents some limitations when applied to bonded pre-stressing tendons. The guided ultrasonic wave experiences loss when penetrating the surrounding medium. The readers are advised to refer to the articles on the details and applications of this method mentioned above.

13.3.3   Magneto-elastic sensor method

Stress measuring technique using the magnetostriction effect utilizes magnetic field distortion by strain of steel itself. Magnetostriction effect was discovered by Joule in 1842 and confirmed by others (i.e. Vilari, 1865). Magnetostriction effects is such that when a ferromagnetic material in the form of a rod is subjected to an alternating magnetic field parallel to its length, the rod undergoes alternate contractions and expansions at a frequency equal to the frequency of applied field. The technique utilizes the magnetic field induced by an electromagnet and the magnetostriction effect, in which the magnetic susceptibility or permeability of a ferromagnetic material changes when subjected to a mechanical stress. This technique was applied to the nuts of rock anchors used in underground powerhouses to infer their axial forces (i.e. Laguerre, 2000; Akutagawa *et al.*, 2008). The readers are advised to refer to the articles by Laguerre *et al.* (2002) and Akutagawa *et al.* (2008) for the details and applications of this method.

13.3.4   Lift-off testing technique

Lift-off testing is also used to check the state of rock anchors and the level of the axial forces. This test may cause anchorage failure or tendon failure. Figure 13.56 shows

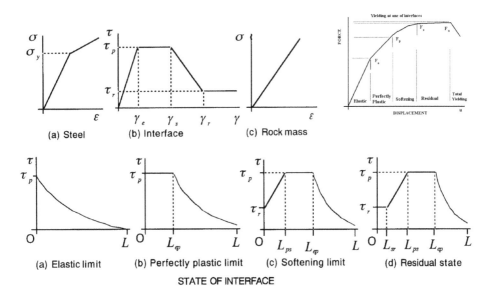

Figure 13.56 Constitutive laws and state of interfaces assumed in the theoretical derivation of equations for pull-out response of rock anchor systems (from Aydan, 1989).

the constitutive laws and the state of interfaces in a rockbolt/rock anchor during a pull-out or lift-off test. Under static loading, the equilibrium equation in the direction of pull-force can be reduced to the following equation (i.e. Aydan, 1989; Aydan *et al.*, 1985–1988).

$$\frac{d\sigma_z}{dz} + \frac{2}{r_b}K_g u_b = 0 \tag{13.37}$$

The specific value of K_g is given by Eq. (13.14).

13.3.4.1 Elastic behaviour

If the anchor and surrounding media behave elastically (Figure 13.56), one can easily get the following (see Aydan, 1989; Aydan *et al.* (1985–1988) for details):

a)   Anchor head displacement

$$u_b = \frac{\sigma_0}{E_b \alpha} \frac{e^{\alpha(L-z)} + e^{-\alpha(L-z)}}{e^{\alpha L} - e^{-\alpha L}} \tag{13.38}$$

b)   Anchor head stress

$$\sigma_z = \sigma_0 \frac{e^{\alpha(L-z)} - e^{-\alpha(L-z)}}{e^{\alpha L} - e^{-\alpha L}} \tag{13.39}$$

c) Shear stress at tendon-grout interface τ_b

$$\tau_b = \frac{\sigma_0 K_g}{E_b \alpha} \frac{e^{\alpha(L-z)} + e^{-\alpha(L-z)}}{e^{\alpha L} - e^{-\alpha L}} \qquad (13.40)$$

where $K_g = \frac{\alpha^2 r_b E_b}{2}$

The displacement, axial stress and shear stress at the yielding of tendon-grout interfaces are obtained as

a) Anchor head displacement

$$u_b = \frac{\tau_p^{bg}}{K_g} \frac{e^{\alpha(L-z)} + e^{-\alpha(L-z)}}{e^{\alpha L} - e^{-\alpha L}} \qquad (13.41)$$

b) Anchor stress (σ_z)

$$\sigma_z = \frac{\tau_p^{bg} E_b \alpha}{K_g} \frac{e^{\alpha(L-z)} - e^{-\alpha(L-z)}}{e^{\alpha L} - e^{-\alpha L}} \qquad (13.42)$$

c) Shear stress along tendon-grout interface τ_b

$$\tau_b = \tau_p^{bg} \frac{e^{\alpha(L-z)} + e^{-\alpha(L-z)}}{e^{\alpha L} - e^{-\alpha L}} \qquad (13.43)$$

where τ_p^{bg} is the peak shear strength of tendon-grout interface.

13.3.4.2 Elasto-plastic behaviour of rock anchor system

In this sub-section, the elasto-plastic behaviour of tendon-grout interface and steel bar are considered and the behaviour of grout and rock are assumed to be elastic. For other cases, please see Aydan (1989).

Region 1: $0 \le z \le L_{p1}$ (L_{p1}: anchor length, and tendon exhibit strain hardening behaviour)

a) Axial displacement u_b

$$u_b = u_0 + \frac{\eta \tau_p^{bg}}{E_T r_b} z^2 - \frac{\sigma_0 - F}{E_T} \qquad (13.44)$$

b) Anchor stress σ_z

$$\sigma_z = \sigma_0 - \frac{2\eta \tau_p^{bg}}{r_b} z \qquad (13.45)$$

c) Shear stress along tendon-grout interface τ_b

$$\tau_b = \eta \tau_p^{bg} \qquad (13.46)$$

The stress at the anchor head is

$$\sigma_0 = \sigma_y + \frac{2\eta\tau_p^{bg}}{r_b} L_{p1} \tag{13.47}$$

Anchor head displacement

$$u_0 = u_b^{ep} + \left(\frac{\sigma_y}{E_b} + \frac{\eta\tau_p^{bg}}{E_T r_b} L_{p1}\right)(L_{p2} - L_{p1}) + \frac{\eta\tau_p^{bg}}{r_b}\left(\frac{1}{E_b} - \frac{1}{E_t}\right) L_{p1}^2$$

$$- \frac{\eta\tau_p^{bg}}{E_b r_b} L_{p2}^2 + \frac{1}{E_t}(\sigma_0 - F) L_{p1} \tag{13.48}$$

Region 2: $L_{p1} \le z \le L_{p2}$ (L_{p2}: Anchor exhibits strain hardening behaviour and tendon-grout interface under flow state)

a) Anchor displacement u_b

$$u_b = \left(\frac{\sigma_y}{E_b} + \frac{\eta\tau_p^{bg}}{E_b r_b} L_{p1}\right)(L_{p2} - z) - \frac{\eta\tau_p^{bg}}{E_b r_b}\left(L_{p2}^2 - z^2\right) + \varepsilon_r \frac{\tau_p^{bg}}{K_g} \tag{13.49}$$

b) Anchor axial stress σ_z

$$\sigma_z = \sigma_y - \frac{2\eta\tau_p^{bg}}{r_b}(z - L_{p1}) \tag{13.50}$$

c) Tendon-anchor interface shear stress τ_b

$$\tau_b = \eta\tau_p^{bg} \tag{13.51}$$

Region 3: $L_{p2} \le z \le L_{p3}$ (L_{p3}: Anchor exhibits strain hardening behaviour and tendon-grout interface exhibits strain-softening behaviour)

a) Anchor displacement u_b

$$u_b = \frac{\tau_p^{bg}}{K_g}\left\{\frac{\varepsilon_r - \varepsilon_s}{1 - \eta}\frac{(A - \eta B)\cos(pz) - (D - \eta C)\sin(pz)}{\sin(p\,(L_{p3} - L_{p2}))}\right\} + \frac{\varepsilon_r - \varepsilon_s}{1 - \eta} \tag{13.52}$$

b) Anchor axial stress σ_z

$$\sigma_z = \frac{p\tau_p^{bg} E_b}{K_g}\frac{\varepsilon_r - \varepsilon_s}{1 - \eta}\left\{\frac{(A - \eta B)\sin(pz) - (D - \eta C)\cos(pz)}{\sin(p\,(L_{p3} - L_{p2}))}\right\} \tag{13.53}$$

c) Tendon-grout interface shear stress τ_b

$$\tau_b = \tau_p^{bg} \left\{ \frac{-(A - \eta B)\cos(pz) + (D - \eta C)\sin(pz)}{\sin(p(L_{p3} - L_{p2}))} \right\} \qquad (13.54)$$

where

$$A = \sin(pL_{p2}); \quad B = \sin(pL_{p3}); \quad C = \cos(pL_{p3}); \quad D = \sin(pL_{p2})$$

Region 4: $L_{p3} \leq z \leq L_{p4}$ (L_{p4}): Tendon-grout interface behaviour is perfectly plastic

a) Anchor displacement u_b

$$u_b = \frac{\tau_p^{bg}}{E_b r_b} z^2 + \frac{1}{L_{p4} - L_{p3}} \left[\left(w_p - \frac{\tau_p^{bg}}{E_b r_b} L_{p4}^2 \right)(z - L_{p3}) \right.$$
$$\left. + \left(\varepsilon_s w_p - \frac{\tau_p^{bg}}{E_b r_b} L_{p3}^2 \right)(L_{p4} - z) \right] \qquad (13.55)$$

b) Anchor axial stress σ_z

$$\sigma_z = \frac{\tau_p^{bg}}{r_b} z + \frac{E_b}{L_{p4} - L_{p3}} \left[-\left(w_p - \frac{\tau_p^{bg}}{E_b r_b} L_{p4}^2 \right) + \left(\varepsilon_s w_p - \frac{\tau_p^{bg}}{E_b r_b} L_{p3}^2 \right) \right] \qquad (13.56)$$

c) Tendon-grout interfaces τ_b

$$\tau_b = \tau_p^{bg} \qquad (13.57)$$

Region 5: $L_{p4} \leq z \leq L$ (L: Anchor length): The interface of tendon-grout and tendon behaves elastically.

a) Anchor displacement u_b

$$u_b = \frac{\tau_p^{bg}}{K_g} \frac{e^{\alpha(2L - L_{p4} - z)} + e^{\alpha(z - L_{p4})}}{e^{2\alpha(L - L_{p4})} + 1} \qquad (13.58)$$

b) Anchor axial stress σ_z

$$\sigma_z = \frac{\tau_p^{bg} \alpha E_b}{K_g} \frac{e^{\alpha(2L - L_{p4} - z)} - e^{\alpha(z - L_{p4})}}{e^{2\alpha(L - L_{p4})} + 1} \qquad (13.59)$$

c) Tendon-grout interface τ_b

$$\tau_b = \tau_p^{bg} \frac{e^{\alpha(2L - L_{p4} - z)} + e^{\alpha(z - L_{p4})}}{e^{2\alpha(L - L_{p4})} + 1} \qquad (13.60)$$

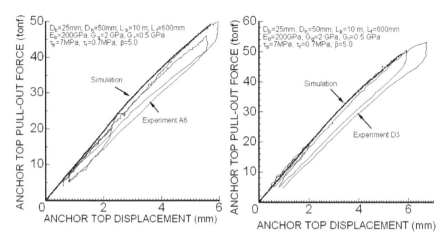

Figure 13.57 Comparison of measured responses of the displacement and applied force of anchor heads for bar-type anchors with theoretical solutions.

At an underground power station, which was constructed about 40 years ago, several lift-off tests were carried out to check the soundness of rock anchors and the value of pre-stress applied at the time of construction. Figure 13.57 and Figure 13.58 show the displacement and applied force of anchor heads for bar-type anchors and cable type anchors, respectively. As noted from the figure, the responses of the bar type anchors are quite close to their original state and their deterioration is almost none despite 40 years' elapse after their installation. On the other hand, the responses of cable type rock anchors are remarkably different from their original state and permanent displacements occur after each cycle of loading. Furthermore, many cable type anchors failed during lift-off experiments. These results also imply that the cable type rock anchors consisting of wires are quite vulnerable to deterioration with time and they may not be qualified as permanent supporting systems.

13.4 ESTIMATION OF FAILURE TIME OF TENDONS

In this section, the failure time of pre-stressed tendons subjected to corrosion and creep is assessed. The long-term strength of materials can be expressed using the following equation (i.e. Aydan and Nawrocki, 1998).

$$\frac{\sigma_{tl}}{\sigma_{ts}} = 1 - b \ln\left(\frac{t}{t_s}\right) \tag{13.61}$$

where t_s is short term testing period (5–20 minutes). Coefficient b is an empirical constant and its value for steel ranges between 0.02 to 0.04.

As many tendons have circular cross-section, the corrosion starts from the outer surface and progresses inward. As a result, the initial cross section area (A_o) of the

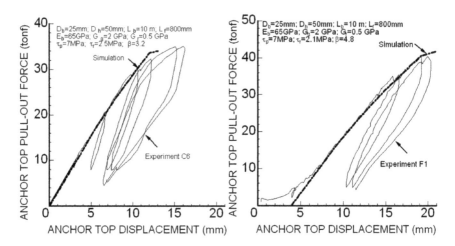

Figure 13.58 Comparison of measured responses of the displacement and applied force of anchor heads for cable-type anchors with theoretical solutions.

tendon would decrease as time progresses. Thus, one can write the following relation for the cross section area for any time normalized by the initial cross section area:

$$\frac{A}{A_o} = \left(\frac{r}{r_0}\right)^2 \tag{13.62}$$

Let us assume that the decrease of the radius of the bar due to corrosion may be estimated from the following equation:

$$r = r_o - \beta t \tag{13.63}$$

If an initial prestress acts on the tendon, the current stress on the tendon may be related to the prestress in the following form provided that the force remains same:

$$\frac{\sigma}{\sigma_o} = \left(\frac{r_o}{r}\right)^2 \tag{13.64}$$

The tendon will rupture when the stress given by Eq. (13.64) becomes equal to the strength given by Eq. (13.61). If we denote the failure time by t_f, it would be specifically obtained as follows:

$$t_f = t_s \exp\left(\frac{1}{b}\left[1 - \frac{\sigma_o}{\sigma_{ts}}\left(\frac{r_o}{r_o - \beta t}\right)^2\right]\right) \tag{13.65}$$

As the surface area of cable type anchors is larger than the bar type rock anchor, Eq. (13.65) implies that the failure time of cable anchors would be shorter than that of the bar type anchors. Table 13.3 is a summary of corrosion rates of steel used in

Table 13.3 Corrosion rates of steel in underground excavations.

	Mine-A Canada	Mine-B Canada	Tunnel Japan	M-UGPH Japan	Mine-E France	Mine-D Australia	Mine-E Australia
pH	7.8	3.4	6.7	6.7	–	7.9	3.0
Rate (mm/year)	0.1–0.5	0.1–0.5	0.02–0.1	0.025	0.1	0.05–0.2	0.1–0.5

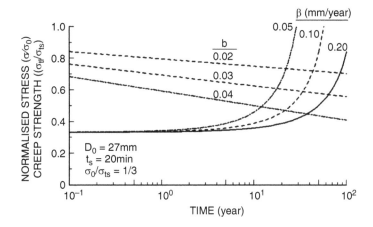

Figure 13.59 Estimation of failure time of bar type rock anchor.

underground excavations in various countries. Figure 13.59 and Figure 13.60 show the estimation of failure time of bar type rock anchors of 27 mm and cable type anchors with 6 mm steel wires for an initial prestress value of 1/3 of their ultimate load bearing capacity.

When the anchor ruptures, the anchor head would be thrown and its ejection velocity can be estimated from a similar procedure described for the ejection velocity of rock fragments during rockburst. The ejection velocity (unit is m/s) can be given as

$$v = 0.0255\sigma_{ct} \tag{13.66}$$

The failure stress (σ_{ct}) would be a fraction of its original tensile stresses. It is very likely to be in the range of 0.6–0.8 times the short-term tensile strength. As the tensile strength of high-grade steel bars is about 1 GPa, the ejection velocity of the anchor head could likely be in the range of 15–20 m/s.

13.5 EFFECT OF DEGRADATION OF SUPPORT SYSTEM

The support system of the caverns consists of rockbolts, rock anchors and arch concrete. The rock anchors installed on sidewalls are expected to perform their functions during the service life of the cavern. The rock anchors may consist of either

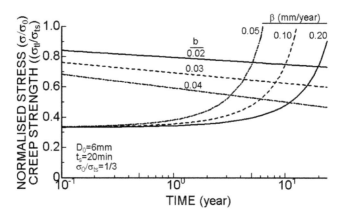

Figure 13.60 Estimation of failure time of cable type rock anchor.

high-strength steel bars and/or high-strength steel wires. The underground environment is always humid and ground water may have corrosive solved chemical substances. Steel is known as a corrosion resistant material. The site investigations by the authors as well as by other researchers indicate that rockbolts and rock anchors may undergo corrosion (Aydan *et al.*, 2008, 2012). Cable-type rockbolts or rock anchors are much more prone to corrosion in view of in-situ pullout experiments. Theoretical estimations for the service life of different type rockbolts and rock anchors under pre-stress based on the experimentally measured corrosion rate confirm this experimental finding. The last computational example is concerned with the effect of the degradation of support system. The excavation of the cavern was simulated first and rock anchors with a 5 m-anchorage length and a pre-stress value 200–300 kN are assumed to be installed. Since blasting technique was used for the excavation of the cavern, it is likely that the rock mass adjacent to the excavation boundary might be damaged. The computation was carried for the following three cases, namely, Case 1: Sound rock with supports, Case 2: Sound rock with degraded supports, Case 3: Rock with damaged zone, whose deformation modulus is 1/10 of the non-damaged rock mass and degraded support. Figure 13.61 illustrates the pattern of rock anchors, the extent of blasting-induced damage zone and selected points for comparisons of displacement. Table 13.4 gives the computed displacements at selected points. As expected, the degradation of support system causes the increases of the displacement field around the cavern. Nevertheless, the amplitude of the increase is not so large. When there is a blasting-induced damage zone around the cavern, the amplitude of the displacement field become much larger. In any case, no plastic zone developed around the cavern. However, it should be noted that these type continuum analyses are valid for global stability of the caverns. Rock mass always has discontinuities in different forms. These pre-existing discontinuities may constitute unstable rock blocks around the cavern. The size of these blocks may be quite large when the cavern height and width are more than 40 m and 25 m, respectively. In such cases, the stabilization of such blocks by support system may become the prime concern (Aydan, 1989; Kawamoto *et al.*, 1991).

Table 13.4 Horizontal and vertical displacement of selected points.

Node No.	Case 1		Case 2		Case 3	
	U (mm)	V (mm)	U (mm)	V (mm)	U (mm)	V (mm)
1781	4.966	−0.343	4.946	−0.330	12.32	−13.33
1807	−13.89	−12.86	−14.03	−12.89	−39.19	−18.40
1815	−8.947	−15.63	−9.049	−15.65	−27.51	−24.69
1826	0.006	−22.15	0.008	−22.17	1.597	−49.14
1839	7.587	−16.12	8.647	−16.08	27.54	−25.03
1848	13.03	11.25	13.24	−11.29	38.84	−15.65

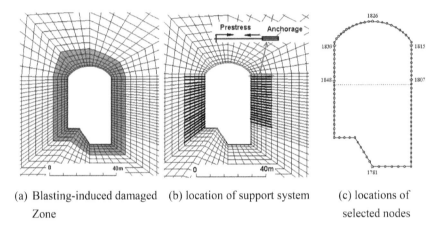

(a) Blasting-induced damaged   (b) location of support system   (c) locations of
    Zone   selected nodes

Figure 13.61 Illustration of blasting-induced damage zone, locations of support system and selected nodes.

13.6 CONCLUSIONS

The assessment of support systems such as rockbolts and rock anchors are very important for the maintenance of existing rock engineering structures. Particularly dynamic loads involve blasting, vehicle traffic and vibrations induced by power generation. As time goes by, support systems may deteriorate as a result of corrosion, micro cracking and motions along rock discontinuities in rock masses.

In the first part of this chapter, several experimental set-ups are used to clarify the response of rockbolts and rock anchors through model tests subjected to dynamic loads through shaking tables. The experiments described in this chapter clearly demonstrated that the dynamic loads may cause additional stretching on support systems utilizing rockbolts and rock anchors due to the permanent deformation along rock discontinuities as a result of active sliding and active toppling type of motions. A theoretical model for the model tests was presented and the computed results are compared with the measured responses. The computed results are quite similar to the measured results both qualitatively and quantitatively although some discrepancies do exist. However, these discrepancies could result from the complexity of actual frictional behaviour of discontinuities as compared with the assumed frictional behaviour in computations. In addition, the fully grouted rockbolts were tested in the model tests under dynamic loads. The experiments clearly showed that the strains of rockbolts crossing discontinuities could be very high and the dynamic loads may cause additional stretching of rockbolts and rock anchors due to the permanent deformation along rock discontinuities as a result of active sliding and active toppling type of motions. These results are in accordance with those from the shaking table tests of mechanically anchored rockbolts and rock anchors.

The non-destructive tests utilizing dynamic responses of support members is very preferable compared to destructive tests. These tests should generally provide some information on the constructional conditions, corrosion and axial force conditions. In the first part of this study, some theoretical and finite element models for the axial and traverse dynamic tests of rockbolt and rock anchors are presented. Then results of numerical tests are explained by considering possible conditions in-situ such as the bonding quality, the existence of corroded parts and pre-stress. The numerical experiments indicate that it is possible to evaluate the length of the rockbolts and rock anchors, the value of the applied pre-stress and the location of corrosion. A nondestructive system was developed using impact waves and various typical situations were experimentally investigated. The non-destructive testing technique based on impact waves is promising. Nevertheless, there is still some room for further advancement of upgrading sensors as well as data-acquisition system.

In addition to the non-destructive testing, lift-off tests provide direct information on the soundness state of rock anchors and their axial load. Nevertheless, this test is expensive and labor-intensive and replacement of rock anchors is necessary if they rupture. Lift-off tests on rock anchors were carried out at an underground power station constructed about 40 years ago, to check the soundness of rock anchors and the value of pre-stress applied at the time of construction. The responses of the bar type anchors are quite close to their original state and their deterioration is almost nil despite 40 years elapsing after their installation. On the other hand, the responses of cable type rock anchors are remarkably different from their original state and permanent displacements occurred after each cycle of loading. These results also imply that the cable type rock anchors consisting of wires are quite vulnerable to deterioration over time and they may not be qualified as permanent supporting systems.

Chapter 14

Dynamics of impacts

The dynamic response of objects during free fall and impact to barriers may have some important implications in various fields such as collision of vehicles in transportation engineering, collision of adjacent structures during earthquakes, standard penetration tests in soil mechanics, and impact craters due to meteorites, anchoring of ships or platforms in marine engineering and bullets and missiles destroying targets. Figure 14.1

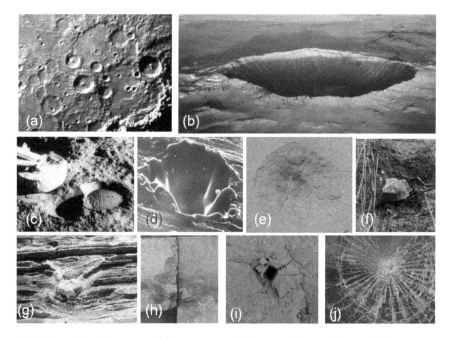

Figure 14.1 Examples of impacts. (a) Moon craters; (b) Arizona Meteor Crater; (c) Crater of Apollo landing pad; (d) Small crater in glass of Apollo Module; (e) Crater on the wall of Victoria and Albert Museum in London; (f) Rock-fall crater caused by 2007 Noto Peninsula Earthquake; (g) Cratering in Volcanic sediments in Hawaii; (h) Collision of bridge deck with abutment by 2008 Wenchuan earthquake; (i) Crater induced by a geologic hammer in Shimajiri mudstone; (j) Broken glass induced during 1999 Kocaeli earthquake. Figures (a) to (d) are from the pictures of images in Smithsonian Air and Space Museum taken by the author during his visits.

illustrates several examples of effects of impacts. In this chapter, the dynamics of impact phenomenon is explained and discussed.

14.1 CRATER FORMATION BY METEORITES AND ITS ENVIRONMENTAL EFFECTS

14.1.1 Dynamics of crater formation by meteorites

Craters due to impacts on the surface of planets and their satellites are well known and there are some excellent books on this topic (i.e. Melosh, 1989, 2011). Such craters are also found on Earth. It is well known that nickel, copper, platinum, gold and palladium ores in the Sudbury basin was caused by the impact of a gigantic large meteorite. It is also well-known that the impact by a gigantic meteorite at Yucatan Peninsula (Mexico) caused the extinction of dinosaurs on Earth 66 million years ago. This crater is named Chicxulub Crater with an estimated diameter of 180 km. The 1908 airburst of Tunguska event (M5) in Siberia caused heavy damage to wooden buildings and blew down 90% of trees within a radius of about 15 km. A similar event took place again in the Chelyabinsk region on Feb. 13, 2013, causing damage to buildings and the surroundings due to airburst (Figure 14.2). The estimated seismic magnitude is about 2.7 according to USGS. The most recent event of meteorite fall occurred near Sarıçiçek village in the Bingöl Province of Turkey.

Figure 14.2 Some views of the meteorite that fell in Chelyabinsk region on Feb. 13, 2013 and its damage (pictures from AP Photo, see also IRIS, 2013).

The cratering process has been well investigated by various researchers (i.e. Melosh, 1989, 2011; Collins *et al.*, 2005; Kenkmann *et al.*, 2011) and Figure 14.3 shows an illustration of impact cratering. The impact mechanism by meteorite is divided into three main stages; a) contact and compression; b) excavation stage; c) modification stage. Depending upon the size of meteoroid, the craters are categorized as simple and complex craters as illustrated in Figure 14.4.

The entry velocity of meteoroids ranges from 11 km/sec to 72 km/sec. However, the speed of a meteoroid rapidly decelerates as it penetrates into increasingly denser portions of the atmosphere. Due to atmospheric drag, the weight of most meteorites ranges from a few kilograms up to about 7,000 kg and their terminal velocity is about 90–180 m/s when they hit the ground.

Kenkmann *et al.* (2011) conducted laboratory experiments with dry and wet sandstone blocks impacted by centimeter-sized steel spheres. They utilized a powerful two-stage light gas gun to achieve impact velocities of up the 5.4 km/s. They concluded that cratering efficiency, ejection velocities, and spall volume are enhanced if the pore space of the sandstone is filled with water. In addition, the crater morphologies differed substantially in both experiments and the vaporization of water upon pressure release significantly contributed to the impact process. Strength and elastic modulus of sandstone used in experiments were 62.4 ± 2.8 MPa and 14.8 ± 1.4 GPa

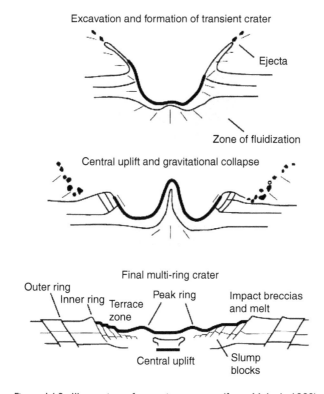

Figure 14.3 Illustration of cratering process (from Melosh, 1989).

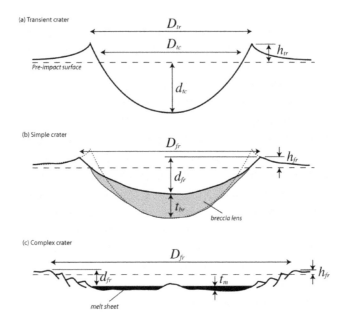

Figure 14.4 Illustration of cratering process (from Collins *et al.*, 2005).

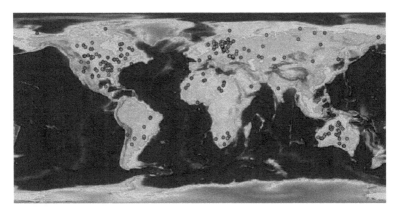

Figure 14.5 Distributions of terrestrial impact structures (from Osinski, 2006).

for the dry, and 47.0 ± 3.7 MPa and 12.1 ± 1.0 GPa for a water-saturated sandstone measured perpendicular to the bedding planes.

Major terrestrial craters found on Earth are well documented (i.e. Osinski, 2006). Figure 14.5 shows the distribution of major craters around the world. However, it should be noted that there may be some hidden large craters in sea-beds or beneath the ice-covered north and south poles, which have gone unnoticed so far.

14.1.2   Effects of meteorites

Effects of terrestrial impacts on the earth may be categorized as airburst, seismic, dust-fall and tsunami. In this sub-section, the state of the art on these aspects is summarized. The readers are advised to consult the quoted references for further information.

Collins *et al.* (2005) summarized some fundamental formulae to estimate the geometry of craters, seismic effects, dust-fall induced by the impact of meteoroids. The kinetic energy (K) of an impactor with mass (m_i) and velocity (v_o) just before atmospheric entry can be written as

$$K = \frac{1}{2}m_i v_o^2 \tag{14.1}$$

If the impactor is generally assumed to be spherical body, its initial mass is given as

$$m_i = \frac{\pi}{6}\rho_i D_o^3 \tag{14.2}$$

Density of the earth's atmosphere is assumed to be decaying outward from the given density at the surface of the earth and it is given by the following equation

$$\rho = \rho_o e^{-z/H} \tag{14.3}$$

The velocity of the impactor as a function of altitude is derived on the basis of exponential decaying of the density of Earth's atmosphere and is given by:

$$v(z) = v_o \exp\left\{ -\frac{3\rho(z)C_D H}{4\rho_i D_o \sin\theta} \right\} \tag{14.4}$$

The estimation of the geometry of craters caused by meteorites is a very complex problem. The following are relations for the diameters of transient, simple and complex craters, its depth and volume of breccias:

Transient Diameter

$$D_{tc} = 1.161 L^{0.78} v_i^{0.44} g_E^{-0.22} \sqrt[3]{\frac{\rho_i}{\rho_t}} \sin\theta \tag{14.5}$$

Simple Crater Diameter

$$D_{fr} = 1.25 D_{tc} \tag{14.6}$$

Volume of Breccias

$$V_{br} = 0.032 D_{tc}^3 \tag{14.7}$$

Depth of Breccias

$$d_{fr} = 0.4 D_{fr}^{0.3} \tag{14.8}$$

Seismic effect is related through the empirical relation given by the Gutenberg-Richter magnitude (M)-energy relation:

$$M = 0.67 \log_{10} E - 5.87 \tag{14.9}$$

where E is assumed to be equal to the kinetic energy of the impactor expressed in Joules. The peak overpressure decreases exponentially with distance from ground zero:

$$p = p_o e^{-\beta r_1} \tag{14.10}$$

where p_o and β are functions of burst altitude given by

$$p_o = 3.14 \times 10^{11} z_b^{-2.6} \tag{14.11}$$

where

$$\beta = 34.87 z_b^{-1.73} \tag{14.12}$$

$$r_1 = \frac{r}{E_{kT}^{1/3}} \tag{14.13}$$

14.2   CRATER FORMATION BY PROJECTILES IN ROCKS

Craters produced by missiles with oblique trajectories and kinetic energies between 2.1×1014 and 81×1014 ergs produce craters 2 to 10 meters across. The craters and their ejecta are typically bilaterally symmetrical about the plane of the trajectory. Crater dimensions are strongly dependent on target material and kinetic energy of the missile. Craters in wet targets are significantly larger than those in dry porous targets (Moore, 1971). The explosion velocity of missiles generally ranges between 71 m/s to 204 m/s.

Bullets or shrapnel have similar characteristics and the size of craters depends upon the characteristics of target material as well as the size of bullets or shrapnel. Figure 14.6 shows several pictures of craters on the wall rocks of Victoria and Albert Museum in London induced by the shrapnel of the bombs. The craters are generally 100–150 mm in diameter having a depth of 15–30 mm and the wall rock is Portland stone, which is a Jurassic aged limestone with a uniaxial compressive strength of about 70–80 MPa. The bullet muzzle velocity is in the range of 1067 m/s to 1524 m/s while the conventional artillery cannons ranges between 914 m/s to 1067 m/s.

14.3   MONITORING OF VIBRATIONS CAUSED BY METEORITES

There is almost no remarkable seismic record of meteorite impacts on Earth. The record reported by Roelofse and Launders (2013) is probably the first seismic record of a meteorite. The record is shown in Figure 14.7 and it was measured by the Mussina seismic station of South African National Seismograph Network. The signal consists of two distinct parts: a signal originating at 20:55:37.59 (UT) with a duration of

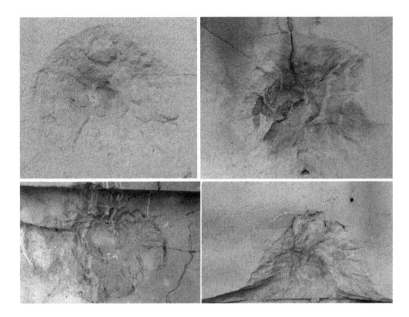

Figure 14.6 Craters on the wall of Victoria and Albert Museum induced by the bomb shells.

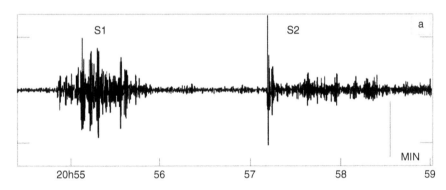

Figure 14.7 Seismic record at the Mussina seismograph by a meteorite (from Roelofse and Launders, 2013).

approximately 77 s (Figure 14.7), which is referred to as S1, and a second signal (S2) commencing almost 138 s after the first at 20:58:11.08 (UT). The second signal's duration is approximately 193.14 s.

As mentioned in the previous sub-section, the recent meteorite impact in Chelyabinsk on Feb. 13, 2013 induced seismic surface waves observed at five seismograph stations of the IRIS/USGS Global Seismographic Network (IRIS, 2013) (Figure 14.8). The nearest station is denoted by ARU, which is about 160 km away from the impact location. In addition, the infrasound detectors installed at USArray stations extending from the midcontinent to the eastern seaboard. The USArray recordings from the meteor show complex arrivals lasting for almost an hour.

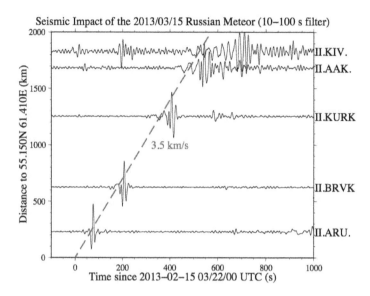

Figure 14.8 Surface wave records at several seismograph stations induced by the Chelyabisnk Meteorite (IRIS, 2013).

The entry mass was 13000 tons with a diameter of about 20 m and a density of 3.3 g/cm³. A 7-m sized hole was discovered in 70-cm thick ice on Lake Chebarkul. Small meteorite fragments were recovered over an area up to 50 m from the impact location. It is estimated that a 200–1,000 kg meteorite would be required to create such a hole. A mass of ≥570 kg was recovered from the lakebed.

NASA had installed some seismometers to observe earthquakes and meteorite impacts in Moon. Since 1970, four stations with seismic sensors have been placed on the moon at Apollo 14, 15, 16, and 17 sites (i.e. Toksöz, 1975; Nakamura et al., 1974; Khan et al., 2013). When combined with the seismic station operating at the Apollo 12 site since November 1969, these have formed an excellent network. The network has detected a large number of moonquakes and meteoroid impacts. In addition to these there have been nine artificial impacts (ascent stage of the lunar module (LM) and Saturn third-stage booster (S4B)) of known energies, impact times, and locations (Toksöz, 1975). The records from meteoroid impacts and lunar module (LM) of Apollo 12 are shown in Figure 14.9. Due to characteristics of the inherent internal structure of the moon (Toksöz et al., 1972, 1974), dissipation of surface waves induced by surface impacts on the Moon takes a very long time.

14.4 FREE-FALL (DROP) EXPERIMENTS

The dynamic response of objects during free-fall and impact may have some important implications in earth science and geo-engineering. In this sub-section, some free-fall (drop) experiments under gravitational environment are presented (Aydan, 2004).

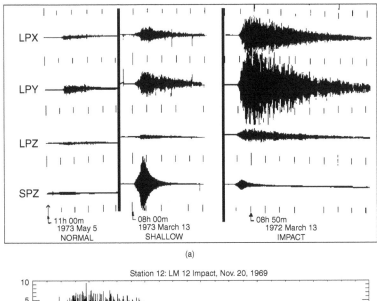

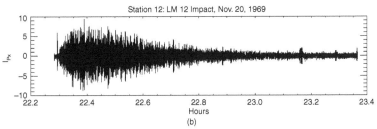

Figure 14.9 Seismic records for the Moon. (a) Comparison of seismic records due to meteoroid impacts with moonquakes (i.e. Nakamura *et al.*, 1974; Toksöz, 1975; Toksöz *et al.*, 1972, 1974); (b) seismic records at Station 12 on Moon by Lunar Module 12 (Lindsay, 2008).

14.4.1   Objects falling onto dry sand layer or a mixture of BaSO₄, ZnO and Vaseline oil

The experimental set-up consists of the cylindrical and spherical objects with given dimensions and a container with dry sand or a mixture of $BaSO_4$, ZnO and Vaseline oil as illustrated in Figure 14.10. The falling object and container are equipped with accelerometers which can measure the acceleration up to 12G. The physical and strength characteristics of sand are given in Table 14.1 and its grain-size distributions are shown in Figure 14.11. During tests, the size of objects and free-fall height were varied and accelerations of the impact object and container were measured. In addition, the embedment depth following the impact was also measured. The objects were concrete, glass, marble and iron. Figure 14.12 shows views of some of the experiments.

(a) Experiments with concrete cylinders

Concrete cylinders had a diameter of 100 mm and height of 200 mm. Their weight was 3750 gf (unit weight: 23.87 kN/m³). They were dropped from a height of 1000 mm on the dry sand layer. Three tests were carried out. Figure 14.13 shows measured

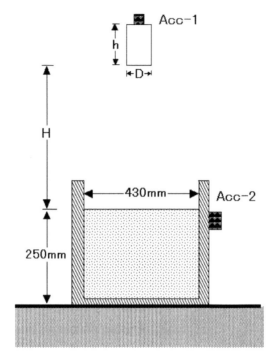

Figure 14.10 Experimental set-up for free-fall and impact tests.

Table 14.1 Physical and mechanical properties of dry sand.

Mean Grain Size (mm)	Dry Unit Weight (kN/m³)	Porosity (%)	Friction Angle (°)
0.46–0.60	14.7	43.73	32–34

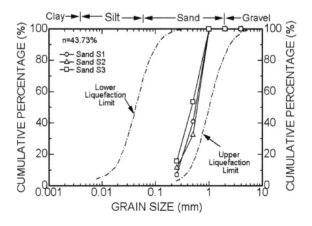

Figure 14.11 Grain size distributions of the sand.

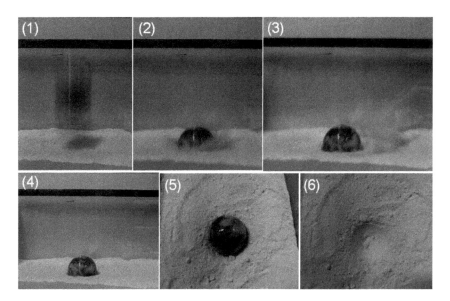

Figure 14.12 Views of a marble sphere during different stages of drop and its crater.

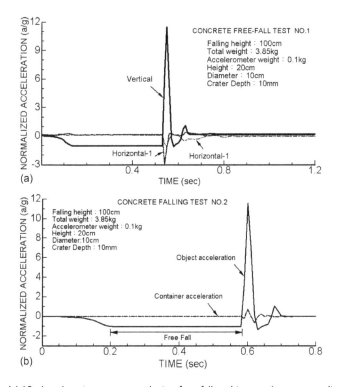

Figure 14.13 Acceleration response during free-fall and impact (concrete cylinders).

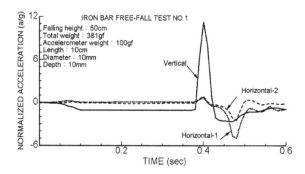

Figure 14.14 Acceleration responses during free-fall and impact (iron bars).

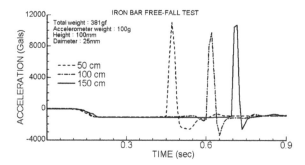

Figure 14.15 Vertical acceleration responses during free-fall and impact (iron bars) (drop height is variable: 50, 100 and 150 cm).

accelerations during one of the experiments. It is of great interest that the maximum accelerations of the concrete cylinders are almost the same; about 11.6–11.7G.

When the object is released, the vertical acceleration reaches gravitational acceleration and it remains the same during free-fall as expected. The maximum ground acceleration occurs during impact and the generated wave form is not symmetric. The ground acceleration at a distance of $R/a = 4.3$ is about 0.6–0.72G and its wave form is symmetric. From the experimental results, it seems that the following expression holds:

$$ma_{max} = (m + M)a_g \tag{14.14}$$

where m and M are the masses of the object and ground permanently displaced, a_{max} and a_g are the accelerations of the object and the ground at the time of impact. The cratering depth was 10 mm and it was almost the same in all experiments.

(b) Experiments with iron cylinders

Iron cylinders had a diameter of 25 mm and height of 100 mm. They were dropped from heights of 500, 1000 and 1500 mm on dry sand layer. Three tests were carried out. Figure 14.14 and Figure 14.15 show measured accelerations during the experiments.

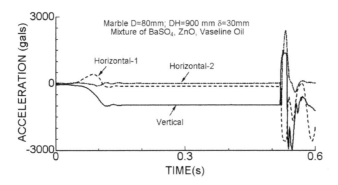

Figure 14.16 Acceleration responses during free-fall and impact (marble sphere).

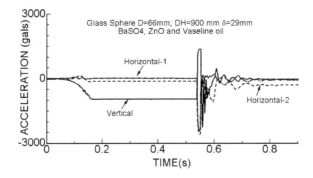

Figure 14.17 Acceleration responses during free-fall and impact (glass sphere).

It is of great interest that the maximum acceleration of the iron cylinders varies with and it is more than 11G and it seems that the vertical accelerations are same for three different drop heights as seen in Figure 14.15. When the object is released, the vertical acceleration reaches gravitational acceleration and it remains the same during free-fall as expected. However, the cables of accelerometer might impose some resistance to the magnitude of the maximum acceleration during free-fall.

(c) Experiments with marble sphere and glass sphere

A series of experiments was carried out on a marble sphere with a 80 mm diameter and a glass-sphere with a diameter of 66 mm. The spheres were dropped from a height of 900 mm on a loosely compacted layer of a mixture of $BaSO_4$, ZnO and Vaseline oil. Accelerations were measured using stand-alone type accelerometers developed by the author. Views of the marble sphere during different stages of free-fall are shown in Figure 14.12. The all stages of cratering depicted by Mellor (1989) is clearly observed in the tests despite the color of base ground is whitish.

Figure 14.16 and Figure 14.17 show the acceleration responses measured during free-fall and impact. As noted from the figure, the vertical acceleration is equal to gravitational acceleration as expected. However, the vertical acceleration becomes

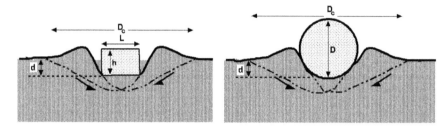

Figure 14.18 Types and parameters of craters created by the impact of falling objects.

reversed is more than 2.4 times the original acceleration. Furthermore, the horizontal accelerations are also remarkably high and 2.3–2.5 times the gravitational acceleration. The penetration depth of ranged between 28–31 mm for spheres. The penetration depth in the loosely compacted layer of a mixture of $BaSO_4$, ZnO and Vaseline oil was remarkably higher than that in dry sandy layer. As a result, the maximum accelerations observed during impact were significantly lower in the loosely compacted layer of a mixture of $BaSO_4$, ZnO and Vaseline oil. This implies that the ground motions caused by the impactor in soft ground are less as compared with those on hard ground.

(d) Cratering

The impact of objects causes cratering in the ground as illustrated in Figure 14.18. The energy dissipated through the bearing failure of ground beneath the object and elastic wave propagation. The energy (E) of the falling object under gravitational field the time of impact may be written in terms of the fall height (H) and the weight (W) of the object as follows:

$$E = W \cdot H \tag{14.15}$$

The energy intensity (e) per unit area (A_c) of the object projected in the direction of movement may be defined as follows:

$$e = \frac{E}{A_c} \tag{14.16}$$

Figure 14.19 shows the relation between the energy intensity and embedment depth for cylindrical steel and tuff samples on dry sand. The falling direction was perpendicular to the longitudinal axis of the samples. It seems that penetration depth normalized by the diameter of the object is influenced by the material type of the objects.

14.4.2   Sand bags falling onto hard-base

The author with his colleagues carried out some impact tests by dropping a 1000 kgf sand bag onto a hard base from a height of 1 m and 1.8 m in relation to the assessment

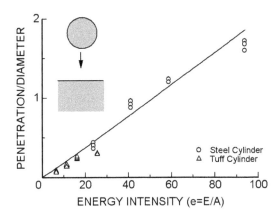

Figure 14.19 The relation between energy intensity and penetration depth.

Figure 14.20 Views of a drop test.

of impact forces on a large underground sub-structure built for the protection of karstic caves by landing airplanes (Aydan and Tokashiki, 2011). A total of 10 accelerometers (6 accelerometers are compact stand-alone type) were used and one of the accelerometers was attached to the falling object and four of the accelerometers were installed in the underground structure, which were 23 m from the drop location. Views of a drop test are shown in Figure 14.20. Five different layout patterns of drop tests were designed and Figure 14.21 shows the acceleration records for Pattern-0.

The maximum acceleration was about 23G when the sand-bag hits the hard-base. However, the maximum acceleration quickly decreases with distance. Taking

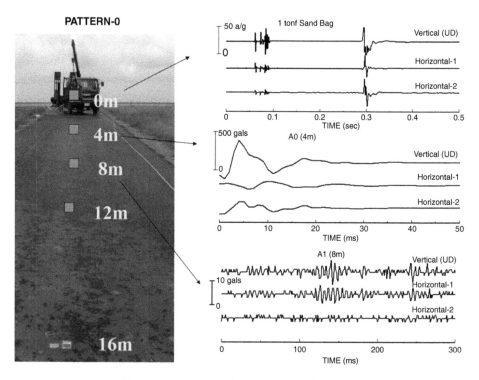

Figure 14.21 Acceleration records measured in a Pattern-0 drop test.

into account possible hemi-spherical wave propagation from the drop location, the measured maximum accelerations were fitted to the following empirical relation.

$$a_{\max} = 20000e^{0.8*W*h}\frac{1}{1+r^3} \tag{14.17}$$

where r, W and h are radial distance (m), drop object weight (kgf) and drop height (m), respectively. The unit of acceleration is gals. Figure 14.22 compares estimations by Eq. (14.17) with measured maximum accelerations.

The measurements for all patterns of instrumentation including measurements in underground structure are plotted in Figure 14.23. Eq. (14.17) was modified as given below:

$$a_{\max} = Ae^{0.8*W*h}\frac{1}{1+r^b} \tag{14.18}$$

where b and A are empirical constants. The value of b ranges between 2 and 3 while the value of A is 20000. When b is 2, it corresponds to cylindrical attenuation.

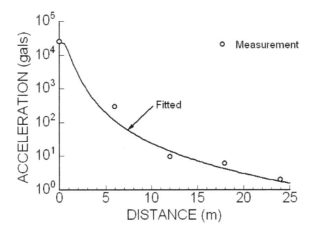

Figure 14.22 Comparison of estimations with measurements for Pattern-0.

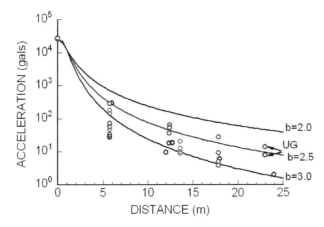

Figure 14.23 Comparison of estimations with measurements.

On the other hand, if the value of b is 3, it corresponds to spherical attenuation. When the measurements shown in Figure 14.23 are taken into account, the best fit was given when the value of b is 2.5.

The drop height was increased from 1 m to 1.85 m and the validity of the attenuation relation given by Eq. (14.18) was checked. Figure 14.24 compares the measurements with the estimations from Eq. (14.18). Despite some scattering of measurements, the empirical relation could be used for estimating maximum accelerations with distance. However, it should be noted that the value of A would probably depend upon the stiffness of the base.

14.4.3   Drop or back-hoe impact test at a bridge foundation site

Some impact tests by a back-hoe bucket were carried out at a bridge foundation site consisting of weak sandy-gravelly limestone formation with the purpose of

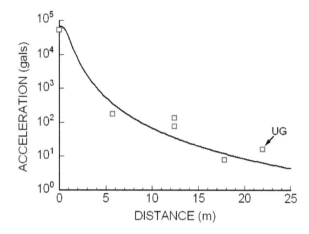

Figure 14.24   Comparison of estimations with measurements (drop height: 1.85 m and the value of b is 2.5).

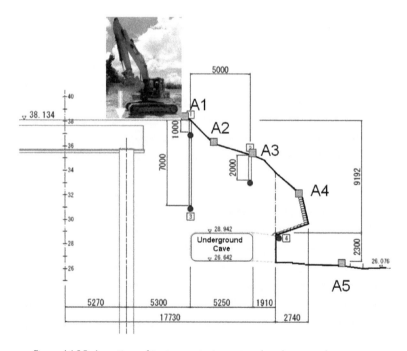

Figure 14.25   Location of instruments, impact and underground structure.

investigating the effect of piles of the abutment of a bridge on a nearby underground structure. Five accelerometers were spaced with a distance of roughly 5 m with a sampling interval of 50 microseconds. Figure 14.25 shows the locations of the underground cave, impact and accelerometers on a cross-section of the construction site. The nearest accelerometer to the impact location is about 5 m while the farthest accelerometer is

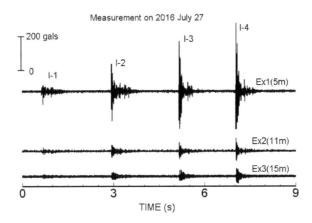

Figure 14.26 Acceleration records due to the impact by the back-hoe bucket.

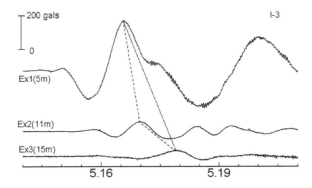

Figure 14.27 A close-up of acceleration records induced by the third impact in Figure 14.26.

about 25 m. Figure 14.26 shows the acceleration records of the devices located at a distance of 5, 11 and 15 m, respectively. As anticipated from previous investigations, the amplitude of accelerations decreases with distance. Although the acceleration of the impact induced by the backhoe bucket was not measured, the overall tendency is similar to that estimated from Eq. (14.18). Furthermore, the P-wave velocity was estimated to be ranging between 670 m/s to 2000 m/s (Figure 14.27). The velocity of the top layer is much higher than that of the below.

14.5 IMPACT OF SLOPE FAILURES

An experimental set-up was created to investigate the effects of impacts induced by slope failures. For this purpose, planar failure was considered and the inclination plane was selected as 30 and 60 degrees. The rock was selected Ryukyu limestone

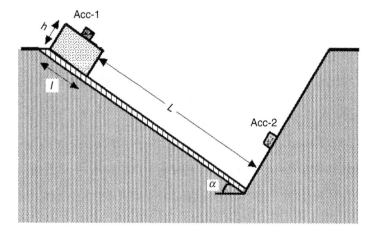

Figure 14.28 Experimental set-up for sliding and impact tests.

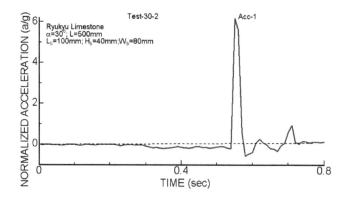

Figure 14.29 Measured acceleration and temperature variations ($\alpha = 30°$).

and planes were saw-cut. The size of sliding rock block was $40 \times 80 \times 100$ mm and a Ryukyu limestone base slab with a thickness of 10 mm fixed on a frame as illustrated in Figure 14.28. The sliding block was equipped with an accelerometer. The accelerometer can measure acceleration up to 12G and they were filtered at 10 Hz. The friction of friction angle of saw-cut surfaces was about 26 degrees. In some experiments, the surface of the stopper was also instrumented with an accelerometer. The travel distance of the block was 500 mm and the acceleration during the sliding stage for a friction resistance can be shown to be

$$a_s = g(\sin \alpha - \cos \alpha \tan \phi) \tag{14.19}$$

where ϕ is friction angle of the plane inclined at an angle α.

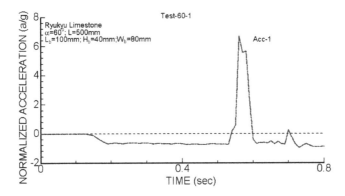

Figure 14.30 Measured acceleration and temperature variations ($\alpha = 60°$).

Three tests were carried out for each inclination of failure plane. Figure 14.29 and Figure 14.30 show measured acceleration response during the experiments for inclination of 30 and 60 degrees, respectively. The acceleration of the block, which can be inferred from Eq. (14.19), remains the same during sliding and it has the maximum value when the block hits the restraining wall. In other words, the acceleration is small and constant if the sliding velocity is constant. The acceleration is quite high when the block movement was restrained and it is more than 6 times the gravitational acceleration. Furthermore, the measured acceleration wave was not symmetric with respect to time axis. Any subsequent peak accelerations are always smaller than that caused by impact unless the block has fallen from the frame. When the inclination of the failure planes was 60 degrees, the period of sliding was shorter. Although the maximum acceleration due to impact for the failure plane inclination of 60 degrees was slightly higher than that for the failure plane inclination of 30 degrees.

It is of great interest that the maximum acceleration occurs only during the impact and it is much smaller during sliding process. In other words, the maximum acceleration must be observed when the motion of the sliding body is restrained. This is a very important conclusion and it has a great implication on the expected ground accelerations during slope failures as well as earthquakes.

The experiments were recently repeated using other type rocks and it was found that the results were almost the same. Ryukyu Limestone plate with a thickness of 20 mm and length of 300 mm was put on the restrainer side. The plate was found to be ruptured at the mid-level for the failure plane inclination of 60 degrees.

14.6 FORMULATION OF IMPACTOR PENETRATION AND ITS APPLICATIONS

14.6.1 Mechanical modeling

Let us consider an impactor with a given mass (m) and geometry is released from at a given height H_f above the ground surface (Figure 14.31). If the air resistance

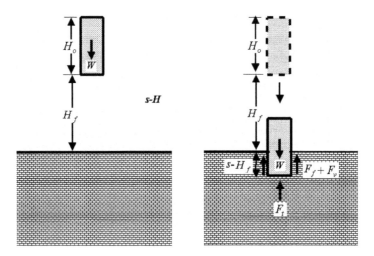

Figure 14.31 Mechanical modeling of penetration of impactor into ground.

is neglected, the dynamic equilibrium equation of the object during freefall may be written as follows

$$\sum F_v = mg - m\frac{d^2s}{dt^2} = 0 \qquad (14.20)$$

where g is gravitational acceleration and s is the distance from release point.

On the other hand, the resistance from the ground against the motion of the object would be imposed following its impact and penetration into the ground, provided that the collision is inelastic (without rebounding). The resistance (F_r) of the ground may consist of three components, namely, tip bearing resistance, friction between the object and ground and velocity dependent resistance. Therefore, one may write the following relation for the dynamic equilibrium equation of the object during impact and penetration

$$\sum F_v = mg - F_r - m\frac{d^2s}{dt^2} = 0 \qquad (14.21)$$

The tip resistance of the ground may be evaluated using the concept of the bearing capacity of foundations and proportional to the mass of the falling object as follows $(s > H_f)$:

$$F_t = mg \cdot q \left(1 + \alpha \frac{s - H_f}{D}\right) \qquad (14.22)$$

where

$$q = \frac{1 + \sin\phi}{1 - \sin\phi} \cdot \frac{\rho_s}{\rho_o}; \quad \alpha = \tan\left(45 - \frac{\phi}{2}\right) \qquad (14.23)$$

The tip resistance may be neglected for sufficiently slender objects. The second term on the right hand side of Eq. (14.23) takes into account the effect overburden following the penetration of the object.

The frictional resistance between the object and surrounding ground may be given as follows

$$F_f = \mu \frac{s - H_f}{H_o} mg \quad \text{for } (s - H_f) < H_o \qquad (14.24)$$

and

$$F_f = \mu mg \quad \text{for } (s - H_f) > H_o \qquad (14.25)$$

The viscous component of the resistance of ground against penetration may also be given in the following form:

$$F_V = mg \cdot \beta v^n \qquad (14.26)$$

where β is viscous coefficient, v is velocity and n is a power coefficient. Thus the total resistance from the ground against impact and penetration would be a sum of the above components as follows:

$$F_r = F_t + F_f + F_v \qquad (14.27)$$

14.6.2 Solution procedure

While Eq. (14.20) is quite easy to solve, Eq. (14.21) would be difficult to solve due to its non-linear characteristics of viscous resistance and cumbersome conditions resulting from the tip bearing resistance and frictional resistance. Therefore, a numerical integration technique would be employed in this study. For a given time step $n+1$, Eqs. (14.20) and (14.21) may be represented in a unified style as follows

$$\left(\frac{d^2 s}{dt^2} \right)_{n+1} = A_r^{n+1} \qquad (14.28)$$

with initial conditions as follows

$$s = 0 \quad \text{and} \quad v = 0 \text{ at } t = 0 \qquad (14.29)$$

The velocity and distance from the release point can be obtained from the following discretised equation using the Feynman-Newton method as follows

$$v_{n+1} = v_n + A_r^{n+1} \Delta t \qquad (14.30a)$$

$$s_{n+1} = s_n + v_{n+1} \Delta t \qquad (14.30b)$$

with

$$v_0 = \frac{1}{2} A_r^1 \Delta t \qquad (14.30c)$$

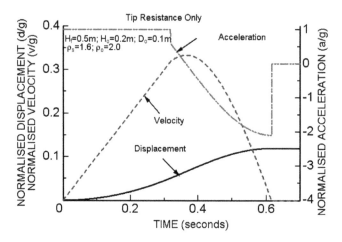

Figure 14.32 Computed responses of the acceleration, velocity and penetration depth of an impactor (CASE 1: Tip resistance only).

14.6.3 Examples

Some computational examples are given herein and their implications are discussed. The physical and geometrical properties of ground, object and freefall height are given in each respective figure. In computations three different cases are analyzed:

CASE 1: Tip resistance only,
CASE 2: Tip and side frictional resistance, and
CASE 3: Tip, side friction and viscous resistance.

Figure 14.32 shows the computed acceleration, velocity and displacement responses for the properties shown in the same figure. In the freefall stage, the acceleration is constant, velocity linearly increases as expected from the integration of Eq. (14.20). After impact, the object starts to penetrate into the ground. As a result, the velocity starts to decrease and the motion of the impactor tends to decelerate. When the velocity of the impactor becomes nil, the motion of the impactor terminates. The computational responses for CASE 2 and CASE 3 are fundamentally similar to those for CASE 1 as shown in Figures 14.33 and Figure 14.34. The biggest difference is that the penetration depth decreases due to the increase of the resistance from the ground.

14.7 WATER SURFACE CHANGES DUE TO IMPACTORS

14.7.1 Tsunami occurrence by meteorite impacts

Following the tsunami by the 2004 Off-Sumatra earthquake and the recent Chelyabinsk meteorite impact in Russia, there is a growing interest in the possibility of tsunami occurrence induced by meteorite impacts (i.e. Weiss *et al.*, 2006; Wünnemann

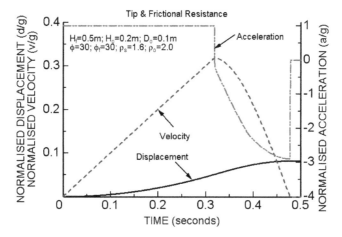

Figure 14.33 Computed responses of the acceleration, velocity and penetration depth of an impactor (CASE 2: Tip + frictional resistance).

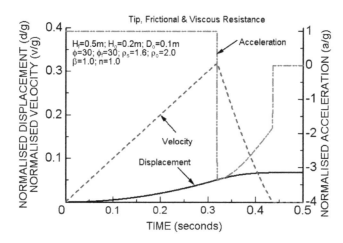

Figure 14.34 Computed responses of the acceleration, velocity and penetration depth of an impactor (CASE 3: Tip + frictional and viscous resistance).

et al., 2010). While there are no historic examples of meteorite impacts to have produced a tsunami, the impact of a meteorite at the end of the Cretaceous Period, about 65 million years ago near the tip of the Yucatan Peninsula of Mexico, produced a tsunami that left deposits all along the Gulf coast of Mexico and the United States (i.e. Ward and Aspaug, 2000; Weiss *et al.*, 2006; Wünnemann *et al.*, 2010). Ward and Asphaug (2000) have simulated the tsunami induced by impact of a meteorite with a diameter of 200 m hitting the sea surface at a velocity of 20 km/s. Figure 14.35 shows their computational results.

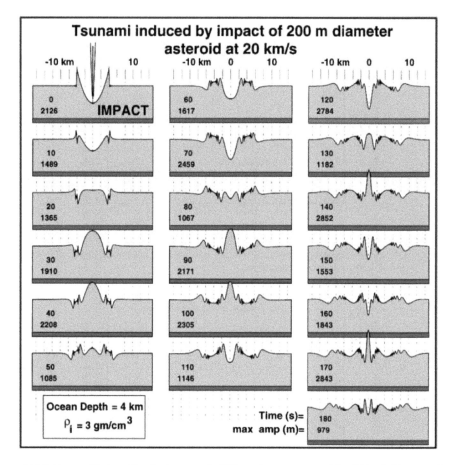

Figure 14.35 Tsunami induced by the impact of a 200-m-diameter asteroid at 20 km/s. The waveforms (shown at 10-s intervals) trace the surface of the ocean over a 30-km cross section that cuts rings of tsunami waves expanding from the impact site at $x = 0$. Maximum amplitude in meters is listed to the left (from Ward and Asphaug, 2000).

14.7.2 Experiments on water-level variations due to impactor in closed water bodies

The disastrous effects of tsunamis are are well known world-wide and tsunamis are generally caused by the earthquakes of normal faulting or thrust faulting types. They may also be caused in closed-water bodies even by the strike-slip faulting. The rockslide into Vaiont dam in 1960 into the reservoir caused a huge tsunami and caused the inundation of a settlement downstream and killed more than 2000 people. The 2008 Iwate-Miyagi intraplate earthquake also caused a huge landslide resulting in a small-scale tsunami in the reservoir of Aratozawa Dam as seen in Figure 14.36 (Aydan, 2015). The landslide acted like an impactor on the reservoir.

An experimental study on the tsunami generation in reservoirs was performed by Aydan (2006). The object was dropped from different heights and induced tsunami

Figure 14.36 A small-scale tsunami induced by the landslide in Aratozawa Dam induced by the 2008 Iwate-Miyagi intraplate earthquake (from Aydan, 2015).

waves were recorded. For this purpose, a cylindrical acrylic container with an internal diameter of 144 mm and water height of 75 mm was used a reservoir and a plastic prismatic block ($50 \times 50 \times 40$ mm) was used as a falling object. The wave height was measured using a water pressure sensor attached to the side-wall of the reservoir and the acceleration of the base was measured in order to evaluate the time of the impact and its vibration on the surroundings (Figure 14.37). The unit weight of the prismatic block was 10.6 kN/m^3. The drop height of the block was selected as 0, 50, 100 and 150 mm. The variation of drop height (h) changes the initial impact velocity (v_{in}) of the falling body, which may be given in the following form by considering the transfer of potential energy into kinetic energy:

$$v_{in} = \sqrt{2gh} \tag{14.31}$$

where g is gravitational acceleration.

Four experiments were carried out by varying the height of drop of the falling object. Figures 14.38 to 14.41 show the time histories of water level variations and acceleration for 10 seconds. The fluctuations of tsunami waves disappeared almost after 10 seconds in an exponential manner. As expected the maximum wave height increased as the drop height increased. The recorded acceleration at the base of the model occurred just after the peak tsunami wave height and remained almost constant thereafter.

The water level of the reservoir increases in proportion to the volume of falling object. The water level increases for the given geometry of the experimental set-up when the fluctuations disappear can be expressed by the following expression:

$$\Delta H = \frac{4 \times a \times b \times c}{\pi D^2} \tag{14.32}$$

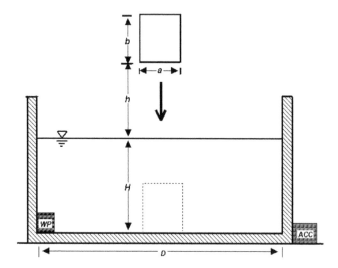

Figure 14.37 An illustration of the experimental set-up.

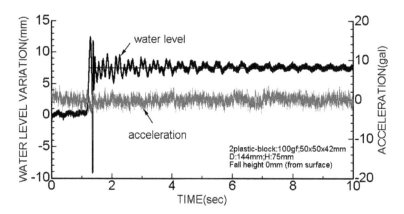

Figure 14.38 Time histories of water height and acceleration for drop height of 0 mm.

The experimental results were consistent with the above relation. Figure 14.42 shows the relation between drop height and induced maximum wave height together with experimental results. The experimental results fit the following relation very well:

$$\Delta h = \left(37.5 + 6.9h^{0.7}\right)\frac{r_1}{r_2} \tag{14.33}$$

where r_1 and r_2 are the equivalent radius of the falling object at the time of impact and the radius of the leading edge of the wave. r_2/r_1 is about 3 in the experiments.

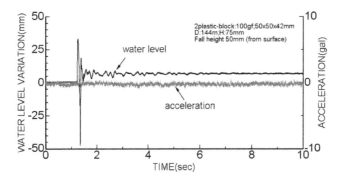

Figure 14.39 Time histories of water height and acceleration for drop height of 50 mm.

Figure 14.43 shows the relation between normalized drop height and induced wave height by the height of the falling object together with experimental results. The experimental results fit the following relation very well:

$$\frac{\Delta h}{b} = \left(1.0 + 2.25 \left(\frac{h}{b}\right)^{0.7}\right) \frac{r_1}{r_2} \tag{14.34}$$

It is also interesting to note that the power of the function remains the same.

14.7.3 Theoretical modeling on water-level variations due to impactor in closed water bodies and its applications

The water-level variations in closed water bodies can be related to the volume change due to the volume change of the impactor. Let us consider a very simple geometry of both closed water body and impactor. Their volumes are given as

$$V_w = H \cdot L \cdot B; \quad V_i = h \cdot l \cdot b \tag{14.35}$$

H, h, L, l, B and b are dimensions, respectively. By considering the mass conservation law, the final water level change (u_f) may be given in the following form

$$u_f = \frac{l}{L} \cdot \frac{b}{B} h \tag{14.36}$$

The vertical force equilibrium relation may be written as

$$\sum F_v = W - F_v - P - m \frac{d^2 s}{dt^2} = 0 \tag{14.37}$$

where

$$W = \rho_w g (H \cdot B \cdot L - u_b \cdot l \cdot b + u \cdot L \cdot B); \quad P = \rho_w g \cdot H \cdot L \cdot B; \quad m = \rho_w H \cdot L \cdot B;$$

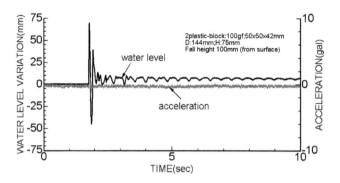

Figure 14.40 Time histories of water height and acceleration for drop height of 100 mm.

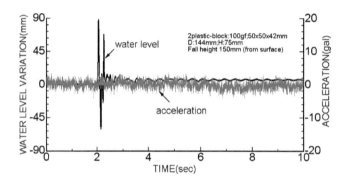

Figure 14.41 Time histories of water height and acceleration for drop height of 150 mm.

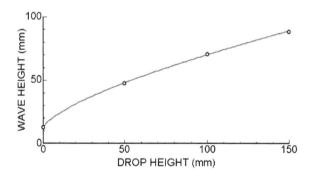

Figure 14.42 Comparison of the empirical relation with experimental results.

ρ_w: density of water; u_b: displacement of water body by impactor; u: water body level change; g: gravitational acceleration.

The rate dependent resistance (F_v) of the water body against motion may be assumed to given in the following form:

$$F_V = \rho_w \cdot H \cdot L \cdot B \cdot g \cdot \beta \cdot v^n \tag{14.38}$$

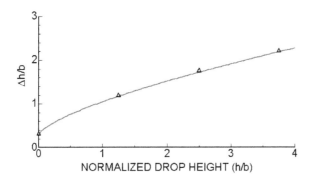

Figure 14.43 Comparison of the empirical relation with experimental results.

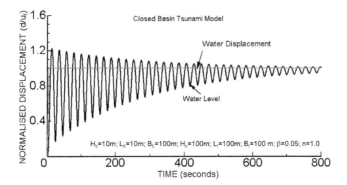

Figure 14.44 Water level fluctuations following the entry of impactor.

where β is viscous coefficient, v is velocity and n is a power coefficient. Thus the following differential equation is obtained for the water level change of closed water body

$$\frac{d^2u}{dt^2} + \beta g \left(\frac{du}{dt}\right)^n + g\frac{u}{H} = g\frac{l}{L} \cdot \frac{b}{B} \cdot \frac{u_b}{H}$$ (14.39)

If the variation of displaced body may be assumed to be

$$u_b = \frac{t}{T_r}u_f$$ (14.40)

where T_r is called "rise time".

Ordinary differential equation given by Eq. (14.39) is non-homogenous and non-linear. Its solution can be obtained through only numerical methods. The procedure described in Sub-section 14.5.2 can be utilized. Figure 14.44 shows computational results for a water body displaced by a prismatic impactor. The entry period of the impactor was assumed to be 5 seconds in this particular example.

Chapter 15

Conclusions

Rock dynamics has become one of the most important topics in the field of rock mechanics and rock engineering. As pointed out in Chapter 1, the spectrum of rock dynamics is very wide and it includes, the failure of rocks, rock masses and rock engineering structures such as rockbursting, spalling, popping, collapse, toppling, sliding, blasting, non-destructive testing, geophysical explorations, science and engineering of rocks and impacts. However, there are very few publications on rock dynamics and many publications are related to rock mechanics and rock engineering under static conditions. The main purpose of the author has been to accomplish a unified and complete treatise on Rock Dynamics and to be a mile-stone in advancing the knowledge in this field and leading to the new techniques for experiments, analytical and numerical modelling as well as monitoring in dynamics of rocks and rock engineering structures.

The fundamental governing equation used in the dynamics of materials and structures is presented in Chapter 2 and it is pointed out that the fundamental equation of motion would be valid irrespective of how rock mass is treated. However, rock mass in nature has discontinuities in the form of tiny cracks in small scale to fault zones in large scale, such discontinuities require treating in an appropriate manner in any type of analyses involving rock masses. The fundamental constitutive models for rocks, discontinuities and rock masses are described in detail and some fundamental features of the numerical methods used in the rock dynamics problems are outlined. This chapter provides fundamentals for anyone interested in Rock Dynamics in a rigorous manner.

The experimental techniques and monitoring equipment are quite important for evaluating the dynamic characteristics of rocks, discontinuities and rock masses. The advances in sensor and monitoring technology enabled scientists and engineers to carry out experiments for understanding of multi-parameter variations including electric potential, electrical resistivity, magnetic field, and acoustic emissions during deformation and fracturing process of geomaterials, which ranges from crystals, fault gouge-like materials to rocks under different loading regimes and environment. These advances also enabled us to measure and to monitor the dynamic responses of geomaterials during fracturing and slippage, which are quite important in geo-science and geo-engineering in association with earthquakes, earthquake engineering, and stability of various engineering structures. In Chapter 3, current available experimental techniques for measuring the dynamic properties in rock mechanics are described and it also points out new directions as to how to deal with actual issues in this field.

The present experimental techniques such as SHBT and cyclic tests are not sufficient enough to determine properties of rocks under different dynamic loading conditions. Furthermore, the dynamic constitutive laws should be more advanced and those used in soil-dynamics are not appropriate for rocks and rock masses. Dynamic shearing induces temperature rises along discontinuities and adjacent rock mass. Temperature rise depends on the dynamic shearing rate, normal stress and frictional properties of discontinuities as well as thermal properties of adjacent rocks. The increase of normal stress, dynamic shearing rate and frictional properties proportionally increase the temperature rise. Particularly normal stress and dynamic shearing rate have great influence on the overall rise of temperature.

The dynamic responses such as acceleration, velocity and displacement of geomaterials during fracturing and slippage under laboratory conditions are presented in Chapter 4. In addition to dynamic measurements of load, displacement responses, it is shown that the multi-parameter responses during experiments are of great importance in geosciences and geoengineering. The implications of stick-slip experiments and observations by infrared imaging techniques provide some insights in the assessment of instability problems such as the rockburst problem. The strong motions and fracture patterns in the vicinity of faults are presented and it is experimentally shown that the ground motions are always higher on the hanging-wall side of the faults. The velocity and displacement responses are obtained through the Erratic Pattern Screening (EPS) integration technique proposed by Aydan and Ohta (2011).

Earthquakes are known to be one of the natural disasters resulting in huge losses of human life and properties as experienced in the recent earthquakes. It is pointed out that the design of structures such as foundations of super-structures such as suspension bridges, nuclear power plants, dams, underground structures such as tunnels, underground power houses, gas and oil storage caverns, and residential and industrial developments must be done according to possible types of earthquakes and associated ground motions. Besides shaking, it was shown that the residual (permanent) relative displacement of the ground does occur. The permanent ground deformation may result from different causes such as faulting, slope failure, liquefaction and plastic deformation induced by ground shaking. The fundamental aspects and features current methods for estimating strong motions and permanent ground deformations are described in Chapter 5. Furthermore, the best procedure to estimate both shaking and permanent straining caused by earthquakes is explained and some practical applications are given.

Many examples of foundation damage by earthquakes are presented in Chapter 6 and it is shown that the foundations of large structures on rock foundations may be damaged by permanent ground deformation resulting from faulting or slope failure. To explain the fundamental mechanisms of damages to foundations, some model experiments, case histories on various type foundations of bridges, buildings, highways, railways, dams, pylons, pipelines are described and possible methods for evaluating the effects of permanent ground deformation of foundations and associated structures are summarized. It is shown that some available techniques are capable of evaluating the potential damage modes of foundations by earthquakes.

As shown in Chapter 7, ground motions are smaller in underground structures compared to those on surface structures. Nevertheless, permanent ground deformations resulting from faulting or slope failure cause tremendous forces on long and/or large structure underground structures such as tunnels, powerhouses and underground storage facilities of oil, gas and nuclear wastes. These forces may cause great damage to underground structures. Therefore, the current seismic design codes must be revised to include these effects. Although observations and studies on various underground openings showed that they are strong against shaking, the existence of discontinuities makes them vulnerable to collapses particularly in case of shallow underground openings. This may have some important implications on areas where shallow abandoned mines, underground shelters and old tunnels exist. Underground openings crossing faults and fracture zones may be enlarged to accommodate relative slips along faults and fracture zones. The lining of the openings should be ductile to accommodate permanent ground deformations at such zones. Furthermore, the brittle linings of the existing underground structures should be lined with ductile thin plates or fiber-reinforced polymers together with rockbolts at fracture and fault zones, where permanent ground deformations may occur.

Chapter 8 is concerned with rock slopes subjected to seismic forces and permanent ground movements. The slope failures induced tremendous damage to infrastructures as well as to residential areas, and they involved not only cut slopes but also natural rock slopes. Compared to the scale of soil slope failures, the scale and the impact of rock slope failures are very large and the form of failure differs depending upon the geological structures of rock mass of slopes. Furthermore, the failure of the rock slope failures may involve both active and passive modes. However, the passive modes are generally observed when the ground shaking is quite large. Besides some empirical methods, the assessment of the stability of natural rock slopes against earthquakes by limiting equilibrium method is possible if they consider possible failure forms. Regarding the use of numerical methods, the equivalent rock mass approach may not be appropriate for rock slopes, and discontinuum-type approaches should be used.

Chapter 9 is concerned with historical structures that are mainly masonry structures, which are composed of blocks made of natural stones, bricks or both, and they are assembled in different patterns with or without mortar. The observations of damage to actual masonry and historical structures, the shaking table experiments, available limiting equilibrium and numerical methods for estimating their responses are explained in Chapter 9. It is shown that the utilization of methods developed in rock mechanics and rock engineering could be effectively used for assessing the stability of such structures under dynamic loading conditions. Furthermore, the multi-parameter monitoring system is found to be quite effective for long-term stability assessment and their dynamic stability under seismic effects.

In Chapter 10, it is shown that the responses might be quite different during the transient phase from those under static assumptions. It is concluded that the loading of samples and structures as well as excavation of rock engineering structures should be treated as a dynamic phenomenon rather than a static problem. The treatment of excavation problem as a dynamic problem clarifies many aspects, which could not be explained by static analyses. For example, the excavation of tunnels through a dynamic-type formulation clearly shows that displacement and stresses are larger than those obtained from static analyses.

Blasting is the most commonly used excavation technique in mining and civil engineering applications. Blasting induces strong ground motions and fracturing of rock mass in rock excavations. Particularly, high ground motions may also induce some instability problems of rock mass and structures nearby. The characteristics of blasting agents, vibration monitoring in open-pit mines, quarries and underground openings are presented in Chapter 11. Furthermore, the negative and positive effects of blasting are also presented and discussed in this chapter. It is shown how to utilize p-wave explorations to assess the average equivalent mechanical properties of rock masses. Flyrocks is another major issue for the safety of people and damage to machinery and equipment when blasting is employed. A procedure is described for estimating the path of flyrocks.

The rockburst phenomenon is one of most dangerous forms of instability in rock engineering. Available methods on this topic are described and some effective monitoring and analysis methods for predicting rockburst are presented in Chapter 12. The fundamentals of various possible methods such as empirical techniques, analytical approaches and various finite element methods are briefly described and their applicability is discussed. Some laboratory tests were carried out on the circular openings excavated in sandstone from Tarutoge Tunnel and Shizuoka intercalated sandstone and shale samples and multi-parameter measurements were done in order to develop some observational and monitoring techniques for predicting rockburst. It is experimentally shown that the combined utilization of parameters such as AE, rock temperature, infrared imaging and electric potential may be a quite effective real-time in-situ monitoring tool for predicting rockburst. Some specific conclusions are:

1   The real-time stability of the tunnel face may be assessed using AE counting system and geo-electric potential (GEP) system during the excavation.
2   When the tunnel face is stable, both AE and GEP activities nearby the tunnel face cease within a short period of time.
3   The AE and GEP activities after initial phase may be indicators that the rock mass surrounding the tunnels is stable
4   The AE and GEP activities have a maximum value when the tunnel face distance is about 0.5D.
5   De-stressing and the use of flexible support system could be quite useful in rock masses prone to rockburst.
6   TBM is not a suitable excavation technique in rock masses prone to rockburst.

Rockbolts and rock anchors are principal support members in underground and surface excavations and they may be subjected to earthquake loading, vibrations induced by turbines, vehicle traffic and long-term corrosion. Chapter 13 is concerned with rockbolts and rock anchors. First, besides seismic loads, some possible sources of vibrations in underground structures are indicated and several examples of measurements were described. Then some theoretical, numerical and experimental studies on rockbolts and rock anchors under shaking are presented. In the remaining part, the fundamentals of non-destructive techniques for the evaluation of the soundness of rockbolts and rock anchors are described and several practical applications of non-destructive technique utilizing impact waves are given and discussed. It is shown that the non-destructive testing using impact waves could be an efficient tool for non-destructive evaluation of soundness of rockbolts and rock anchors. Lift-off tests on rock anchors at an underground power station constructed about 40 years ago were performed to check the soundness of rock anchors and the value of pre-stress applied at the time of construction. The responses of the bar type anchors are quite close to their original state and their deterioration is almost nil despite 40 years' elapse after their installation. On the other hand, the responses of cable type rock anchors are remarkably different from their original state and permanent displacements occurred after each cycle of loading. These results also imply that the cable type rock anchors consisting of wires are quite vulnerable to deterioration with time and they may not be qualified as permanent supporting systems.

Chapter 14 is concerned with impact phenomena observed in collision of vehicles in transportation engineering, collision of adjacent structures during earthquakes, standard penetration tests in soil mechanics, and impact craters due to meteorites, anchoring of ships or platforms in marine engineering and bullets and missiles destroying targets. The state of art on the impacts of meteorites and their effects are presented and it is shown that the impacts by projectiles such as bullets and missiles and their effects are quite similar to those of the meteorites except their size. The dynamic responses of objects during free-fall and impact are described in several laboratory tests. Furthermore, empirical and analytical formulations are presented and several examples of practical applications are given. As expected the maximum wave height increased as the drop height increased. The recorded acceleration at the base of the model occurred just after the peak tsunami wave height and remained almost constant thereafter. Furthermore, impact induced tsunami issue is discussed and some experimental and analytical formulations are described in relation to water level variations in closed water bodies such as lakes and reservoirs.

References

Abrahamson, N.A. & Somerville, P.G. (1996) Effects of the hanging wall and footwall on ground motions recorded during the Northridge earthquake. *Bulletin of the Seismological Society of America*, 86(1B), 593–599.

Adachi, K., Iwata, N., Kiyota, R., Aydan, Ö. & Tokashiki, N. (2016) Model experiment on seismic stability of discontinuous rock slope and numerical simulation. In: *EUROCK2016, Ürgüp*. pp. 629–633.

Ak, H., Iphar, M., Yavuz, M. & Konuk, A. (2009) Evaluation of ground vibration effect of blasting operations in a magnesite mine. *Soil Dynamics and Earthquake Engineering*, 29(4), 669–676.

Akutagawa, S., Arimura, Y., Nakamura, E., Sakurai, S., Baba, S. & Mori, S. (2008) A new evaluation method of PS-anchor force by using magnetic sensor and its application in a large underground powerhouse cavern (in Japanese). *JSCE, Division F*, 64(4), 413–430.

Amberg, R. (1983) Design and construction of the Furka base tunnel. *Rock Mechanics and Rock Engineering*, 16(4), 215–231.

Ambraseys, N.N. (1988) Engineering seismology. *Earthquake Engineering and Structural Dynamics*, 17, 1–105.

AP Photo (2013) *Pictures of Chelyabinsk Meteorite*. Associate Press Photo. Available from: http://www.apimages.com/.

Arıoğlu, E., Arıoğlu, B. & Girgin, C. (1999) Stability problems in tunnels excavated in massif-brittle rocks (in Turkish). *Mühendislik Jeolojisi Bülteni*, 27, 69–78.

Asakura, T. & Sato, Y. (1998) Mountain tunnels damage in the 1995. *Hyogo-ken Nanbu Earthquake* 39(1), 9–16. Railway Technical Research Institute (RTRI).

Asakura, T., Shiba, Y., Sato, Y. & Iwatate, T. (1996) Mountain tunnels performance in 1995 Hyogo-ken Nanbu Earthquake. In: *Special Report of the 1995 Hyogo-ken Nanbu Earthquake, Committee of Earthquake Engineering, JSCE*.

Attewell, P.B., Farmer, I.W. & Haslam, D. (1965) Prediction of ground vibration from major quarry blasts. *Mining and Mineral Engineering*, 1, 621–626.

Aydan, Ö. (1986) Stability of slope and shallow underground openings in discontinuous rock mass. In: *Nagoya University, Dept. of Geotechnical Engineering, Interim Report*. 16 p. (unpublished).

Aydan, Ö. (1987) *Approximate Estimation of Plastic Zones About Underground Openings*. Nagoya University, Department of Geotechnical Engineering, (unpublished interim report). 7 p.

Aydan, Ö. (1989a) *The Stabilization of Rock Engineering Structures by Rockbolts*. Doctorate Thesis. Nagoya University. 204 p.

Aydan, Ö. (1989b) *The Stabilization of Rock Engineering Structures by Rockbolts*. Doctorate Thesis. Nagoya University. 205 p.

Aydan, Ö. (1994) The dynamic shear response of an infinitely long visco-elastic layer under gravitational loading. *Soil Dynamics and Earthquake Engineering*, 13, 181–186.

Aydan, Ö. (1995a) Mechanical and numerical modelling of lateral spreading of liquified soil. In: *The 1st Int. Conf. on Earthquake Geotechnical Engineering, IS-TOKYO'95, Tokyo*. pp. 881–886.

Aydan, Ö. (1995b) The stress state of the earth and the earth's crust due to the gravitational pull. In: *The 35th US Rock Mechanics Symposium, Lake Tahoe*. pp. 237–243.

Aydan, Ö. (1997a) Dynamic uniaxial response of rock specimens with rate-dependent characteristics. In: *SARES'97*. pp. 322–331.

Aydan, Ö. (1997b) Seismic characteristics of Turkish earthquakes. In: *Turkish Earthquake Foundation, TDV/TR 97-007*. 41 p.

Aydan, Ö. (1998a) A simplified finite element approach for modelling the lateral spreading of liquefied ground. In: *The 2nd Japan-Turkey Workshop on Earthquake Engineering, Istanbul*.

Aydan, Ö. (1998b) Analysis of masonary structures by finite element method. In: *Prof. Dr. Rifat Yarar Symposium, TDV*. pp. 141–150.

Aydan, Ö. (2000) A stress inference method based on GPS measurements for the directions and rate of stresses in the earth' crust and their variation with time. In: *Yerbilimleri, No. 22*. pp. 223–236.

Aydan, Ö. (2001) Comparison of suitability of submerged tunnel and shield tunnel for subsea passage of Bosphorus. *Geological Engineering Journal*, 25(1), 1–17.

Aydan, Ö. (2003a) The inference of crustal stresses in Japan with a particular emphasis on Tokai region. In: *Int. Symp. on Rock Stress, Kumamoto*. pp. 343–348.

Aydan Ö. (2003b) Actual observations and numerical simulations of surface fault ruptures and their effects engineering structures. In: *The Eight U.S.-Japan Workshop on Earthquake Resistant Design of Lifeline Facilities and Countermeasures Against Liquefaction. Technical Report, MCEER-03-0003*. pp. 227–237.

Aydan, Ö. (2003c) An experimental study on the dynamic responses of geomaterials during fracturing. *Journal of School of Marine Science and Technology, Tokai University*, 1(2), 1–7.

Aydan, Ö. (2004a) Dynamic responses of objects during free fall and impacts. In: *Tokai University, Dept. of Marine Civil Eng., Interim Report*. 8 p.

Aydan, Ö. (2004b) Dynamic responses of objects during free fall and impacts. In: *Tokai University, Dept. of Marine Civil Eng., Interim Report*. 5 p.

Aydan, Ö. (2004c) Implications of GPS-derived displacement, strain and stress rates on the 2003 Miyagi-Hokubu earthquakes. In: *Yerbilimleri, No.30*. pp. 91–102.

Aydan, Ö. (2006a) An experimental study on the tsunami generation in reservoirs by falling objects. In: *Tokai University, Dept. of Marine Civil Eng., Interim Report*. 7 p.

Aydan Ö. (2006b) Geological and seismological aspects of Kashmir Earthquake of October 8, 2005 and geotechnical evaluation of induced failures of natural and cut slopes. *Journal of Marine Science and Technology, Tokai University*, 4(1), 25–44.

Aydan, Ö. (2006c) The possibility of earthquake prediction by global positioning system (GPS). *Journal of School of Marine Science and Technology*, 4(3), 77–89.

Aydan, Ö. (2007a) The inference of seismic and strong motion characteristics of earthquakes from faults with a particular emphasis on Turkish earthquakes. In: *The 6th National Earthquake Engineering Conference of Turkey, Istanbul*. pp. 563–574.

Aydan, Ö. (2007b) Inference of seismic characteristics of possible earthquakes and liquefaction and landslide risks from active faults (in Turkish). In: *The 6th National Conference on Earthquake Engineering of Turkey, Istanbul*, Vol. 1. pp. 563–574.

Aydan, Ö. (2008) Seismic and tsunami hazard potentials in Indonesia with a special emphasis on Sumatra Island. *Journal of the School of Marine Science and Technology, Tokai University*, 6(3), 19–38.

Aydan, Ö. (2010) Numerical modelling of heat flow in porous rock masses and its application to Armutlu geothermal area. *Bulletin of Engineering Geology*, 30, 1–16.

Aydan, Ö. (2011a) Some issues in tunnelling through rock mass and their possible solutions. In: *First Asian Tunnelling Conference, ATS11-15*. pp. 33–44 (Invited lecture).

Aydan, Ö. (2011b) Some issues in tunnelling through rock mass and their possible solutions. In: *First Asian Tunnelling Conference, ATS11*. pp. 33–44.

Aydan, Ö (2011c) Some issues in tunnelling through rock mass and their possible solutions. In: *Proc. First Asian Tunnelling Conference, ATS-15*. pp. 33–44.

Aydan, Ö. (2012a) The utilization of non-destructive testing method for the evaluation of soundness of rockbolts and rock anchors used in rock engineering structures. In: *Elastic wave Inspection Technology Association (EITAC), Invited Lecture*. 110 p.

Aydan, Ö. (2012b) Ground motions and deformations associated with earthquake faulting and their effects on the safety of engineering structures. In: Meyers, R. (ed.) *Encyclopedia of Sustainability Science and Technology, Springer*. pp. 3233–3253.

Aydan, Ö. (2012c) The inference of physico-mechanical properties of soft rocks and the evaluation of the effect of water content and weathering on their mechanical properties from needle penetration tests. In: *ARMA 12-639, 46th US Rock Mechanics/Geomechanics Symposium, Chicago, Paper No. 639*. 10 p. (on CD).

Aydan, Ö. (2013a) In-situ stress inference from damage around blasted holes. *Journal of Geosystem Engineering, Taylor and Francis*, 16(1), 83–91.

Aydan, Ö. (2013b) Inference of contemporary crustal stresses from recent large earthquakes and its comparison with other direct and indirect methods. In: *6th International Symp. on In Situ Rock Stress (RS2013), Sendai, Paper No. 1051*. 8 p.

Aydan, Ö. (2015) Some considerations on a large landslide at the left bank of the Aratozawa Dam caused by the 2008 Iwate-Miyagi intraplate earthquake. *Rock Mechanics and Rock Engineering, Special Issue on Deep-Seated Landslides*, 49(6), 2525–2539.

Aydan, Ö. (2016) The state of art on large cavern design for underground powerhouses and some long-term issues. In: *Lehr/Wiley Encyclopedia Energy: Science, Technology and Applications, John Wiley and Sons, Chp45*. pp. 465–487.

Aydan, Ö. & Amini, M.G. (2009) An experimental study on rock slopes against flexural toppling failure under dynamic loading and some theoretical considerations for its stability assessment. *Journal of Marine Science and Technology, Tokai University*, 7(2), 25–40.

Aydan, Ö. & Daido, M. (2002) An experimental study on the seepage induced geo-electric potential in porous media. *Journal of School of Marine Science and Technology, Tokai University*, 55, 53–66.

Aydan, Ö. & Geniş, M. (2004) Surrounding rock properties and openings stability of rock tomb of Amenhotep III (Egypt). In: *ISRM Regional Rock Mechanics Symposium, Sivas*. pp. 191–202.

Aydan, Ö. & Geniş, M. (2008) Assessment of dynamic stability of an abandoned room and pillar underground lignite mine (in Turkish). In: *Turkish National Bulletin of Rock Mechanics, TNSRM, Ankara*.

Aydan, Ö. & Geniş, M. (2010) Rockburst phenomena in underground openings and evaluation of its counter measures. *Turkish Journal of Rock Mechanics*, 17, 1–62.

Aydan, Ö. & Geniş, M. (2014) A numerical study on the response and stability of abandoned lignite mines in relation to the excavation of a large underground opening below. In: *The 14th*

International Conference of International Association for Computer Methods and Advances in Geomechanics (IACMAG), Kyoto, Japan.

Aydan, Ö. & Hamada, M. (1992) The site investigation of the Erzincan (Turkey) Earthquake of March 13, 1992. In: *The 4th Japan-US Workshop on Earthquake Resistant Design of Lifeline Facilities and Countermeasures Against Soil Liquefaction, Honolulu.* pp. 17–34.

Aydan, Ö. & Hamada, M. (2006) Damage to civil engineering structures by Oct. 8, 2005 Kashmir earthquake and recommendations for recovery and reconstruction. *Journal of Disaster Research*, 1(3), 1–9.

Aydan, Ö. & Kawamoto, T. (1992) Stability of slopes and underground openings against flexural toppling and their stabilization. *Rock Mechanics and Rock Engineering*, 25(3), 143–165.

Aydan, Ö. & Kawamoto, T. (2000a) Assessing mechanical properties of rock masses by RMR rock classification method. In: *Proc. of GeoEng 2000 Symposium, Sydney, Paper No. OA0926.*

Aydan, Ö. & Kawamoto, T. (2000b) The assessment of mechanical properties of rock masses through RMR rock classification system. In: *GeoEng2000, Melbourne, OA926.*

Aydan, Ö. & Kawamoto, T. (2004) The damage to abandoned lignite mines caused by the 2003 Miyagi-Hokubu earthquake and some considerations on its causes. In: *3rd Asian Rock Mechanics Symposium, Kyoto.* pp. 525–530.

Aydan, Ö. & Kim, Y. (2002) The inference of crustal stresses and possible earthquake faulting mechanism in Shizuoka Prefecture from the striations of faults. *Journal of the School of Marine Science and Technology, Tokai University*, 54, 21–35.

Aydan, Ö. & Kumsar, H. (1997) A site investigation of Dinar earthquake of October 1, 1995. In: *Turkish Earthquake Foundation, TDV/DR 97-003, Istanbul, Turkey.* 115 p.

Aydan, Ö. & Kumsar, H. (2002) Dinamik Yükler altında şevlerin kama yenilmesi sırasında mekanik davranışlarının modellenmesi üzerine deneysel ve kuramsal bir yaklaşım. In: *KAYAMEK'2002, VI. Bölgesel Kaya Mekaniği Sempozyumu.* pp. 235–243.

Aydan, Ö. & Kumsar, H. (2005) Features of the 2003 Buldan earthquakes and an evaluation of damaged roman underground tombs in Yenice (Denizli). In: *Bulleting of Engineering geology, Istanbul, No. 20.* pp. 61–73.

Aydan, Ö. & Kumsar, H. (2010) An experimental and theoretical approach on the modeling of sliding response of rock wedges under dynamic loading. *Rock Mechanics and Rock Engineering*, 43(6), 821–830.

Aydan, Ö. & Nawrocki, P. (1998) Rate-dependent deformability and strength characteristics of rocks. In: *Proceedings of Symposium on the Geotechnics of Hard Soils-Soft Rock, Napoli, Vol. 1.* pp. 403–411.

Aydan, Ö. & Ohta, Y. (2006) The characteristics of ground motions in the neighbourhood of earthquake faults and their evaluation. In: *Symposium on the Records and Issues of Recent Great Earthquakes in Japan and Overseas, EEC-JSCE, Tokyo.* pp. 114–120.

Aydan, Ö. & Ohta, Y. (2011) A new proposal for strong ground motion estimations with the consideration of characteristics of earthquake fault. In: *Seventh National Conference on Earthquake Engineering, Istanbul, 30 May–3 June 2011, Paper No 65.* pp. 1–10.

Aydan, Ö. & Tokashiki, N. (2010) *The Off-Okinawa Island Earthquake of February 27, 2010.* Available from: http://www.jsce.or.jp/library/eq_repo/Vol3/14/20100227okinawa_report.pdf.

Aydan, Ö. & Tokashiki, N. (2011) A comparative study on the applicability of analytical stability assessment methods with numerical methods for shallow natural underground openings. In: *The 13th International Conference of the International Association for Computer Methods and Advances in Geomechanics, Melbourne, Australia.* pp. 964–969.

Aydan, Ö. & Ulusay, R. (2002) A back analysis of the failure of a highway embankment at Bakacak during the 1999 Düzce-Bolu earthquake. *Environmental Geology*, 42, 621–631.

Aydan, Ö., Ichikawa, Y. & Kawamoto, T. (1985) Load bearing capacity and stress distributions in/alongrockbolts with inelastic behaviour of interfaces. In: *The 5th Int. Conf. on Num. Meths. in Geomechanics, Nagoya*, Vol. 2. pp. 1281–1292.

Aydan, Ö., Ichikawa, Y. & Kawamoto, T. (1986) Reinforcement of geotechnical engineering structures by grouted rockbolts. In: *Int. Symp. Engng. Complex Rock Formation, Beijing.* pp. 183–189.

Aydan, Ö., Kyoya, T., Ichikawa, Y. & Kawamoto, T. (1987) Anchorage performance and reinforcement effect of fully grouted rockbolts on rock excavations. In: *The 6th Int. Congress on Rock Mechanics, ISRM, Montreal*, Vol. 2. pp. 757–760.

Aydan, Ö., Kyoya, T., Ichikawa, Y., Kawamoto, T., Ito, T. & Shimizu, Y. (1988) Three-dimensional simulation of an advancing tunnel supported with forepoles, shotcrete, steel ribs and rockbolts. In: *The 6th Int. Conf. on Num. Meths. in Geomechanics, Innsbruck*, Vol. 2. pp. 1481–1486.

Aydan, Ö., Shimizu, Y. & Ichikawa, Y. (1989) The effective failure modes and stability of slopes in rock mass with two discontinuity sets. *Rock Mechanics and Rock Engineering*, 22(3), 163–188.

Aydan, Ö., Ichikawa, Y. & Kawamoto, T. (1990) Numerical modelling of discontinuities and interfaces in rock mass. In: *The 4th Japan Computational Mechanics Symposium, JUSE.* pp. 40–46.

Aydan, Ö., Ichikawa, Y., Shimizu, Y. & Murata, K. (1991) An integrated system for the stability of rock slopes. In: *The 5th Int. Conf. on Computer Methods and Advances in Geomechanics, Cairns*, Vol. 1. pp. 469–465.

Aydan, Ö., Shimizu, Y. & Kawamoto, T. (1992a) The stability of rock slopes against combined shearing and sliding failures and their stabilisation. *Int. Symp. on Rock Slopes, New Delhi.* pp. 203–210.

Aydan, Ö., Shimizu, Y. & Kawamoto, T. (1992b) The reach of slope failures. In: *The 6th Int. Symp. Landslides, ISL 92, Christchurch*, Vol. 1. pp. 301–306.

Aydan, Ö., Akagi, T. & Kawamoto, T. (1993) Squeezing potential of rocks around tunnels; theory and prediction. *Rock Mechanics and Rock Engineering*, 26(2), 137–163.

Aydan, Ö., Akagi, T., Okuda, H. & Kawamoto, T. (1994) The cyclic shear behaviour of interfaces of rock anchors and its effect on the long term behaviour of rock anchors. In: *Int. Symp. on New Developments in Rock Mechanics and Rock Engineering, Shenyang.* pp. 15–22.

Aydan, Ö., Shimizu, Y. & Karaca, M. (1994) The dynamic and static stability of shallow underground openings in jointed rock masses. In: *The 3rd Int. Symp. on Mine Planning and Equipment Selection, Istanbul.* pp. 851–858.

Aydan, Ö., Akagi, T., Ito, T., Ito, J. & Sato, J. (1995) Prediction of deformation behaviour of a tunnel in squeezing rock with time-dependent characteristics. *Numerical Models in Geomechanics NUMOG V.* pp. 463–469.

Aydan, Ö., Komura, S.S. & Ebisu, T.K. (1995) A unified design method for rock anchor foundations of super-high pylons. In: *The Int. Workshop on Rock Foundation of Large-scaled Structures, ISRM, Tokyo.* pp. 279–284.

Aydan, Ö., Akagi, T. & Kawamoto, T. (1996a) The squeezing potential of rock around tunnels: Theory and prediction with examples taken from Japan. *Rock Mechanics and Rock Engineering*, 29(3), 125–143.

Aydan, Ö., Mamaghani, I.H.P & Kawamoto, T. (1996b) Application of discrete finite element method (DFEM) to rock engineering. In: *North American Rock Mechanics Symp. Montreal*, Vol. 2. pp. 2039–2046.

Aydan, Ö., Sezaki, M. & Yarar, R. (1996c) The seismic characteristics of Turkish Earthquakes. In: *The 11th World Conf. on Earthquake Eng., CD-2, Paper No: 1270.*

Aydan, Ö., Shimizu, Y. & Kawamoto, T. (1996d) The anisotropy of surface morphology and shear strength characteristics of rock discontinuities and its evaluation. In: *NARMS'96, Montreal.* pp. 1391–1398.

Aydan, Ö., Tokashiki, N. & Seiki, T. (1996e) Micro-structure models for porous rocks to jointed rock mass. In: *APCOM'96,* Vol. 3. pp. 2235–2242.

Aydan, Ö., Ulusay, R., Kumsar, H. & Ersen, A. (1996f) Buckling failure at an open-pit coal mine. In: *EUROCK'96.* pp. 641–648.

Aydan, Ö., Üçpırtı, H. & Kumsar, H. (1996g) The stability of a rock slope having a visco-plastic sliding surface. *Rock Mechanics Bulletin (Kaya Mekaniği Bülteni),* 12, 39–49.

Aydan, Ö., Kumsar, H. & Sakoda, S. (1997a) Model wedge tests and re-assessment of limiting equilibrium methods for wedge sliding. In: *Proceedings of Rock Mechanics and Evironmental Geotechology RMEG, Chonging University Press, China.* pp. 261–266.

Aydan, Ö., Shimizu, Y., Akagi, T. & Kawamoto, T. (1997b) Development of fracture zones in rock masses. In: *Int Symp. on Deformation and Progressive Failure in Geomechanics, IS-NAGOYA.* pp. 533–538.

Aydan, Ö., Ulusay, R. & Kawamoto, T. (1997c) Assessment of rock mass strength for underground excavations. In: *Proc. of the 36th US Rock Mechanics Symp., New York.* pp. 777–786.

Aydan, Ö., Ulusay, R., Hasgür, Z. & Taşkın, B. (1999a) A site investigation of Kocaeli Earthquake of August 17, 1999. In: *Turkish Earthquake Foundation – Earthquake Report, TDV-DR08-49, Istanbul.* 180 pp.

Aydan, Ö., Ulusay, R., Hasgür, Z. & Hamada, M. (1999b) The behaviour of structures built in active fault zones in view of actual examples from the 1999 Kocaeli and Chi-chi Earthquakes. In: *ITU-IAHS International Conference on the Kocaeli Earthquake 17 August 1999: A Scientific Assessment and Recommendations for Re-building, Istanbul.* pp. 131–142.

Aydan, Ö., Seiki, T., Shimizu, Y. & Hamada, M. (2000a) Some considerations on rock slope failures due to earthquakes. In: *Chonquing-Waseda Joint Seminar on Chi-Chi Earthquake.*

Aydan, Ö., Ulusay, R., Kumsar, H. & Tuncay, E. (2000b) Site investigation and engineering evaluation of the Düzce-Bolu Earthquake of November 12, 1999. In: *Turkish Earthquake Foundation, TDV/DR 09-51.* pp. 220.

Aydan, Ö., Ulusay, R., Kumsar, H. & Tuncay, E. (2000c) Site investigation and engineering evaluation of the Düzce-Bolu Earthquake of November 12, 1999. In: *Turkish Earthquake Foundation, TDV/DR 095-51.* pp. 307.

Aydan, Ö., Minato, T. & Fukue, M. (2001a) An experimental study on the electrical potential of geomaterials during deformation and its implications in Geomechanics. In: *38th US Rock Mech. Symp., Washington,* Vol. 2. pp. 1199–1206.

Aydan, Ö., Tokashiki, N., Shimizu, Y. & Mamaghani, I.H.P. (2001b) A stability analysis of masonry walls by Discrete Finite Element Method (DFEM). In: *10th IACMAG Conference, Austin.* pp. 1625–1628.

Aydan, Ö., Geniş, M., Akagi, T. & Kawamoto, T. (2001c) Assessment of susceptibility of rockbursting in tunnelling in hard rocks. In: *Int. Symp. on Modern Tunnelling Science and Technology, IS-KYOTO,* Vol. 1. pp. 391–396.

Aydan, Ö., Bilgin, H.A. & Aldas, U.G. (2002a) The dynamic response of structures induced by blasting. In: *Int. Workshop on Wave Propagation, Moving load and Vibration Reduction, Okayama, Japan, Balkema.* pp. 3–10.

Aydan, Ö., Ito, T., Akagi, T., Watanabe, H. & Tano, H. (2002b) An experimental study on the electrical potential of geomaterials during fracturing and sliding. In: *Korea-Japan Joint Symposium on Rock Engineering, Seoul, Korea, July.* pp. 211–218.

Aydan, Ö., Kumsar, H. & Ulusay, R. (2002c) How to infer the possible mechanism and characteristics of earthquakes from the striations and ground surface traces of existing faults. *JSCE, Earthquake and Structural Engineering/Earthquake Engineering*, 19(2), 199–208.

Aydan, Ö., Tokashiki, N. & Harada, A. (2002d) An experimental study on the dynamic responses and stability of retaining walls of masonry type. In: *Korea-Japan Joint Symposium on Rock Engineering, Seoul, Korea, July*. pp. 769–776.

Aydan, Ö., Daido, M., Ito, T., Tokashiki, N. & Kawamoto, T. (2003a) Development of experimental and numerical techniques for predicting rockburst and their applications. In: *35th Japan Rock Mechanics Symposium, Tokyo*. pp. 305–311.

Aydan, Ö., Ogura, Y., Daido, M. & Tokashiki, N. (2003b) A model study on the seismic response and stability of masonary structures through shaking table tests. In: *Fifth National Conference on Earthquake Engineering, Istanbul, Turkey, Paper No: AE-041(CD)*.

Aydan, Ö., Tokashiki, N., Ito, T., Akagi, T., Ulusay, R. & Bilgin, H.A. (2003c) An experimental study on the electrical potential of non-piezoelectric geomaterials during fracturing and sliding. In: *9th ISRM Congress, South Africa*. pp. 73–78.

Aydan, Ö., Daido, M., Owada, Y., Tokashiki, T. & Ohkubo, K. (2004) The assessment of rock bursting in rock engineering structures with a particular emphasis on underground openings. In: *3rd Asian Rock Mechanics Symposium, Kyoto*, Vol. 1. pp. 531–536.

Aydan, Ö., Daido, M., Tano, H., Tokashiki, N. & Ohkubo, K. (2005a) A real-time multi-parameter monitoring system for assessing the stability of tunnels during excavation. In: *ITA Conference, Istanbul*. pp. 1253–1259.

Aydan, Ö., Kinbara, T., Uehara, F. & Kawamoto, T. (2005b) A numerical analysis of non-destructive tests for the maintenance and assessment of corrosion of rockbolts and rock anchors. In: *35th Japan Rock Mechanics Symposium, Tokyo*. pp. 371–376.

Aydan, Ö., Ogura, Y., Daido, Y., Tokashiki, N., Irabu, S. & Takara, S. (2005c) Re-assessment of the seismic response and stability of stone masonry structures of shuri castle through shaking table tests. In: *Proceedings of International Symposium on Industrial Minerals and Building Stones, September 15–18, Istanbul, Turkey*. pp. 109–117.

Aydan, Ö., Sakamoto, A., Yamada, N., Sugiura, K. & Kawamoto, T. (2005d) A real time monitoring system for the assessment of stability and performance of abandoned room and pillar lignite mines. In: *Post Mining 2005, Nancy*.

Aydan, Ö., Tokashiki, N. & Mamaghani, I.H.P. (2005e) Numerical analyses of discontinuous rock masses using Discrete Finite Element Method. In: *Special Issue on the Frontier Line of Simulation of Disaster Prevention Analyses and Environment, Chishitsu and Chosa, No. 4*. pp. 8–15 (in Japanese).

Aydan, Ö., Daido, M., Ito, T., Tano, H. & Kawamoto, T. (2006a) Prediction of post-failure motions of rock slopes induced by earthquakes. In: *4th Asian Rock Mechanics Symposium, Singapore, Paper No. A0356 (on CD)*.

Aydan, Ö., Daido, M., Tano, H., Nakama, S. & Matsui, H. (2006b) The failure mechanism of around horizontal boreholes excavated in sedimentary rock. In: *50th US Rock Mech. Symp., Paper No. 06-130*.

Aydan, Ö., Daido, M., Ito, T., Tano, H. & Kawamoto, T. (2006c) Instability modes of abandoned lignite mines and the assessment of their stability in long-term and during earthquakes. In: *3rd Asian Rock Mechanics Symposium, Singapore, on CD*.

Aydan, Ö., Daido, M., Tokashiki, N., Bilgin, A.H. & Kawamoto, T. (2007a) Acceleration response of rocks during fracturing and its implications in earthquake engineering. In: *11th ISRM Congress, Lisbon*, Vol. 2. pp. 1095–1100.

Aydan, Ö., Sugiura, K. & Suzumura, Y. (2007b) A quick report on Mie-ken Kameyama earthquake of April 15, 2007. In: *Japan Society of Civil Engineers, Earthquake Disaster Investigation Comittee*. 35 p. Available from: http://www.jsce-int.org/.

Aydan, Ö., Tokashiki, N. & Sugiura, K. (2008a) Characteristics of the 2007 Kameyama Earthquake with some emphases on unusually high strong ground motions and the collapse of Kameyama Castle wall. *Journal of the School of Marine Science and Technology*, 7(1), 83–105.

Aydan, Ö., Tsuchiyama, S., Kinbara, T., Uehara, F., Tokashiki, N. & Kawamoto, T. (2008b) A numerical analysis of non-destructive tests for the maintenance and assessment of corrosion of rockbolts and rock anchors. In: *The 12th International Conference of International Association for Computer Methods and Advances in Geomechanics (IACMAG), Goa, India*. pp. 40–45.

Aydan, Ö., Ulusay, R. & Atak, V.O. (2008c) Evaluation of ground deformations induced by the 1999 Kocaeli earthquake (Turkey) at selected sites on shorelines. *Environmental Geology, Springer Verlag*, 54, 165–182.

Aydan, Ö., Hamada, M., Itoh, J. & Ohkubo, K. (2009a) Damage to civil engineering structures with an emphasis on rock slope failures and tunnel damage induced by the 2008 Wenchuan Earthquake. *Journal of Disaster Research*, 4(2), 153–164.

Aydan, Ö., Kumsar, H., Toprak, S. & Barla, G. (2009b) Characteristics of 2009 l'Aquila earthquake with an emphasis on earthquake prediction and geotechnical damage. *Journal Marine Science and Technology, Tokai University*, 9(3), 23–51.

Aydan, Ö., Ohta, Y. & Hamada, M. (2009c) Geotechnical evaluation of slope and ground failures during the 8 October 2005 Muzaffarabad earthquake in Pakistan. *Journal Seismology*, 13(3), 399–413.

Aydan, Ö., Ohta, Y., Hamada, M., Ito, J. & Ohkubo, K. (2009d) The characteristics of the 2008 Wenchuan Earthquake disaster with a special emphasis on rock slope failures, quake lakes and damage to tunnels. *Journal of the School of Marine Science and Technology, Tokai University*, 7(2), 1–23.

Aydan, Ö., Kumsar, H., Toprak, S. & Barla, G. (2010a) Characteristics of 2009 l'Aquila earthquake with an emphasis on earthquake prediction and geotechnical damage. *Journal Marine Science and Technology, Tokai University*.

Aydan, Ö., Ohta, Y., Geniş, M., Tokashiki, N. & Ohkubo, K. (2010b) Response and stability of underground structures in rock mass during earthquakes. *Rock Mechanics and Rock Engineering*, 43(6), 857–875.

Aydan, Ö., Ohta, Y. & Tano, H. (2010c) Multi-parameter response of soft rocks during deformation and fracturing with an emphasis on electrical potential variations and its implications in geomechanics and geoengnineering. In: *The 39th Rock Mechanics Symposium of Japan, Tokyo*. pp. 116–121.

Aydan, Ö., Ohta, Y., Daido, M., Kumsar, H., Genis, M., Tokashiki, N., Ito, T. & Amini, M. (2011) Chapter 15: Earthquakes as a rock dynamic problem and their effects on rock engineering structures. In: Zhou, Y. & Zhao, J. (eds.) *Advances in Rock Dynamics and Applications*. Boca Raton, FL, CRC Press, Taylor and Francis Group. pp. 341–422.

Aydan, Ö., Tokashiki, N. & Geniş, M. (2012a) Some considerations on yield (failure) criteria in rock mechanics ARMA 12-640. In: *46th US Rock Mechanics/Geomechanics Symposium, Chicago, Paper No. 640*, 10 p. (on CD).

Aydan, Ö., Uehara, F. & Kawamoto, T. (2012b) Numerical study of the long-term performance of an underground powerhouse subjected to varying initial stress states, cyclic water heads, and temperature variations. *International Journal of Geomechanics, ASCE*, 12(1), 14–26.

Aydan, Ö., Ulusay, R. & Kumsar, H. (2012c) Site investigation and engineering evaluation of the Van earthquakes of October 23 and November 9, 2011. In: *Turkish earthquake Foundation (TDV), TDV/DR 015-92.* 136p+T54.

Aydan, Ö., Ulusay, R., Hamada, M. & Beetham, D. (2012d) Geotechnical aspects of the 2010 Darfield and 2011 Christchurch earthquakes of New Zealand and geotechnical damage to structures and lifelines. *Bulletin of Engineering Geology and Environment*, 71(4), 637–662.

Aydan, Ö., Geniş, M. & Bilgin, H.A. (2013a) The effect of blasting on the stability of benches and their responses at Demirbilek open-pit mine. *Environmental Geotechnics ICE.*

Aydan, Ö., Geniş, M., Tokashiki, N. & Tano, H. (2013b) Some considerations on the appropriate dimension in the numerical analysis of geoengineering structures. In: *APCOM2013-ISCM, Paper No 1828, Singapore.* 8 p.

Aydan, Ö., Sato, A. & Yagi, M. (2013c) The inference of geo-mechanical properties of soft rocks and their degradation from needle penetration tests. *Rock Mechanics and Rock Engineering, Springer.*

Aydan, Ö., Ulusay, R. & Tokashiki, N. (2013d) A new rock mass quality rating system: Rock Mass Quality Rating (RMQR) and its application to the estimation of geomechanical characteristics of rock masses. *Rock Mechanics and Rock Engineering*, 47(4), 1255–1276.

Aydan, Ö., Geniş, M. & **Bilgin**, A.H. (2014a) Effect of blasting on the bench stability at the Demirbilek open-pit mine. *Environmental Geotechnics, Institute of Civil Engineering, London*, 1(4), 240–248. doi:10.1680/envgeo.13.00018.

Aydan, Ö., Manav, H., Yaoita, T. & Yagi, M. (2014b) Multi-parameter thermo-dynamic response of minerals and rocks during deformation and fracturing. In: *8th Asian Rock Mechanics Symposium, Sapporo.* pp. 817–826.

Aydan, Ö., Fuse, T. & Ito, T. (2015a) An experimental study on thermal response of rock discontinuities during cyclic shearing by Infrared (IR) thermography. In: *Proc. 43rd Symposium on Rock Mechanics, JSCE.* pp. 123–128.

Aydan, Ö., Imazu, M., Soya, M. & Ideura, H. (2015b) The possibility of Infrared Camera Thermography for assessing the real-time stability of underground excavations. In: *ITA WTC 2015 Congress and 41st General Assembly, Dubrovnik, Croatia.* 10 p.

Aydan, Ö., Imazu, M., Soya, M. & Ideura, H. (2016a) The dynamic response of the Taru-Toge tunnel during Blasting. In: *ITA WTC 2016 Congress and 42nd General Assembly, San Francisco, USA.* 10 p.

Aydan, Ö., Tano, H., Ideura, H., Asano, A., Takaoka, H., Soya, M. & Imazu, M. (2016b) Monitoring of the dynamic response of the surrounding rock mass at the excavation face of Tarutoge Tunnel, Japan. In: *EUROCK2016, Ürgüp.*

Aydan, Ö., Tomiyama, J., Matsubara, H., Tokashiki, N. & Iwata, N. (2017) The characteristics of damage to rock engineering structures induced by the 2016 Kumamoto earthquakes. In: *14th Domestic Rock Mechanics Symposium, Kobe* (in Press).

Backman, M.E. & Goldsmith, W. (1978) Mechanics of penetration of projectiles into targets. *International Journal of Engineering Science*, 16(1), 1–99.

Badawy, A. (2001) Status of the crustal stress in Egypt as inferred from earthquake focal mechanisms and borehole breakouts. *Tectonophysics*, 343, 49–61.

Bagde, M.N. & Petros, V. (2005) Fatigue properties of intact sandstone samples subjected to dynamic uniaxial cyclical loading. *International Journal of Rock Mechanics and Mining Sciences*, 42(2), 237–250.

Barton, N.R. (1973) Review of a new shear strength criterion for rock joints. *Engineering Geology*, 7, 287–332.

Barton, N. & Bandis, S.C. (1987) Rock joint model for analyses of geological discontinue. In: *Proc. Int. Symp. on Constitutive Laws for Engineering Materials: Theory and Applications, Elsevier, Amsterdam*. pp. 993–1002.

Barton, N., Harvik, L., Christianson, M. &Vik, G. (1986) Estimation of joint deformations, potential leakage and lining stresses for a planned urban road tunnel. In: *Int. Symp. On Large Rock Caverns*. pp. 1171–1182.

Baudendistel, M., Malina, H. & Müller, L. (1970) Einfluss von Discontinuitaten auf die Spannungen und Deformationen in der Umgebung einer Tunnelröhre. *Rock Mechanics*, 2, 17–40.

Beard, M.D. & Lowe, M.J.S. (2003) Non-destructive testing of rock bolts using guided ultrasonic waves. *International Journal of Rock Mechanics & Mining Sciences*, 40, 527–536.

Berberian, M., Jackson, J.A., Qorashi, M., Khatib, M.M., Priestley, K., Talebian, M. & Ghafuri-Ashtiani, M. (1999) The 1997 May 10 Zirkuh (Qa'enat) earthquake (Mw 7.2): Faulting along the Sistan suture zone of eastern Iran. *Geophysical Journal International*, 136(3), 671–694.

Bernard, R.S. (1977) *Empirical Analysis of Projectile Penetration in Rock*. United States. 23 p.

Betournay, M.C., Gorski, B., Labrie, D., Jackson, R. & Gyenge, M. (1991) New considerations in the determining of Hoek and Brown material constants. In: *Proc. 7th Int. Cong. on Rock Mechanics, Aachen*, Vol. 1. pp. 195–200.

Bieniawski, Z.T. (1974) Geomechanics classification of rock masses and its application in tunnelling. In: *Third Int. Congress on Rock Mechanics, ISRM, Denver, IIA*. pp. 27–32.

Bieniawski, Z.T. (1989) *Engineering Rock Mass Classifications*. New York, Wiley.

Birkimer, D.L. (1971) A possible fracture criterion for the dynamic tensile strength of rock. In: Clark, G.B. (ed.) *Proc. 12th U.S. Symp. Rock Mech. Maryland, Baltimore*. pp. 573–590.

Blake, W. (1972) Rock-burst mechanics. In: *Colorado School of Mines Quarterly*. 67. No. 1, Golden. Colorado. 64 p.

Bosman, J.D. & Malan, D.F. (2000) Time dependent deformation of tunnels in deep hard rock mines. *News Journal, ISRM*, 6(2), 8–9.

Borcherdt, R.D. (1994) Estimates of site-dependent response spectra for design (methodology and justification). *Earthquake Spectra*, 10, 617–654.

Bowden, F.P. & Leben, L. (1939) The nature of sliding and the analysis of friction. *Proceedings of the Royal Society of London*, A169, 371–391.

Bowden, F.P. & Tabor, D. (1950) *The Friction and Lubrication of Solids*. Oxford, Clarenden Press.

Brace, W.F. & Byerlee, J.D. (1966) Stick-slip as a mechanism for earthquakes. *Science*, 153, 990–992.

Buys, B.J., Heyns, P.S. & Loveday, P.W. (2009) Rock bolt condition monitoring using ultrasonic guided waves. *The Journal of the Southern African Institute of Mining and Metallurgy*, 108, 95–105.

Campbell, K.W. (1981) Near source attenuation of peak horizontal acceleration. *Bulletin of Seismological Society of America*, 71(6), 2039–2070.

CEDC (2008) *A Quick Report on the Strong Motion Data of 5.12 Wenchuan Earthquake*. Available from: http://www.smsd-iem.net.cn/.

Chaki, S. & Bourse, G. (2009) Guided ultrasonic waves for non-destructive monitoring of the stress levels in prestressed steel strands. *Ultrasonics*, 49, 162–171.

Chang, T.-Y., Cotton, F., Tsai, Y.-B. & Angelier, J. (2004) Quantification of hanging-wall effects on ground motion: Some insights from the 1999 Chi-Chi Earthquake. *Bulletin of Seismological Society of America*, 94(6), 2186–2197.

Cho, S.H., Ogata, Y. & Kaneko, K. (2003) Strain-rate dependency of the dynamic tensile strength of rock. *International Journal of Rock Mechanics and Mining Sciences*, 40(5), 763–777.

Christensen, R.J. (1971) *Dynamic Properties of Rocks*. MS Thesis. Salt Lake City, UT, University of Utah.

Christensen, R.J., Swanson, S.R. & Brown, W.S. (1972) Split-Hopkinson-bar tests on rock under confining pressure. *Experimental Mechanics*, 12, 508–513.

Cohee, B.P., Somerville, P.G. & Abrahamson, N.A. (1991) Simulated ground motions for hypothesized MW=8 subduction earthquake in Washington and Oregon. *Bulletin of Seismological Society of America*, 81, 28–56.

Collings, G.S., Melosh, J.J. & Marcus, R.A. (2005) Earth impact effects program: A web-based computer program for calculating the regional environmental consequences of a meteoroid impact on Earth. *Meteoritics & Planetary Science*, 40(6), 817–840.

Costin, L.S. & Holcomb, D.J. (1981) Time dependent failure of rock under cyclic loading. *Tectonophysics*, 79(3–4), 279–296.

Crotty, J.M. & Wardle, L.J. (1985) Boundary integral analysis of piece-wise homogenous media with structural discontinuities. *International Journal of Rock Mechanics and Mining Sciences*, 22(6), 419–427.

Crouch, S.L. & Starfield, A.M. (1983) *Boundary Element Methods in Solid Mechanics*. London, Allen & Unwin.

Cundall, P.A. (1971a) A computer model for simulating progressive, large-scale movements in blocky rock systems. In: *Proc. Int. Symp. Rock Fracture, 11-8, Nancy, France.*

Cundall, P.A. (1971b) *The Measurement and Analysis of Acceleration in Rock Slopes*. Ph.D. Thesis. London, University of London (Imperial College).

CWB (1999) Free field strong-motion data from the 921 Chi-Chi earthquake. In: *Seismological Center, Central Weather Bureau, Taipei, Taiwan.*

DAD-ERD (1995) Available from: http://www.deprem.gov.tr.

Davies, E. & Hunter, S.C. (1963) The dynamic compression testing of solids by the method of the split Hopkinson pressure bar. *Journal of the Mechanics and Physics of Solids*, 11, 155–179.

DERMAS (1993) A collaborative research on the dynamic resistance of masonry structures between Tokai University, Waseda University, Turkish Earthquake Foundation, Princeton University and National Earthquake Center of USA (unpublished report). 30 p.

Desai, C.S., Zaman, M.M., Lightner, J.G. & Siriwardane, H.J. (1984) Thin layer element for interfaces and joints. *International Journal for Numerical and Analytical Methods in Geomechanics*, 8, 19–43.

Detournay, E. (1983) *Two Dimensional Elasto-Plastic Analysis of Deep Cylindrical Tunnel Under Non-Hydrostatic Loading*. PhD Thesis. Dept. of Civil Eng. and Min. Eng., University of Minnesota.

Detzlhofer, H. (1969) Gebirgswassereinflüsse beim stollenbau. *Rock Mechanics*, 1, 207–240.

Donath, F.A. (1964) Strength variation and deformational behavior in anisotropic rock. In: Judd, W.R. (ed.) *State of Stress in the Earth's Crust*. New York, NY, Elsevier. pp. 281–297.

Donath, F.A. & Fruth, L.S. (1971) Dependence of strain rate effects on deformation mechanism and rock type. *Journal of the Geological Society*, 79, 343–371.

Dowding, C.H. (1985) *Blast Vibration Monitoring and Control*. Englewood Cliffs, NJ, Prentice-Hall.Dowding, C.H. & Rozen, A. (1978) Damage to rock tunnels from earthquake shaking. *Journal of the Geotechnical Engineering Division*, ASCE, GT2, 175–191.

Drucker, D.C. & Prager, W. (1952) Soil mechanics and plastic analysis for limit design. *Quarterly of Applied Mathematics*, 10(2), 157–165.

Egger, P. (1983) A new development in the base friction technique. In: *Coll. Phys. Geomech. Models, ISMES, Bergamo*. pp. 67–87.

Eringen, A.C. (1961) Propagations of elastic waves generated by dynamical loads on a circular cavity. *Journal of Applied Mechanics, ASME*, 28(2), 218–212.

Eringen, A.C. (1980) *Mechanics of Continua*. John Wiley & Sons Ltd (1st), 520 p. 2nd edition. Huntington, New York, Robert E. Krieger Publ. Co.

Eringen, A.C. (1980) *Mechanics of Continua*. 2nd edition. New York, NY, Krieger Pub. Co.

Eshghi, S. & Zare, M. (2003) Bam (SE Iran) earthquake of 26 December 2003, Mw6.5: A Preliminary Reconnaissance Report. Available form: http://www.iiees.ac.ir/.

Follansbee, P.S. & Frantz, C. (1983) Wave propagation in the split Hopkinson pressure bar. *ASME Journal of Engineering Materials and Technology*, 105, 61–66.

Forrestal, M.J. (1986) Penetration into dry porous rock. *International Journal of Solids and Structures*, 22(12), 1485–1500.

Forrestal, M.J. & Luk, V.K. (1988) Dynamic spherical cavity-expansion in a compressible elastic-plastic solid. *Journal of Applied Mechanics, Transactions ASME*, 55(2), 275–279.

Fukushima, K., Yuji Kanaori, K. & Fusanori Miura, F. (2010) Influence of fault process zone on ground shaking of inland earthquakes: Verification of Mj=7.3 Western Tottori Prefecture and Mj=7.0 West Off Fukuoka Prefecture earthquakes, southwest Japan. *Engineering Geology*, 116, 157–165.

Gee, L.S. (1996) Real-time seismology at UC Berkeley; the Rapid Earthquake Data Integration Project. *Bulletin of the Seismological Society of America*, 86, 1037–1106.

Geniş, M. (2002) *Evaluation of Dynamic Response and Stability of Shallow Underground Openings in Discontinuous Rock Masses Using Model Tests*. Doctorate Thesis. Zonguldak, Zonguldak Karaelmas University.

Geniş, M. & Aydan, Ö. (2002) Evaluation of dynamic response and stability of shallow underground openings in discontinuous rock masses using model tests. In: *Korea-Japan Joint Symposium on Rock Engineering, Seoul, Korea, July*. pp. 787–794.

Geniş, M. & Aydan, Ö. (2007) Static and dynamic stability of a large underground opening. In: *Proceedings of 2nd Symposium on Undergound Excavations for Transportation*. pp. 317–326.

Geniş, M. & Aydan, Ö. (2008) Assessment of dynamic response and stability of an abandoned room and pillar underground lignite mine. In: *The 12th International Conference of International Association for Computer Methods and Advances in Geomechanics (IACMAG), Goa, India*. pp. 3899–3906.

Geniş, M. & Gerçek, H. (2003) A numerical study of seismic damage to deep underground openings. In: *ISRM 2003-Technology Roadmap for Rock Mechanics, 10th ISRM Congress, South African Institute of Mining and Metallurgy*. pp. 351–355.

Geniş, M., Tokashiki, N. & Aydan, Ö. (2009) The stability assessment of karstic caves beneath Gushikawa Castle remains (Japan), In: *EUROCK2009 ISRM Regional Sym., Dubrovnik*. pp. 449–454.

Genis, M., Aydan, Ö. & Derin, Z. (2013) Monitoring blasting-induced vibrations during tunnelling and its effects on adjacent tunnels. In: *Proc. of the 3rd Int. Symp. on Underground Excavations for Transportation, Istanbul*. pp. 210–217.

Gerçek, H. (1996) Special elastic solutions for underground openings. In *Milestones in Rock Engineering – The Bieniawski Jubilee Collection, Balkema, Rotterdam*. pp. 275–290.

Gerçek, H. & Geniş, M. (1996) A comparative study of the effect of in situ stress field on the stability of underground openings. In: *Proceedings EUROCK'96 ISRM Symposium on Prediction and Performance in Rock Mechanics and Rock Engineering, Vol. 2, Balkema, Rotterdam*. pp. 869–874.

Gerçek, H. & Geniş, M. (1999) Effect of anisotropic in situ stresses on the stability of underground openings. In: *Proceedings of the 9th International Congress on Rock Mechanics, ISRM, Vol. 1, Balkema, Rotterdam*. pp. 367–370.

Ghaboussi, J. (1988) Fully deformable discrete analysis using a finite element approach. *Computer and Geotechnical*, 5, 175–195.

Ghaboussi, J., Wilson, E.L. & Isenberg, J. (1973) Finite element for rock joints and interfaces. *Journal of the Soil Mechanics and Foundations Engineering Division, ASCE, SM10*, 99, 833–848.

Goldsmith, W., Sackman, J.L. & Ewert, C. (1976) Static and dynamic fracture strength of Barre granite. *The International Journal of Rock Mechanics and Mining Sciences*, 13, 303–309.

Goodman, R.E., Taylor, R. & Brekke, T.L. (1968) A model for the mechanics of jointed rock. *Journal of the Soil Mechanics and Foundations Engineering Division, ASCE, SM3*, 94, 637–659.

Graves, R. (1996) Simulating seismic wave propagation in 3D elastic media using staggered-grid finite differences. *Bulletin of the Seismological Society of America*, 86(4), 1091–1106.

Gray, G.T. (2000) Classic split-Hopkinson pressure bar testing. In: *ASM Handbook Vol 8, Mechanical Testing and Evaluation, vol. 8. ASM Int, Materials Park, OH*. pp. 462–476.

Grimal, A.N. (1988) *History of Ancient Egypt*. Blackwell, None Stated.

Grimstad, E. (1999) Experiences from excavation under high rock stress in the 24.5 km long Laerdal tunnel. In: *International Conf. on Rock Engineering Techniques for Site Characterisation. Bangalore, India*. pp. 135–146.

GSI (Geographical Survey Institute of Japan) (2004) *Mid Niigata Prefecture Earthquakes in 2004*. Available from: http://www.gsi.go.jp/.

Hagedorn, H., Stadelmann, R. & Husen, S. (2008) Gotthard base tunnel rock burst phenomena in a fault zone, measuring and modelling results. In: *World Tunnel Congress 2008 – Underground Facilities for Better Environment and Safety – India*. pp. 419–430.

Haimson, B.C. (1974) Mechanical behavior of rock under cyclic loading. In: *Proceedings of the 3rd Congress of the International Society for Rock Mechanics, Part A. Advances in Rock Mechanics: Report of Current Research, Sept 1–7, 1974*. Washington, DC, Denver, National Academy of Sciences. pp. 373–387.

Haimson, B.C. & Fairhurst, C. (1970) Some bit penetration characteristics in pink Tennessee marble. In: *Proc. 12th US Rock Mechanics Symp*. pp. 547–559.

Hamada, M. & Aydan, Ö. (1992) A report on the site investigation of the March 13 Earthquake of Erzincan, Turkey. In: *ADEP, Association for Development of Earthquake Prediction*. 86 p.

Hao, H. (2002) *Characteristics of Non-Linear Response of Structures and Damage of RC Structures to High Frequency Blast Ground Motion*. Okayama, Wave 2002.

Hartzell, S.H. (1978) Earthquake aftershocks as Green's functions. *Geophysical Research Letters*, 5, 1–4.

Hartzell, S.H. & Heaton, T.H. (1983) Inversion of strong ground motion and teleseismic waveform data for fault rupture history of the 1979 Imperial Valley, California earthquake. *Bulletin of the Seismological Society of America*, 73, 1553–1583.

Hashimoto, S., Miwa, K., Ohashi, M. & Fuse, K. (1999) Surface soil deformation and tunnel deformation caused by the September 3, 1998, Mid-North Iwate Earthquake. In: *7th Tohoku Regional Convention, Japan Society of Engineering Geology*.

Hendron, A.J. (1977) Engineering of rock blasting on civil projects. In: *Structural and Geotechnical Mechanics*. Englewood Cliffs, NJ, Prentice-Hall.

Henke, A. (2005) Tunnelling in Switzerland: From long tradition to the longest tunnel in the world. In: *World Long Tunnels*. pp. 57–70.

Hestholm, S. (1999) Three-dimensional finite difference viscoelastic wave modelling including surface topography. *Geophysical Journal International*, 139, 852–878.

Heuze, F.E. (1990) An overview of projectile penetration into geological materials, with emphasis on rocks. *International Journal of Rock Mechanics and Mining Science & Geomechanics Abstracts*, 27(1), 1–14.

Hirth, G. & Tullis, J. (1994) The brittle-plastic transition in experimentally deformed quartz aggregates. *Journal of Geophysical Research*, 99, 11731–11747.

Hoek, E. & Bray, J.W. (1981) *Rock Slope Engineering*. 3rd edition. London, Inst. Mining and Metallurgy.

Hoek, E. & Brown, E.T. (1980) Empirical strength criterion for rock masses. *Journal of the Geotechnical Engineering Division, ASCE*, 106(GT9), 1013–1035.

Hoek, E. & Brown, E.T. (1997) Practical estimates of rock mass strength. *International Journal of Rock Mechanics and Mining Sciences*, 34(8), 1165–1186.

Hopkinson, B. (1914) A method of measuring the pressure produced in the detonation of high explosives or by the impact of bullets. *Philosophy Transactions of the Royal Society (London)* A, 213, 437–456.

Hutchings, L. & Viegas, G. (2012) Application of empirical green's functions in earthquake source, wave propagation and strong ground motion studies. In: D'Amico, S. (ed.) *Earthquake Research and Analysis, New Frontiers in Seismology, Chapter 3*. In Tech Publishing. pp. 87–140.

Ikeda, K. (1970) A classification of rock conditions for tunnelling. In: *Proceedings of the 1st Int. Congr. on Engineering Geology, IAEG, Paris*. pp. 1258–1265.

Ikeda, T., Konagai, K., Kamae, K., Sato, T. & Takase, Y. (2016) Damage investigation and source characterization of the 2014 Northern Part of Nagano Prefecture earthquake. *Journal of Structural Mechanics and Earthquake Engineering*, 72(4), I_975–I_983.

Imazu, M., Ideura, H. & Aydan, Ö. (2014) A monitoring system for blasting-induced vibrations in tunneling and its possible uses for the assessment of rock mass properties and in-situ stress inferences. In: *Proc. of the 8th Asian Rock Mechanics Symposium, Sapporo*. pp. 881–890.

Imazu, M., Soya, M., Ideura, H. & Aydan, Ö. (2015) A monitoring system for blasting-induced vibrations in tunneling and its possible uses for the assessment of rock mass properties and in-situ stress inferences. In: *ITA WTC 2015 Congress and 41st General Assembly, Dubrovnik, Croatia*.

Ingliss, C.E. (1913) Stresses in a plate due to the presence of cracks and sharp corners. In: *Trans. Inst. Nav. Archit., London*, Vol. 55. pp. 219–241.

Inoma, H. (1981) Rock burst in the Kanetsu tunnel. *Journal of the Japan Society of Engineering Geology (in Japanese)*, 22(3), 286–295.

Irikura, K. (1983) Semi-empirical estimation of strong ground motions during large earthquakes. *Bulletin of the Disaster Prevention Research Institute (Kyoto University)*, 33, 63–104.

IRIS (2013) Teachable Moments. M0.0 Meteor Explosion, Russia: Friday, February 15, 2013 at 03:20:26 UTC. Available from: https//www.iris.edu/130215Russia.

Ishihara, K. (1985) Stability of natural deposits during earthquakes. In: *Procs. of 11th Int. Conf. on Soil Mechanics and Foundation Engineering*, Vo.1. pp. 321–376.

Itasca (2005) *FLAC3D-Fast Lagrangian Analysis of Continua-User Manual (Dynamic Option) (Version 2.1)*. Minneapolis, Itasca Consulting Group Inc.

Ito, T., Aydan, Ö. & Akagi, T. (2004) An experimental study on the electrical potential response of rocks during creep and cyclic loading. In: *3rd Asian Rock Mechanics Symposium, Kyoto*, Vol. 2. pp. 881–884.

Ivanović, A. & Neilson, R.D. (2008) Influence of geometry and material properties on the axial vibration of a rock bolt. *International Journal of Rock Mechanics and Mining Sciences*, 45, 941–951.

Iwata, N., Adachi, K., Takahashi, Y., Aydan, Ö., Tokashiki, N. & Miura, F. (2016) Fault rupture simulation of the 2014 Kamishiro Fault Nagano Prefecture Earthquake using 2D and 3D-FEM. EUROCK2016. In: *Ürgüp*. pp. 803–808.

Jaeger, J.C. (1959) The frictional properties of joints in rock. *Geology Pura Applied*, 43, 148–158.

Jaeger, J.C. & Cook, N.G.W. (1969) *Fundamentals of Rock Mechanics*. 1st edition. London, Methuen. p. 593.

Jaeger, J.G. & Cook, N.G.W. (1979) *Fundamentals of Rock Mechanics*. 3rd edition. London, Chapman and Hall.

Jafari, M.K., Pellet, F., Boulon, M. & Hosseini, K.A. (2004) Experimental study of mechanical behavior of rock joints under cyclic loading. *Rock Mechanics and Rock Engineering*, 37(1), 3–23.

Japan Geotechnical Society (2000) *Report on 1999 Kocaeli, Turkeyand Chi-Chi Taiwan Earthquakes*.

Japan Society of Civil Engineers. Archives of structural damage by the 1923 Great Kanto Earthquake. Available from: http://www.jsce.or.jp.

Jiang, Q., Feng, X.T., Xiang, T.B. & Su, G.S. (2010) Rockburst characteristics and numerical simulation based on a new energy index: A case study of a tunnel at 2,500 m depth. *Bulletin of Engineering Geology and the Environment*, 69(3), 381–388.

JMA (2007) Available from: http://www.seisvol.kishou.go.jp/eq/ (in Japanese).

Joule, J. (1842) On a new class of magnetic forces. *Annals of Electricity, Magnetism, and Chemistry*, 8, 219–224.

Joyner, W.B. & Boore, D.M. (1981) Peak horizontal acceleration and velocity from strong motion records from the 1979 Imperial Valley California Earthquake. *Bulletin of the Seismological Society of America*, 71(6), 2011–2038.

Kaiser, P.K., Jesenak, P. & Brummer, R.K. (1993) Rockburst damage potential assessment. In: *Int. Symp. on Assessment and prevention of Failure Phenomena in Rock Engineering, Istanbul*. pp. 591–596.

Kaiser, P.K., McCreath, D.F. & Tannant, D.D. (1996) *Canadian Rockburst Support Handbook*. Sudbury, Geomechanics Research Centre, Laurentian University.

Kamae, K., Irikura, K. & Pitarka, A. (1998) A technique for simulating strong ground motion using hybrid Green's function. *Bulletin of the Seismological Society of America*, 88, 357–367.

Kamai, T., Wang, W.N. & Shuzui, H. (2000) The Slope failure disaster induced by the Taiwan Chi-Chi earthquake of 21 September 1999. *Slope Failure News*, 13, 8–12.

Kana, D.D., Chowdhury, A.H., Hsiung, S.M., Ahola, P.A., Brady, B.H.G. & Philip, J. (1991) Experimental techniques for dynamic shear testing of natural rock joints. In: *Proc. 7th ISRM Congress*. pp. 519–525.

Kanai, K. & Tanaka, T. (1951) Observations of earthquake motion at different depths of the earth. *Bulletin of the Earthquake Research Institute, Tokyo University*, 28, 107–113.

Karaca, M. & Egger, P. (1993) Failure phenomena around shallow tunnels in jointed rock. In: Pasamehmetoglu et al. (eds.) *Assessment and Preventions of Failure Phenomena in Rock Engineering*. Rotterdam, Balkema. pp. 381–388.

Karaca, M., Egger, P., Aydan, O. & Sezaki, M. (1995) Mechanics of failure around shallow tunnels in jointed rock. In: Rossmanith, H.P. (ed.) *Mechanics of Jointed and Faulted Rock*. Rotterdam, Balkema.

Kastner, H. (1961) *Statik des Tunnel- and Stollenbaues ("Design of Tunnels").* 2nd edition. New York, NY, Springer-Verlag.

Kawakami, H. (1984) Evaluation of deformation of tunnel structure due to Izu-Oshima Kinkai earthquake of 1978. *Earthquake Engineering & Structural Dynamics,* 12(3), pp. 369–383.

Kawamoto, T., Aydan, Ö. & Tsuchiyama, S. (1991) A consideration on the local instability of large underground openings. In: *Int. Conf., GEOMECHANICS'91, Hradec.* pp. 33–41.

Kawamoto, T., Kyoya, T. & Aydan, Ö. (1994) Numerical models for rock reinforcement by rockbolts. In: *Int. Conf. on Computer Methods and Advances in Geomechanics, IACMAG, Morgantown,* Vol. 1. pp. 33–45.

Keefer, D.K. (1984) Slope failures caused by earthquakes. *Geological Society of American Bulletin,* 95, 406–421.

Kenkmann, T., Wunnemann, K., Deutsch, A., Poelchau, M.H., Schafer, F. & Thoma, K. (2011) Impact cratering in sandstone: The MEMIN pilot study on the effect of pore water. *Meteoritics & Planetary Science,* 46, 890–902.

Kesimal, A., Ercikdi, B. & Cihangir, F. (2008) Environmental impacts of blast-induced acceleration on slope instability at a limestone quarry. *Environmental Geology,* 54(2), 381–389.

Khan, A., Pommier, A., Neumann, G.A. & Mosegaard, K. (2013) The lunar moho and the internal structure of the Moon: A geophysical perspective. *Tectonophysics,* 609, 331–352.

KIK-NET (2004) Available from: http://www.bosai.go.jp/KIK-NET.

KiK-Net (2007) Available from: http://www.kik.bosai.go.jp/.

King, G.C.P., Stein, R.S. & Lin, J. (1994) Static stress changes and the triggering of earthquakes. *Bulletin of the Seismological Society of America,* 84(3), 935–953.

Kirsch, G. (1898) Die theorie der elastizitat und die bedürfnisse der festigkeitslehre. *Veit Vereines Deutscher Ingenieure,* 42, 797–807.

K-NET & KiK-Net (2007, 2008) *Digital Acceleration Records of Earthquakes Since 1998.* Available from: http://www.k-net.bosai.go.jp/ and http://www.kik.bosai.go.jp/.

K-net (2007) Available from: http://www.k-net.bosai.go.jp/.

Kreyszig, E. (1983) *Advanced Engineering Mathematics.* New York, NY, John Wiley & Sons.

Kobayashi, R. (1970) On mechanical behaviours of rocks under various loading-rates. *Rock Mechanics in Japan,* 1, 56–58.

Koide, H., Hoshino, K., Nishimatsu, Y. & Koizumi, S. (1988) In situ stress measurements in the Kanto-Tokai region of Japan for the prediction of earthquakes. In: *The Second Int. Symp. on Field Measurements in Geomechanics, Kobe.* pp. 125–134.

Koketsu, K., Fujiwara, H. & Ikegami, Y. (2004) Finite-element simulation of seismic ground motion with a voxel mesh. *Pure and Applied Geophysics,* 161(11–12), 2463–2478.

Kolsky, H. (1949) An investigation of the mechanical properties of materials at very high rates of loading. Proceedings of the Physical Society of London, B62, 676–700.

Kolsky, H. (1953) *Stress Waves in Solids.* Oxford, Clarendon Press.

Komada, H. & Hayashi, M. (1980) Earthquake observation around the site of undergound power station. In: *CRIEPI Report, E379003, Central Research Institute of Electric Power Industry, Japan.* pp. 1–34.

Kumsar, H. & Aydan, Ö. (1998) Kaya şevlerindeki kama tipi kaymaların dinamik duraylılığının model deneyler ile incelenmesi. *Kaya Mekaniği Bülteni (Bulletin of Rock Mechanics, TRMS), in Turkish,* 14, 29–39.

Kumsar, H. & Aydan, Ö. (2014) Response and stability of rock wedges under dynamic loading. In: *8th Asian Rock Mechanics Symposium, Sapporo.*

Kumsar, H., Aydan, Ö. & Ulusay, R. (2000) Dynamic and static stability of rock slopes against wedge failures. *Rock Mechanics and Rock Engineering,* 33(1), 31–51.

Kuno, H. (1935) The geologic section along the Tanna. In: *Tunnel Bull of the Earthquake Research Inst, Univ of Tokyo*, Vol. 14. pp. 92–103.

Kutter, H.K. & Fairhurst, C. (1971) On the fracture process in blasting. *International Journal of Rock Mechanics Mining Sciences*, 8, 181–202.

Laguerre, L., Aime, J.C. & Brissaud, M. (2002) Magnetostrictive pulse-echo device for nondestructive evaluation of cylindrical steel materials using longitudinal guided waves. *Ultrasonics*, 39, 503–514.

Laguerre, L., Christian, J. & Brissaud, M. (2000) Generation and detection of elastic waves with magnetoelastic device for the nondestructive evaluation of steel cables and bars. In: *Proceedings, 15th World Conference on Nondestructive Testing, 15–21 October, Roma, Italy*.

Lee, S.M., Park, B.S. & Lee, S.W. (2004) Analysis of rockbursts that have occurred in a waterway tunnel in Korea. In: *Paper 3B 24 – SINOROCK2004 Symposium, International Journal of Rock Mechanics and Mining Sciences, CD-ROM, Elsevier Ltd*, Vol. 41.

Lin, C.J. & Chai, J.F. (2008) Reconnaissance report on the China Wenchuan Earthquake My 12, 2008. In: *NCREE Newsletter*, Vol. 3, No. 3, Sep. 2008.

Lindsay, H. (2008) *ALSEP Apollo Lunar Surface Experiments Package 19 Nov 1969–30 Sept 1977*. Available from: https://www.hq.nasa.gov/alsj/HamishALSEP.html.

Logan, J.M. & Handin, J. (1972) Triaxial compression testing in intermediate strain rate. In: Clark, G.B. (ed.) *Dynamic Rock Mechanics*. USA, Port City Press. pp. 167–194.

Luk, V.K., Forrestal, M.J., *et al.* (1991) Dynamic spherical cavity expansion of strain-hardening materials. *Transactions of the ASME. Journal of Applied Mechanics*, 58(1), 1–6.

Lundborg, N. (1981) The probability of flyrock damages. In: *Swedish Detoni Research Foundation, Stockholm, D.S. 5*. 39 pp.

Mamaghani, I.H.P., Baba, S., Aydan, O.G. & Shimizu, Y. (1994) Discrete finite element method for blocky systems. In: *Int. Conf. Computer Meth. Adv. Geomech*. pp. 843–850.

Mamaghani, I.H.P., Aydan, Ö. & Kajikawa, Y. (1999) Analysis of masonry structures under static and dynamic loading by discrete finite element method. *Journal of Structural Mechanics and Earthquake Engineering*. Japan Society of Civil Engineers, JSCE, No. 626/I-48, pp. 1–12.

Masuda, K., Mizutani, H. & Yamada, I. (1987) Experimental study of strain-rate dependence and pressure dependence of failure properties of granite. *Journal of Physics of the Earth*, 35, 37–66.

Matsuda, T. (1975) The magnitude and periodicity of earthquakes from active faults. *JISHIN*, 28, 269–283 (in Japanese).

McLamore, R. & Gray, K.E. (1967) The mechanical behaviour of anisotropic sedimentary rocks. In: *Trans. Am. Soc. Mech. Engrs. Series B*. pp. 62–76.

Mellor, M. & Hawkes, I. (1971) Measurement of tensile strength by diametrical compression of discs and annuli. *Engineering Geology*, 5, 173–225.

Melosh, H.J. (1989) *Impact Cratering: A Geologic Process*. New York, NY, Oxford University Press. 245 pp.

Melosh, H.J. (2011) *Planetary Surface Processes*. Cambridge, Cambridge University Press. 500 pp.

Mindlin, R.D. (1949) Compliance of elastic bodies in contact. *Journal of Applied Mechanics*, 16, 259–268.

Mitri, H.S., Scoble, M.J. & McNamara, K. (1988) Numerical studies of destressing mine pillars in highly stressed rock. In: *Department of Mining and Metallurgical Engineering. McGill University, Montreal, QC*. pp. 50–56.

Misawa, Y., Aydan, Ö. & Hamada, M. (1993) A consideration on the failure of Mt. Mayuyama in 1792 from rock mechanics view point. In: *Proc. Int. Symp. on Assessment and Prevention of Failure Phenomena in Rock Engineering, Istanbul, Turkey*. pp. 871–877.

Mizutani, H., Ishido, T., Yokokura, T. & Ohnishi, S. (1976) Electrokinetic phenomena associated with earthquakes. *Geophysical Research Letters*, 3(7), 365–368.

Mogi, K. (1967) Effect of the intermediate principal stress on rock failure. *Journal of Geophysical Research*, 72(20), 5117–5131.

Moore, H.J. (1971) Craters produced by missile impacts. *Journal of Geophysical Research*, 76(23), 5750–5755.

Murai, I. & Matsuda, T. (1975) The earthquake of 1975 in the Central Part of Oita Prefecture. In: *Geological Report on Damage and Ground Disturbance with Special Reference to Quaternary faults. Bulletin of Earthquake Research Institute*, Vol. 50. pp. 303–327.

Muskhelishvili, N.I. (1953) Some Basic Problems of the Mathematical Theory of Elasticity. Groningen, Noordhoff.

Nakamura, Y., Dorman, J., Duennebier, R., Ewing, M., Lammlein, D. & Latham, G. (1974) High-frequency lunar teleseismic events. In: *Proceedings of Lunar and Planetary Science Conference 5th*. pp. 2883.

Nakano, T. & Ohta, Y. (2008) Non-linear dynamic response analysis of bridge crossing earthquake fault rupture plane. In: *The 14th World Conference on Earthquake Engineering October 12–17, 2008, Beijing, China*.

Nasseri, B.M.H., Rao, K.S. & Ramamurthy, T. (2003) Anisotropic strength and deformational behaviour of Himalayan Schists. *International Journal of Rock Mechanics and Mining Sciences*, 40(1), 3–23.

Nasu, N. (1931) Comparative studies of earthquake motions above ground and in a tunnel. *Bulletin of Earthquake Research Institute, Tokyo University*, 9, 454–472.

Nebert, K. (1960) Stratigraphy and tectonics of coal-bearing Neogene sediments in the west and north of Tavşanlı. *Bulletin of the Mineral Research and Exploration Institute (MTA) of Turkey*, 54, 7–35 (in Turkish).

Newmark, N. (1965) Effects of earthquakes on dams and embankments. *Geotechnique*, 15(2), 139–160.

Ngo, D. & Scordelis, A.C. (1967) Finite element analysis of reinforced concrete beams. *Journal of ACI*, 64, 152–163.

Northwood, T.D., Crawford, R. & Edwards, A.T. (1963) Blasting vibrations and building damage. *The Engineer*, 215, 973–978.

Ohnishi, Y., Sasaki, T. & Tanaka, M. (1995) Modification of the DDA for elasto-plastic analysis with illustrative generic problems. In: *35th US Rock Mechanics Symp. Lake Tahoe*. pp. 45–50.

Ohta, K. (1987) Failure mechanism of Mt. Mayuyama and Tsunami (in Japanese). *Chikyu*, 9(4), 214–220.

Ohta, Y. (2011) *A Fundamental Research on the Effects of Ground Motions and Permanent Ground Deformations Neighborhood Earthquake Faults on Civil Engineering Structures (in Japanese)*. Doctorate Thesis. Graduate School of Science and Technology, Tokai University. 272 p.

Ohta, Y. & Aydan, Ö. (2004) An experimental study on ground motions and permanent deformation nearby faults. *Journal of the School of Marine Science and Technology*, 2(3), 1–12.

Ohta, Y. & Aydan, Ö. (2009) An experimental and theoretical study on stick-slip phenomenon with some considerations from scientific and engineering viewpoints of earthquakes. *Journal of the School of Marine Science and Technology*, 8(3), 53–67.

Ohta, Y. & Aydan, Ö. (2010) The dynamic responses of geomaterials during fracturing and slippage. *Rock Mechanics and Rock Engineering*, 43(6), 727–740.

Ohta, Y., Aydan, Ö. & Tokashiki, N. (2008) The dynamic response of rocks during fracturing and its implications in geo-engineering and earth science. In: *5th Asian Rock Mechanics Symposium (ARMS5), Tehran*. pp. 965–972.

Ohta, Y., Aydan, Ö. & Yagi, M. (2014) Laboratory model experiments and case history surveys on response and failure process of rock engineering structures subjected to earthquake faulting. In: *8th Asian Rock Mechanics Symposium, Sapporo*. pp. 842–852.

Ohta, Y., Nakamura, M., Daido, M. & Aydan, Ö. (2004) The ground vibrations caused by fracturing, faulting and blasting. In: *3rd Asian Rock Mechanics Symposium, Kyoto*, Vol. 2. pp. 923–926.

Okada, Y. (1992) Internal deformation due to shear and tensile faults in a half-space. *Bulletin of the Seismological Society of America*, 82(2), 1018–1040.

Okamoto, S. (1973) *Introduction to Earthquake Engineering*. Tokyo, University of Tokyo Press.

Olsson, W.A. (1991) The compressive strength of Tuff as a function of strain rate from 10^6 to 10^3/sec. *International Journal of Rock Mechanics and Mining Sciences*, 8(1), 115–118.

Ortlepp, W.D. (2000) Observation of mining-induced faults in an intact rock mass at depth. *International Journal of Rock Mechanics and Mining Sciences*, 37(1–2), 423–436.

Ortlepp, W.D. & Stacey, T.R. (1994) Rockburst mechanisms in tunnels and shafts. *Tunnelling and Underground Space Technology*, 9(1), 59–65.

Osinski, G.R. (2006) The geological record of meteorite impacts. In: *ESASP06*. 12 p.

Owen, D.R.J. & Hinton, E. (1980) *Finite Element in Plasticity: Theory and Practice*. Swansea, Pineridge Press Ltd.

Panet, M. (1969) Quelques problèmes de mécanique des roches posés par le tunnel du Mont Blanc. In: *Bul. Liaison Labo. Routiers P. et Ch., N. 42*. pp. 115–145.

Park, S. & Elrick, S. (1998) Predictions of shear-wave velocities in Southern California using surface geology. *Bulletin of the Seismological Society of America*, 88, 677–685.

Perzyna, P. (1966) Fundamental problems in viscoplasticity. *Advances in Applied Mechanics*, 9(2), 244–368.

Pietruszczak, S. & Mroz, Z. (1981) Finite element analysis of deformation of strain-softening materials. *International Journal for Numerical Methods in Engineering*, 17, 327–334.

Pierre, B. (2008) Thermography and thermal stress analysis: New developments and trends. In: *Cedip Infrared Systems-France, 2008.2 Cedip User Meeting*.

Pitarka, A. (1999) 3D elastic finite-differences modeling of seismic motion using staggered grids with nonuniform spacing. *Bulletin of the Seismological Society of America*, 89(1), 54–68.

Popova, O.P., *et al.* (2013) Chelyabinsk airburst, damage assessment, meteorite recovery and characterization. *Science*, 342, 1069–1073.

Prentice, C. & Ponti, D. (1997) Coseismic deformation of the Wrights tunnel during the 1906 San Francisco earthquake: A key to understanding 1906 fault slip and 1989 surface ruptures in the southern Santa Cruz Mountains, California. *Journal of Geophysical Research*, 102, 635–648.

Rashid, Y.R. (1968) Ultimate strength analysis of prestresses concrete pressure vessels. *Nuclear Engineering and Design*, 7, 334–344.

Read, R.S., Chandler, N.A. & Dzik, E.J. (1998) In situ strength criteria for tunnel design in highly-stressed rock masses. *International Journal of Rock Mechanics and Mining Sciences*, 35(3), 261–278.

Reilinger, R.E., Ergintav, S., Burgmann, R., McClusky, S., Lenk, O., Barka, A., Gürkan, O., Hearn, L., Feigl, K.L., Çakmak, R., Aktug, B., Özener, H. & Toksöz, M.N. (2000) Coseismic and postseismic fault slip for the 17 August 1999, M = 7.5, Izmit, Turkey Earthquake. *Science*, 289, 1519–1524.

Roelofse, F. & Saunders, I. (2013) A first report on meteor-generated seismic signals as detected by the SANSN. *South African Journal of Science*, 109(5/6), 6.

Rojat, F., Labiouse, V., Kaiser, P.K. & Descoeudres, F. (2009) Brittle rock failure in the Steg lateral adit of the Lötschberg base tunnel. *Rock Mechanics Rock Engineering*, 42, 341–359.

Romano, F., Trasatti, E., Lorito, S., Piromallo, C., Piatanesi, A., Ito, Y., Zhao, D., Hirata, K., Lanucara, P. & Cocco, M. (2014) Structural control on the Tohoku earthquake rupture process investigated by 3D FEM, tsunami and geodetic data. *Science Reports*, 4, 5631.

Rozen, A. (1976) *Response of Rock Tunnels to Earthquake Shaking*. MSc Thesis in Civil Engineering. Massachusetts Institute of Technology.

Sakurai, T. (1999) A report on the earthquake fault appearing in the Tanna tunnel under construction by North-Izu Earthquake 1930 (in Japanese). *Journal of the Japan Society of Engineering Geology*, 39(6), 540–544.

Sato, R. (1989) *Handbook on Parameters of Earthquake Faults in Japan*. Tokyo, Kajima Pub. Co. (in Japanese).

Savin, G.N. (1961) *Stress Concentrations Around Holes*. Oxford, Pergamon.

Seiki, T. & Aydan, Ö. (2003) Deterioration of Oya tuff and its mechanical property change as building stone. In: *Proc. of Int. Symp. on Industrial Minerals and Building Stones, Istanbul, Turkey*. pp. 329–336.

Semenza, E. & Ghirotti, M. (2000) History of the 1963 Vaiont slide: The importance of geological factors. *Bulletin of Engineering Geology and the Environment*, 59, 87–97.

Serdengecti, S. & Boozer, G.D. (1961) The effects of strain rate and temperature on the behaviour of rocks subjected to triaxial compression. In: Hartman, H.L. (ed.) *Proc. 4th Symp. Rock Mech., Penn State Univ.* pp. 83–97.

Sezaki, M., Aydan, Ö., Ichikawa, Y. & Kawamoto, T. (1990) Mechanical properties of rock mass for the pre-design of tunnels by NATM using a rock mass data-base (in Japanese). *Journal of Civil Engineers of Japan, Construction Division*, 421-VI-13, 125–133.

Sharma, S. & Judd, W.R. (1991) Underground opening damage to underground facilities. *Engineering Geology*, 30, 263–276.

Shaw, I. (2000) *Oxford History of Ancient Egypt*. The Oxford University Press, ISBN 0-19-815034-2.

Shi, G.H. (1988) *Discontinuous Deformation Analysis: A New Numerical Model for the Statics and Dynamics of Block Systems*. Ph.D. Thesis. Department of Civil Engineering, University of California, Berkeley. 378 p.

Shimokawa, M., Oda., S. & Kizawa, T. (1977) An investigation from rock-burst phenomenon in Dai-Shimizu Tunnel (in Japanese). In: *5th Domestic Rock Mechanics Conference*, Vol. 5. pp. 79–84.

Shockey, D.A., Petersen, C.F., Curran, D.R. & Rosenberg, J.T. (1973) Failure of rock under high rate tensile loads. In: Hardy, H.R. Jr & Stefanko, R. (eds.) *Proc. 14th Symp. Rock Mech., Penn State Univ., Pennsylvania*. pp. 709–738.

Siskind, D.E., Stagg, M.S., Koop, J.W. & Dowding, C.H. (1980) Structure response and damage produced by ground vibration from surface mine blasting. In: *United States Bureau of Mines, Report of Investigations, No. 8507*.

Smith, P.D. & Hetherington, J.G. (1994) *Blast and Ballistic Loading of Structures*. Great Britain, Butterworth-Heinemann Ltd.

Somerville, P.G., Sen, M. & Cohee, B. (1991) Simulation of strong ground motion recorded during the 1985 Michoacan, Mexico an Valparaiso, Chile earthquakes. *Bulletin of the Seismological Society of America*, 81, 1–27.

Somerville, P.G., Smith, N., Graves, R. & Abrahamson, N. (1997) Modification of empirical strong ground motion attenuation relations to include the amplitude and duration effects of rupture directivity. *Seismic Research Letter*, 68, 199–222.

Stacey, T.R. (1981) A simple extension strain criterion for fracture of brittle rock. *International Journal of Rock Mechanics and Mining Sciences*, 18, 469–474.

Stein, R.S. (January 2003) Earthquake conversations. *Scientific American*, 288(1), 72–79.

Stowe, R.L. & Ainsworth, D.L. (1968) Effect of rate loading on strength and Young's modulus of elasticity of rock. In: Cray, K.E. (ed.) *Proc. 10th U.S. Symp. Rock Mech., Austin, Texas.* pp. 3–34.

Sugito, M., Furumoto, Y. & Sugiyama, T. (2000) Strong motion prediction on rock surface by superposed evolutionary spectra. In: *12th World Conference on Earthquake Engineering, 2111/4/A, CD-ROM.*

Tang, C. (1997) Numerical simulation of progressive rock failure and associated seismicity. *International Journal of Rock Mechanics and Mining Sciences*, 34 (2), 249–261.

Tang, C.L., Hu, J.C., Lin, W.L., Angelier, J., Lu, C.Y., Chan, Y.C. & Chu, H.T. (2009) The Tsaoling landslide triggered by the Chi-Chi earthquake, Taiwan: Insights from a discrete element simulation. *Engineering Geology*, 106, 1–19.

Tannant, D.D., Brummer, R.K. & Yi, X. (1995) Rockbolt behaviour under dynamic loading: Field tests and modelling. *International Journal of Rock Mechanics and Mining Sciences & Geomechanics Abstracts*, 32(6), 537–550.

Tano, H., Aydan, Ö., Ulusay, R., Tokashiki, N., Ito, T., Akagi, T. & Yoshimura, S. (2003a) The assessment of the performance of historical rock structures for preservation and restoration. In: *International Symposium on Industrial Minerals and Building Stones, Istanbul.* pp. 371–379.

Tano, H., Kumsar, H., Aydan, Ö. & Ulusay, R. (2003b) Assessment on Babadağ landslide using by simple field measuring systems. In: Tano, H. & Aydan, Ö. (eds.) *Int. Colloquium on Instrumentation and Monitoring of Landslides and earthquakes in Japan and Turkey, Koriyama.* pp. 1–10.

Tano, H., Abe, T. & Aydan, Ö. (2005) The development of an in-situ AE monitoring system and its application to rock engineering with particular emphasis on tunneling. In: *ITA Conference, Istanbul.* pp. 1245–1252.

TEC-JSCE (Tunnel Engineering Committee, JSCE) (2005) Report of the 2004 Mid Niigata Prefecture Earthquake. *JSCE* (in Japanese).

Terzaghi, K. (1946) Rock defects and loads on tunnel supports. In: Proctor, R.V. & White, T.L. (eds.) *Rock Tunneling with Steel Supports*, Vol. 1. Youngstown, OH. pp. 17–99.

Terzaghi, K. (1960) Stability of steep slopes on hard, unweathered rock. *Geotechnique*, 12, 251–270.

Thoenen, J.R. & Windes, S.L. (1942) Seismic effects of quarry blasting. In: *U.S. Bureau of Mines Bulletin.* p. 442.

Thomson, W. (Lord Kelvin) (1853) On the dynamical theory of heat. *Transactions of the Royal Society of Edinburgh*, 20, 261–283.

Toda, S., Stein, R.S. & Sagiya, T. (2002) Evidence from the 2000 Izu Islands swarm that seismicity is governed by stressing rate. *Nature*, 419, 58–61.

Tokashiki, N. & Aydan, Ö. (2003) Characteristics of Ryukyu Limestone and its utilization as a building stone in historical and modern structures. *International Symposium on Industrial Minerals and Building Stones, Istanbul.* pp. 311–318.

Tokashiki, N. & Aydan, Ö. (2010a) Kita-Uebaru natural rock slope failure and its back analysis. *Environmental Earth Sciences*, 62(1), 25–31.

Tokashiki, N. & Aydan, Ö. (2010b) The stability assessment of overhanging Ryukyu limestone cliffs with an emphasis on the evaluation of tensile strength of Rock Mass. *Journal of Geotechnical Engineering, JSCE*, 66(2), 397–406.

Tokashiki, N. & Aydan, Ö. (2011) A comparative study on the analytical and numerical stability assessment methods for rock cliffs in Ryukyu Islands. In: *The 13th IACMAG Conference, Melbourne*. pp. 663–668.

Tokashiki, N., Aydan, Ö., Mamaghani, I.H.P. & Kawamoto, T. (1997) The stability of a rock block on an incline by discrete finite element method (DFEM). In: *Proceedings of the Ninth International Conference on Computer Methods and Advances in Geomechanics, Wuhan, China*, Vol. 1. pp. 523–528.

Tokashiki, N., Aydan, Ö., Shimizu, Y. & Mamaghani, I.H.P. (2001) A stability analysis of masonry walls by discrete finite element method. In: *Proceedings of the Tenth International Conference on Computer Methods and Advances in Geomechanics, Tucson, Arizona, USA*, Vol. 2. pp. 1625–1628.

Tokashiki, N., Aydan, Ö. & Shimizu, Y. (2002) Some considerations on the stability of masonary structures: Numerical analyses and experiments of masonary structures. *Shitatei*, 22, 32–37 (in Japanese).

Tokashiki, N., Aydan, Ö., Daido, M. & Akagi, T. (2006) An experimental and analytical study on the dynamic stability of masonry structures. In: *35th Japan Rock Mechanics Symposium, Tokyo*. pp. 115–120.

Tokashiki, N., Aydan, Ö., Daido, M. & Akagi, T. (2007) Experiments and analyses of seismic stability of masonry retaining walls. *Rock Mechanics Symposium, JSCE*. pp. 115–120.

Tokashiki, N., Aydan, Ö. & Jeong, G.C. (2008) Stone masonary historical structures in Ryukyu Islands and possible remedial measures. In: *2008, International Symposium on Conservation Science for Cultural Heritage, Seoul*.

Tokashiki, N., Aydan, Ö. & Geniş, M. (2014) Experimental studies on the dynamic response and stability of some historical masonry structures in Ryukyu Archipelago. In: *8th Asian Rock Mechanics Symposium, Sapporo*.

Toki, K. & Miura, F. (1985) Simulation of a fault rupture mechanism by a two-dimensional finite element method. *Journal of Physics of the Earth*, 33, 485–511.

Toksöz, M.N. (1975) Lunar and planetary seismology. *Reviews of Geophysics and Space Physics*, 13(3), 306–311.

Toksöz, M.N., Press, F., Anderson, K., Dainty, A., Latham, G., Ewing, M., Dorman, J., Lammlein, D., Nakamura, Y., Sutton, G. & Duennebier, F. (1972) Velocity structure and properties of the lunar crust. *Moon*, 4, 490.

Toksöz, M.N., Dainty, A.M., Solomon, S.C. & Anderson, K. (1974) Structure of the Moon. *Reviews of Geophysics*, 12, 539.

Toper, A.Z., Kabongo, K.K., Stewart, R.D. & Daehnke, A. (2000) The mechanism, optimization and effects of preconditioning. *Journal of the Southern African Institute of Mining and Metallurgy*, 100, 7–16.

Tripathy, G.R. & Gupta, I.D. (2002) Prediction of ground vibrations due to construction blasts in different types of rock. *Rock Mechanics and Rock Engineering*, 35(3), 195–204.

Tsai, Y.B. & Huang, M.W. (2000) Strong ground motion characteristics of the Chi-Chi Taiwan earthquake of September 21, 1999. In: *2000 NCHU-Waseda Joint Seminar on Earthquake Engineering, July 17–18, Taichung*. pp. 1/1–32.

Tsukahara, H. & Ikeda, R. (1987) Hydraulic fracturing stress measurements and in-situ stress field in the Kanto-Tokai area, Japan. *Tectonophysics*, 135, 329–345.

Tsuneishi, Y., Ito, T. & Kano, K. (1978) Surface faulting associated with the 1978 Izu-Oshima-Kinkai earthquake. In: *Bulletin of the Earthquake Research Institute, University of Tokyo*, Vol. 53. pp. 649–674.

Ueta, K., Miyakoshi, K. & Inoue, D. (2001) Left-lateral deformation of head-race tunnel associated with the 2000 western tottori earthquake. *Journal of the Seismological Society of Japan*, 54(2), 547–556.

Ulusay, R., Aydan, Ö. & Hamada, M. (2002) The behavior of structures built on active fault zones: Examples from the recent earthquakes of Turkey. *Structural Engineering/Earthquake Engineering*, 19(2), 149–167.

Villari, E. (1865) Change of magnetization by tension and by electric current. *Annual Review of Physical Chemistry*, 126, 87–122.

Vuillemeur, F., Teuscher, P. & Beer, R. (1997) The Lotscherberg base tunnel. *Tunnelling and Underground Space*, 12(3), 361–368.

Wald, D.J., Quitoriano, V., Heaton, T.H., Kanamori, H., Scrivner, C.W. & Worden, C.B. (1999) TriNet "ShakeMaps": Rapid generation of instrumental ground motion and intensity maps for earthquakes in southern California. *Earthquake Spectra*, 15, 537–556.

Wald, D., Somerville, P.G., EERI, M. & Burdick, L.J. (1998) The Whitter Narrows, California earthquake of October 1, 1997 – Simulation of recorded accelerations. *Earthquake Spectra*, 4, 139–156.

Wang, W.L., Wang, T.T., Su, J.J., Lin, C.H., Seng, C.R. & Huang, T.H. (2001) Assessment of damage in mountain tunnels due to the Taiwan Chi-Chi earthquake. *Tunneling and Underground Space Technology*, 16, 133–150.

Watanabe, H., Tano, H., Aydan, Ö., Ulusay, R., Tuncay, E., Bilgin, H.A. & Seiki, T. (2003) The measurement of the in-situ stress state by Acoustic Emission (AE) method in weak rocks. In: *Proc. of the 3rd International Symposium on Rock Stress, RS KUMAMOTO '03, November 2–4, Kumamoto, Japan.* pp. 389–394.

Watanabe, H., Aydan, Ö. & Imazu, M. (2013) An integrated study on the stress state of the vicinity of Mt. Kuriko. In: *The 6th International Symposium on In-Situ Rock Stress (SENDAI).* pp. 831–838.

Ward, S.N. & Asphaug, E. (2000) Asteroid impact tsunami: A probabilistic hazard assessment. *Icarus*, 145, 64–78.

Waversik, W.R. & Fairhurst, C. (1970) A study of brittle rock fracture in laboratory compression experiments. *International Journal of Rock Mechanics and Mining Sciences*, 7, 561–575.

Weber, W. (1830) Über die spezifische Warme fester Körper insbesondere der Metalle. *Annual Review of Physical Chemistry*, 96, 177–213.

Weiss, R., Wünnemann, K. & Bahlburg, H. (2006) Numerical modeling of generation, propagation and run-up of tsunamis caused by oceanic impacts: Model strategy and technical solutions. *Geophysical Journal International*, 167, 77–88.

Wells, D.L. & Coppersmith, K.J. (1994) New empirical relationship among magnitude, rupture length, rupture width, rupture area, and surface displacement. *Bulletin of the Seismological Society of America*, 84(4), 974–1002.

Wu, C., Hao, H., Lu, Y. & Zhou, Y. (2003) Characteristics of stress waves recorded in small-scale field blast tests on a layered rock-soil site. *Geotechnique*, 53(6), 587–599.

Wu, J.H., Lin, J.S. & Chan, C.S. (2008) Dynamic discrete analysis of an earthquake-induced large-scale landslide. *International Journal of Rock Mechanics and Mining Sciences*, 46(2), 397–407.

Wünnemann, K., Collins, G.S. & Weiss, R. (2010) Impact of a cosmic body into earth's ocean and the generation of large tsunami waves: Insight from numerical modeling. *Reviews of Geophysics*, 48(4), RG4006.

Wyllie, D.C. & Mah, C.W. (2004) *Rock Slope Engineering: Civil and Mining.* 4th edition. London, Taylor and Francis.

Yashiro, K., Kojima, Y. & Shimizu, M. (2007) Historical earthquake damage to tunnels in Japan and case studies of railways tunnels in the 2004 Niigata-ken Chuetsu earthquake. *QR of RTRI*, 48(3), 136–141.

Yavuz, H., Tufekci, K., Kayacan, R. & Cevizci, H. (2013) Predicting the dynamic compressive strength of carbonate. Rocks from quasi-static properties. *Experimental Mechanics*, 53(3), 367–376.

Yoshida, J., Yoshinaka, R., Sasaki, T. & Osada, M. (2014) Study on dynamic properties of rock discontinuity using dynamic direct shear test machine. In: *ARMS8, Sapporo.*

Zachmanoglou, E.C. & Thoe, D.W. (1986) *Introduction to Partial Differential Equations with Applications.* New York, NY, Dover Pub. Inc.

Zeng, Y. (1994) A composite source model for computing realistic synthetic strong ground motions. *Geophysical Research Letters*, 21, 725–728.

Zhang, C., Feng, X.T., Zhou, H., Qiu, S. & Wu, W. (2012) Case histories of four extremely intense rockbursts in deep tunnels. *Rock Mechanics and Rock Engineering*, 45, 275–288.

Zienkiewicz, O.C. & Pande, G.N. (1977) Time-dependent multi-laminate model of rocks – A numerical study of deformation and failure of rock masses. *International Journal for Numerical and Analytical Methods in Geomechanics*, 1, 219–247.

Zoback, M.D., Tsukahara, H. & Hickman, S.H. (1980) Stress measurements at depth in the vicinity of the San Andreas fault: Implications for the magnitude of shear stress at depth. *Journal of Geophysical Research*, 85(B11), 6157–6173.

Subject index

ISRM Book Series

Book Series Editor: Xia-Ting Feng

ISSN: 2326-6872

Publisher: CRC Press/Balkema, Taylor & Francis Group

1. Rock Engineering Risk
   Authors: John A. Hudson & Xia-Ting Feng
   2015
   ISBN: 978-1-138-02701-5 (Hbk)

2. Time-Dependency in Rock Mechanics and Rock Engineering
   Author: Ömer Aydan
   2016
   ISBN: 978-1-138-02863-0 (Hbk)

3. Rock Dynamics
   Author: Ömer Aydan
   2017
   ISBN: 978-1-138-03228-6 (Hbk)

Milton Keynes UK
Ingram Content Group UK Ltd.
UKHW052028141024
449569UK00017B/735